Founded in 1972, the Institute for Research on Public Policy is an independent, national, nonprofit organization.

IRPP seeks to improve public policy in Canada by generating research, providing insight and sparking debate that will contribute to the public policy decision-making process and strengthen the quality of the public policy decisions made by Canadian governments, citizens, institutions and organizations.

IRPP's independence is assured by an endowment fund established in the early 1970s.

Fondé en 1972, l'Institut de recherche en politiques publiques (IRPP) est un organisme canadien, indépendant et sans but lucratif.

L'IRPP cherche à améliorer les politiques publiques canadiennes en encourageant la recherche, en mettant de l'avant de nouvelles perspectives et en suscitant des débats qui contribueront au processus décisionnel en matière de politiques publiques et qui rehausseront la qualité des décisions que prennent les gouvernements, les citoyens, les institutions et les organismes canadiens.

L'indépendance de l'IRPP est assurée par un fonds de dotation établi au début des années 1970.

IRPP

Institute for Research on Public Policy

Institut de recherche en politiques publiques

GEOPOLITICAL INTEGRITY

EDITED BY HUGH SEGAL

Printed in Canada
Dépôt légal 2005

Bibliothèque nationale du Québec

National Library of Canada
Cataloguing in Publication Data

Geopolitical integrity / edited by Hugh D. Segal.

Includes bibliographical references.
ISBN 0-88645-189-2

1. National security—Canada. 2. Canada—Armed Forces.
3. Canada—Military policy. 4. Canada—Foreign relations—1945-.
5. Combined operations (Military science)
I. Segal, Hugh II. Institute for Research on Public Policy.

UA600.G46 2004 355'.033071 C2004-904710-8

Copy Editing
Brian McIntyre
Timothy Niederman

Proofreading
Wendy Dayton

Grid Design
Studio Duotone inc.

Production and Cover Design
Anne Tremblay

Cover photograph courtesy of Canadian Department of National Defence

Published by
The Institute for Research on Public Policy (IRPP)
L'Institut de recherche en politiques publiques
1470 Peel Street, Suite 200
Montreal, Quebec H3A 1T1
Tel: 514 985 2461 Fax: 514 985-2559
email: irpp@irpp.org www.irpp.org

Contents

Preface i

1 Introduction 1
Hugh Segal

2 Geopolitical Integrity for Canada 9
Hugh Segal

3 Force Structure or Forced Structure?
The 1994 White Paper on Defence
and the Canadian Forces in the 1990s 37
Sean M. Maloney

4 A National Security Framework for Canada 81
W.D. Macnamara and Ann M. Fitz-Gerald

5 Canada and Military Coalitions:
Where, How and with Whom? 105
Douglas Bland

6 The Canadian Forces and the Doctrine of Interoperability:
The Issues 151
Danford W. Middlemiss and Denis Stairs

7 Sailing in Concert:
The Politics and Strategy of Canada-US Naval Interoperability 183
Joel J. Sokolsky

8 Multinational Land Force Interoperability:
Meeting the Challenge of Different Cultural Backgrounds
in Chapter 6 Peace Support Operations 217
Ann M. Fitz-Gerald

9 Missed Opportunity II: The Canadian Forces, Force Structuring and Operation Iraqi Freedom ... 245
Sean M. Maloney

10 Military and Postconflict Security: Implications for American, British and Other Allied Force Planning and for Postconflict Iraq ... 271
Ann M. Fitz-Gerald

11 On Being an Ally: Why Virtue Is Not Reward Enough ... 293
Thomas S. Axworthy

12 Between a Rock and a Soft Place: The Geopolitics of Canada-US Security Relations ... 299
Joel J. Sokolsky

Notes on Contributors ... 333

Preface

The foreign and defence policy challenges Canada faces embrace many variables and pressures. The breadth of subjects covered in this volume reflects the work not only of the authors but also of dedicated fellows and staff at the IRPP who have been engaged in and supportive of the many different aspects of the research program of which Geopolitical Integrity is a part. The papers collected here were originally published under the IRPPs National Security and Military Interoperability research program, which began in June 2001. It has now been broadened to encompass Canada in the World and the nexus between defence, diplomacy and development.

The Board of Directors of the IRPP has been committed to this project from the outset and has offered thoughtful comment and counsel in a fashion that has enriched the entire series of research-focused publications and symposia in the area of national security pursued by the IRPP since 2001.

Current and former IRPP fellows such as Don Macnamara, Joel Sokolsky, Ann Fitz-Gerald and Doug Bland have not only been contributors to the research program in many ways but have also served as key advisers to the Institute on security issues. Denis Stairs, who chairs the Research Committee of the board, has been a constant source of strength, inspiration and sage counsel. The research agenda has also been strengthened by collaboration with scholars and students from the Royal Military College of Canada, the University of Manitoba, Queen's University, Dalhousie University, Université de Montréal, St. Lawrence University, the American Defence University and the Royal Military College of Science at Shrevenham in the United Kingdom.

Many students attended and participated in the symposia and events the IRPP's research program has generated so far. Two, in particular, have made important contributions to this volume: John Haffner, a law student at McGill University, contributed substantially to research and early drafts on the broad issue of geopolitical integrity; and Erin Dyck, a political science student, also at McGill, helped in the editing process. The IRPP staff was a critical part of the effort overall — especially Suzanne Ostiguy McIntyre, who, as vice-president, operations, oversees all production activity — as were the scholars who reviewed, on an anonymous basis, each of the papers in this publication.

The IRPP is a research organization. While scholars and participants may argue for various policy prescriptions, our core purpose as an institute is to provide a range of well-considered, research-based policy options that inform and enrich the debate and the ensuing decision-making process of our democracy. Not all the views on related issues will coincide in this volume. In fact, articulate and well-researched cases are made by scholars and practitioners with opposing views on what constitutes a reasonable level of geopolitical integrity for Canada with regard to Canada's resource profile, geography, ethnic makeup and history. We have a profound respect, as a matter of course, for the advocacy groups who work so hard on various national security issues, both those on the Right and those on the Left, just as we do for those who advocate in the economic and social spheres. All are part of the larger debate and of the multivariant database that fuels the marketplace of ideas. This volume is intended as a constructive and informative contribution to that very marketplace of ideas, focusing explicitly on what genuine geopolitical integrity could mean for Canada's security and defence priorities.

Hugh Segal
President
Institute for Research on Public Policy
Montreal, December, 2004

1

Introduction

Hugh Segal

This collection consolidates a portion of IRPP's contribution to a topic of critical importance in Canada: the nature of a renewed commitment to our military, our national security and to our engagement in the world. The papers chosen come from both the IRPP's National Security and Interoperability research series, initiated in June 2001, and from presentations delivered at various IRPP seminars and symposia that have dealt with Canada's global and hemispheric priorities and the foreign and defence policy options and challenges that those priorities suggest.

In the second paper, I propose a framework for "Geopolitical Integrity for Canada." *Geopolitical integrity*, here, means cohesiveness: Canadian foreign, defence, humanitarian and trade policies should come together in a coherent manner. The term is also used in a secondary sense, to imply an ethical responsibility: Canadian rhetoric at home ought to be consistent with our actions abroad, and both our words and our actions should express our values as a country.

Much of the debate about Canadian military capacity can be traced back to the gap between the military capabilities required to fulfill the recommendations and undertakings proposed in the *1994 White Paper on Defence* and current military strength. The Canadian Forces press on valiantly forward despite troop burnout and widespread equipment deficiencies. In "Force Structure or Forced Structure: The 1994 White Paper and the Canadian Forces in the 1990s," Sean Maloney of the Department of War Studies at the Royal Military College looks back to assess the 1994 White Paper and to discuss how Canada's defence policy and force structure should have developed over the past 10 years.

Maloney's exercise is of more than historical interest. The 1994 White Paper was designed as an interim policy during the first years of the post-Cold War period. In 1994, many looked to the United Nations as the primary vehicle to bring stability to an unsettled world. In Canada, utopian internationalists asserted that the Canadian Forces should eschew "warfighting" and restrict their activities to peacekeeping through the UN; such views were countered by pragmatic realists, who maintained that the future was uncertain and that the forces should maintain a spectrum of capabilities. The utopian view was strongly advanced by the Department of Foreign Affairs and International Trade (DFAIT) and prevailed; consequently, detailed force structuring was eliminated from the draft documents. The result, says Maloney, was the 1994 White Paper.

Although the world has changed quite dramatically since 1994, the White Paper nevertheless continues to guide Canadian Forces policy. It is evident, however, that the struggles of the Canadian Forces today are the direct result of the failure of the White Paper and the policies based on it to deal adequately with force structuring and equipment issues.

According to Maloney, a prioritization exercise should have been undertaken to determine the force structure best suited to implementing Canadian national security policy. Such a structure should have included two clear streams — one dedicated to providing surveillance and security within Canada, the other relating to forces abroad, whose deployment should have been based on "the magnitude of threats."

One reason that the patchwork, ad hoc conception of defence from the 1994 White Paper still persists in 2004 is that Canada lacked a framework for articulating its national security strategy. In "A National Security Framework for Canada," Ann Fitz-Gerald, director of the Centre for Managing Security in Transitional Societies at Cranfield University in the UK, joins forces with retired Air Force Brigadier General W.D. Macnamara, a fellow at the Queen's Centre for International Relations, to examine what the absence of such a strategy framework has meant for defence policy in Canada.

Fitz-Gerald and Macnamara begin by clarifying key terms of reference for strategic defence planning. Next, they explore critical linkages between a broader definition of national security and Canada's core values and interests as expressed in various defence and foreign affairs policy documents. The authors make explicit the connection between these values and interests and the tools and instruments that ought to preserve them, including defence and military power. Much of the thought process reflected in this paper is also reflected in the government's National Security Policy issued in the spring of 2004.

The commitment-capability gap introduced by Maloney and the need for a strategic framework as advanced by Fitz-Gerald and Macnamara are both underlined and expanded as themes in "Canada and Military Coalitions: Where, How and with Whom?" by Douglas Bland, chair of the Defence Management Studies Program at Queen's University. Bland's central theme is the "inescapable relationship between means and ends in foreign policy" as it bears on the question of Canadian involvement in international coalitions. He quotes Michael Ignatieff, both at the beginning and end of his paper, to underscore his central message: "If we will the ends, we had better will the right means. For the means we select may betray our ends." Bland applies the means-end challenge to Canada's present and future military capabilities and addresses what those capabilities imply for a foreign policy based on acting through coalitions.

Canada's ability to play a reasonable and responsible role in international affairs has diminished substantially from 1956 — when Canada initiated and commanded the United Nations Emergency Force in the Middle East — to 1996, when a similar effort was unsustainable in Zaire. Bland attributes this decline to several factors but mainly to the continuous depreciation of the place of military capabilities within Canadian foreign policy; that is, to the failure of *means*. He notes, "Lester Pearson might have won the Nobel Peace Prize for his initiative, but he knew beforehand that Canada had the wherewithal to accomplish what he had proposed."

If Canada is to regain or even maintain its ability to further its interests by acting through like-minded coalitions, Bland argues, our political leaders must develop a national strategy tailored to this purpose. As with other contributors to this monograph, Bland argues that such a strategy must be interdepartmental; it should include discussions of where, when and with whom Canada should act; and it should account for the "machinery of government" needed to plan and control such enterprises.

Increasingly, coalitions rely on the military concept of "interoperability." In the sixth paper, "The Canadian Forces and the Doctrine of Interoperability: The Issues," political scientists Danford Middlemiss and Denis Stairs of the Centre for Foreign Policy Studies at Dalhousie University argue that Canada needs to have more public debate about interoperability, as it is a doctrine that is quickly becoming the central principle of our defence policy.

Interoperability is the "ability of systems, units or forces to provide services to and accept services from other systems, units or forces and to use the services so exchanged to enable them to operate effectively together," and it is a way for Canada to make a difference if it wants to engage its military in global affairs. The doctrine of interoperability is a strategy that permits "lesser powers" like Canada to have their own discernable impact and visibility in theatres of combat by enabling them to make focused contributions in ways that complement the resources of their allies, most of all those of the United States.

Middlemiss and Stairs start with an informal June 1999 Department of National Defence (DND) document on the future of Canadian defence that included signal references to interoperability. From this document and subsequent policy developments, the authors conclude that "Canada is moving towards greater interoperability with the United States in a manner that is far more comprehensive and rapidly paced than most Canadians realize."

Where Middlemiss and Stairs discuss the doctrine of interoperability in its broad outlines, Joel Sokolsky, dean of arts and professor of political science at the Royal Military College, focuses on naval interoperability in particular in his paper, "Sailing in Concert: The Politics and Strategy of Canada-US Naval Interoperability."

Although the tradition of the Canadian Navy cooperating closely with its principal allies goes back to the beginning of the twentieth century, since 1994 the Canadian Navy has sought to enhance interoperability with the United States Navy, and navies from other like-minded nations, as a matter of explicit foreign and defence policy and strategy. At its most advanced, naval interoperability allows maritime forces from a number of different countries to function together seamlessly as a single unit.

Sokolsky points out that, just as the Canadian Navy now seeks to operate seamlessly with the US Navy, the United States is now also promoting interoperability with its allies, including Canada. But Sokolsky's description of the American perspective recalls an observation from Middlemiss's and Stairs's paper, that interoperability has a different significance for the dominant power than it does for lesser powers.

For the United States, interoperability is one of many tools it employs to maintain its dominant global position. Thus, while Washington encourages and welcomes them, Canadian naval contributions are unlikely to afford Ottawa a greater voice or leverage in larger political or strategic decisions. Sokolsky argues that this lack of leverage is where the true issue lies regarding the effect of interoperability on Canadian sovereignty. Canada is always free to choose whether or not to sail in concert with the US Navy, but once Canadian naval forces have committed themselves as part of a seamless larger operation, Ottawa's commitment does have consequences.

For the Americans, this is as it must be. Otherwise, Canadian interoperability with the US Navy would not be to America's advantage, and the US would not favour the doctrine. In this sense, there will always be tension between Canadian foreign policy strategy and the politics of interoperability with the US Navy, but Sokolsky is optimistic that this tension is a limited and familiar one for Canada and that it can be managed with close political oversight.

Ann Fitz-Gerald takes the question of interoperability back to land in her paper "Multinational Land Force Interoperability: Meeting the Challenge of

Different Cultural Backgrounds in Chapter 6 Peace Support Operations." She sets out to examine how efforts to enhance military interoperability — by Canada as well as its allies — have actually worked in recent multinational military interventions. Fitz-Gerald begins by noting an important but neglected feature of many such interventions: that members of the local population, who alone will determine whether there is a lasting peace, "are now able to view the behavioural conduct and operational effectiveness of the peacekeeping forces due to the close proximity in which they operate." If there are "disparate national approaches" among the peacekeepers, the locals will know it, and the international forces will lose both credibility and effectiveness. Thus multinational cohesion — as perceived by local actors — is a key factor in determining the effectiveness of multinational military interventions.

Fitz-Gerald focuses on case studies in Haiti and Bosnia to identify implications for interoperability and multinational training, with specific recommendations for Canada. The experiences in those countries show that disparities in peace-keeping conduct among multinational troops can lead local groups to "re-evaluate their allegiance to the international effort," and such groups "may easily fall vulnerable once again to the forces that ignited the conflict in the first place."

Lastly, Fitz-Gerald focuses on the Canadian contribution to these multinational operations. She examines how political decisions taken in Ottawa have affected the performance of the Canadian Forces and discusses measures that could be implemented to restore local confidence in Canada's position as a front-line peacekeeping partner. She ends by emphasizing that the doctrine of interoperability needs to take account of issues on the ground if it is to be truly effective: "[W]hile much of the literature stresses the importance of trade partnerships, the shared defence industrial base and advanced weapons systems, complementarity at the tactical and operational levels must also be included in the equation."

In "Missed Opportunity II: The Canadian Forces, Force Structuring and Operation Iraqi Freedom" (originally published as "Are we Really Just Peacekeepers? The Perception versus the Reality of Canadian Military Involvement in the Iraq War"), Sean Maloney assesses the proposition that it is only "peacekeeping" that defines the Canadian Forces. He looks specifically at what a Canadian deployment to Iraq, with the US, the UK, Australia and others, might have encompassed.

Maloney lays the groundwork for showing how Canada missed an opportunity in Iraq by examining the military forces Canada contributed to other wars in the 1990s, the principles underlying those commitments, the force structure as it existed in 2002-03 and, therefore, what options existed for a possible deployment to Iraq.

In the early 1990s, there was a push in Canada to expand the conception of peacekeeping beyond its original definition, instead of using a more flexible typology that would acknowledge how our forces have actually been deployed across the spectrum of conflict — in ways that fall well outside traditional forms of UN peacekeeping or humanitarian assistance.

Since 1990, Canadian Forces have been deployed in Iraq War Phase I, from 1990-91; in Haiti in 1993; in Contingency Plan COBRA in Croatia and Bosnia in 1995; in Iraq War Phase II, from 1991-2002; in Kosovo from 1999-2000; and in Afghanistan from 2001-02. By examining case studies on these deployments, Maloney establishes that it is not enough for Canadian Forces to show up to be part of a coalition; our forces must be deployed on the "principle of saliency" — a principle that recalls discussions of interoperability in other papers. Canada must be able to provide a unique capability that no other nation can bring to the table, or use the forces committed in an orthodox or unusual way, or be prepared to accept missions that no other nation in the coalition would accept.

Over the course of the 1990s, argues Maloney, the leadership of the Canadian Forces came to realize that the forces are capable of making militarily significant contributions to diverse alliance and coalition war efforts. He then returns directly to his central question, of how Canada could have participated in the third phase of the Iraqi war. Through a detailed, fact-based exposition of the options that were available, Maloney is able to conclude that "[t]here can be little doubt that Canada had the military means to contribute in an effective and salient fashion to [Operation] Iraqi Freedom in 2003."

Ann Fitz-Gerald provides yet another analysis occasioned by Iraq, "Military and Postconflict Security: Implications for American, British, and Other Allied Force Planning and for Postconflict Iraq." The paper takes postconflict Iraq as a timely point of departure for a broader discussion of postconflict security and especially of the military's role in security sector reform (SSR).

As Fitz-Gerald points out, SSR is a subject that has received considerable attention and debate in the development community and government, and while there is now a theoretical understanding that effective SSR requires broad, holistic approaches, "the way in which these concepts have been translated into practice remains unclear at worst and disparate at best." Fitz-Gerald points out that transitioning societies have comprehensive security needs: if the emergency response phase of a conflict is not properly integrated with the wider agenda, the conflict can be reignited very quickly.

To help address such comprehensive security needs, Fitz-Gerald identifies the key actors and activities within a broad conception of the security sector. She gives an overview of the main program and policy areas for SSR and discusses the potential contribution of international military forces to each of these areas as part of a continuum of actors.

Tom Axworthy, a former principal secretary to Pierre Trudeau and the current executive director of the Historica Foundation, contributes a view of what the duties and obligations of an ally really are in "On Being an Ally: Why Virtue Is Not Reward Enough." Axworthy presents a broad and integrated image of what a country's moral obligations, military capacities and global responsibilities must perforce embrace. He sets a high bar for any Canadian administration to address.

The final paper, "Between a Rock and a Soft Place: The Geopolitics of Canada-US Security Relations," by Joel Sokolsky, reflects on the implications for Canada of new US trends in both security and foreign policy. Without the slightest bit of complacency, Dr. Sokolsky outlines his prescription as to what should be the basis for future Canadian foreign and defence policy priorities.

2

Geopolitical Integrity for Canada

Hugh Segal

Introduction

On September 12, 2001, as the world recoiled from the devastating airplane attacks in the United States, a writer in the French newspaper *Le Monde* captured the global mood of sympathy and solidarity by declaring "We are all Americans, all New Yorkers," a deliberate parallel to John F. Kennedy's famous 1962 declaration on his visit to a besieged Berlin that he, too, was a Berliner. On the political level, article 5 of the 1949 NATO Washington Treaty was invoked for the first time,[1] which holds that an attack on one alliance member is an attack on all. Democratic nations came together as one.

Three years or so after 9-11, however, we are far from the sense of collective resolve and solidarity that surfaced at that time. The outpouring of global sympathy has settled into tough policy questions for America and its allies about how best to negotiate the insecurities and anxieties of a less certain world.

Not surprisingly, in the eyes of some, Canada now finds itself in an uneasy middle space between opposing European and American attitudes. On the one hand, many Canadians have questions about the place of due process and multilateralism in the new American campaign. On the other hand, we still identify ourselves with a sense of global responsibility; and as its closest neighbour we can relate very strongly and immediately to the shock that 9-11 represented to America. What is unclear for many Canadians is exactly where Canada fits into this new world, and what principles ought to govern the form and extent of our engagement in it.

As we struggle to answer those questions, the reality remains that the world has become increasingly interconnected and unpredictable in terms of security risks — even for Canada. Canadians can no longer content themselves in thinking they are immune from global dangers simply by sewing maple leaves on their backpacks. When the October 2002 bomb exploded in Bali, taking more than 200 lives, it emerged that Australians had been targets as much as Americans (88 Australians died). More recently, almost 200 people died and 1,500 were wounded in the bombings in Madrid. If Australia and Spain are targets, why not Canada?

Auditor-General Sheila Fraser recently delivered a report to Parliament pointing to "gaping holes in Canada's security net."[2] Around the same time that Ms. Fraser submitted her report, the Royal Canadian Mounted Police (RCMP) charged a Foreign Affairs contract worker in an alleged international terrorism conspiracy.[3] And beyond these telltale indications of vulnerability, we must not forget that in the 9-11 attacks, "immense destruction was achieved with weapons no more exotic than...box cutters."[4] Any possible target we can imagine within Canadian borders, a terrorist can imagine, too.

Of course, terrorist organizations are not the only threat to global security. The former US special envoy for negotiations with North Korea, Charles "Jack" Pritchard, recently announced that 8,000 spent fuel rods have gone missing there, possibly "reprocessed for weapons grade plutonium," and he pointed out that this discovery, in early January 2004, represents only "the most recent failure in a string of serious North Korea-related intelligence failures."[5] Also, in that region, there is little trust between Japan, South Korea and China, and Taiwanese independence continues to be a volatile issue for China, so instability is quite possible and even likely in Asia in the coming decades.

Other global threats include "extended transnational criminal networks, rings of organized crime, regions ruled by non-state actors and warlords, and continuous violations of human rights" in myriad forms.[6] And threats can be overlapping: if a state tyrant, for example, sells a weapon of mass destruction to an underground terrorist organization, it would then fall beneath the radar of a state-based inspections program. Mohamed ElBaradei, the director-general of the International Atomic Energy Agency (IAEA, the UN atomic weapons watchdog), has recently described an underground "Wal-Mart of private-sector proliferation" of nuclear-related material and equipment.[7] Justifiable anxieties about Iran's nuclear and geopolitical intentions are commonly held in European, Middle Eastern and North American capitals.

There is no doubt that if a terrorist attack were launched in Toronto or Vancouver, the debate over Canadian security would change dramatically. But surely it is infinitely better to act proactively so that we never have to contend with such a tragedy on our own soil, and so our domestic security initiatives are not motivated by regret and manifest failure. The primary meaning of national defence in the context of the asymmetrical threats after September 11, 2001, therefore, is that we must now have the rapidly deployable capacity to engage terrorists — and their source promoters and funders — in their homelands, not ours. Concurrently, the primary focus of our broader foreign policy should be to promote a stable and humane global order that advances social justice and democracy throughout the world as the best antidote to terror and security threats.

We also have a special neighbourhood duty to homeland security with our American allies, one that no other American ally can meet on our behalf. Underestimating the importance of our neighbourhood responsibility to continental homeland security would be a serious strategic misjudgment. The nature of an enemy that is diffused abroad, and likely represented in both Canada and the US, requires us to make the security of our population at home a compelling priority — not just at airports, but at all public places where multiple innocent civilian targets are vulnerable to attack. In this sense, the growing interconnectedness of the world is a threat; but it is also an opportunity.

This reflection envisions a Canadian foreign policy and defence framework for this new world, a view that is grounded in empirical realities and rooted in Canada's vital interests, commitments and values. This framework is one of geopolitical integrity for Canada.

What Is Meant by Geopolitical Integrity?

The primary sense of "integrity" here is coherence and unity: elements of our foreign, defence, humanitarian and trade policies should come together in a coherent manner. On a secondary level, the term suggests an ethical element, seeking to ensure that our words are consistent with our actions, and that both our words and actions express our values as a country.

Geopolitical integrity means, therefore, that the values that drive our way of life at home should be present in our activities and initiatives abroad. Before any real review of defence policy, we must know what our vital interests and values are and we must be comfortable with the commitments they necessitate. Vital interests are threatened when events or trends are deemed to threaten the survival or security of the nation. When these threats are serious enough (e.g., a direct attack on national territory, or a direct threat to national political or economic well-being, or to access to raw materials, etc.), the nation must be prepared to counter by resorting to strong measures, including military action. There is a hierarchy to these values and interests, one that should preclude the "tyranny of small differences" that sometimes distracts allies from their common goals.

The United States, Canada and the UK have differences in terms of favourite national sports, forms of government, voting and party systems, as well as views on health care, private schooling and gun control; but these differences are overshadowed by our shared commitments to democracy, the separation of church and state, respect for diversity and the rule of law. Our common values primarily embrace individual freedoms and social justice, with tolerance and respect for diversity emerging from the core values of democracy, freedom and equality of opportunity. To them we need to add that core Canadian value of civility. Absolutely essential to democracy itself, it underpins our tolerance of the differences in our society and in others.

Democracy, freedom and equality of opportunity form the core of our respective constitutions and social and political infrastructures. That the UK is economically to the left on some issues from the US, or that the Canadian health system works differently from those of both the US and the UK are important domestic policy distinctions, but they do not reflect core differences on the broader questions where we share common beliefs with other democracies in Europe and Asia.

Canadian embassies are now instructed to promote multiculturalism, federalism and respect for bilingualism and diversity. While these ideas are positive and constructive, they are reflective of minor aspects of Canadian society and are not unique to Canada. Germany and Australia have much to share on federalism; Switzerland and Russia cope with either more official languages or more diverse communities than does Canada. Unique Canadian achievements are better found in the ways in which our democratic institutions, and our individual and collective freedoms, have adapted to the need to integrate different regions and nationalities into a truly civil, multinational democracy.

Much of Canada's social and fiscal infrastructure, however imperfect, has been designed explicitly, or been modified, to reflect our regional and multinational realities in a way that seeks to promote not just a civil society but, more importantly, a society that is truly civil. That civility — which is reflected in everything from refugee policy and student loan programs to the Charter of Rights and Freedoms and official languages laws, plus so much more — is what makes Canada unique in the world.

Canada's brand of democracy protects the core freedoms and rights that are so vital to the life Canadians have built for themselves at home and hold up as an example for others less fortunate abroad. Some call it hopelessly naive and unsophisticated, but Canadians believe that Arab women should have the right to vote and drive, that Muslims in Bosnia deserve human rights, and that Israeli children should be able to go to school without being blown up. Canadians believe that 800,000 Rwandans should not have been butchered and that girls in

Afghanistan have the right to be educated. And while the redoubt of the isolationist is to ask just how much Canadians are prepared to contribute or sacrifice — from their own taxes or the money allocated to health care — for any of the positive outcomes, the answer can only be given in the context of a democratic framework that reflects our mix of core values and vital interests.

Canada pursued its vital interests when it proposed (through the St-Laurent government and then-secretary of state for external affairs Pearson) a United Nations emergency force at Suez to find a way for the British, French and Israelis to stand down from a confrontation with Egypt, and perhaps Russia. Our vital interest there, beyond the free movement for world shipping through the Suez Canal, was keeping the cold war between East and West from producing a proximate thermonuclear risk. Not all of those things that violate our core values necessarily impact our vital interests. But the first rule of defence and national security is that threats should be addressed as far away from one's own population and territory as possible. We share this vital interest with the Americans, whether our government is to the left of theirs or their government is to the right of ours, whether we approve of their approach to gun control or they disagree with our approach to the medical use of marijuana. We also share common economic and geographic space with America, and our respective vital interests embrace the safety and security of the citizens of both countries of that space.

Thus a clear foreign policy statement should not begin with normative platitudes about international development or peaceful co-existence. A foreign policy statement that reflects Canadian leadership would stipulate those activities abroad that Canada will not tolerate, the values and vital interests it thinks are at stake, and the prerogatives Canada is prepared to assert in defence of those values and vital interests. Threats to our national security, efforts at ethnic cleansing or genocide, state or non-state commitments to eradicate nations or cultures or other religious or cultural groups, the threat of weapons of mass destruction and violence against civilian populations all need to be factored into a Canadian framework for foreign policy engagement. And while the military capacity to enable Canadian reach and impact is vital, also vital are the abilities to disarm post-conflict factions, to rebuild societies' infrastructures, to promote democracy and to enhance social and economic self-sufficiency.

When close allies are threatened or attacked, Canadian engagement cannot be equivocal. When Canadians are at risk at home or abroad, we must have the capacity to respond without dependence on others. Geopolitical integrity requires that we practise abroad what we say and believe at home about core values and vital interests, with both our diplomatic policy and our military and security capacity on the ground. The extent to which a country allows that gap to grow — between what it says and believes and what it is prepared to do — determines the degree of international influence that country deserves. The dilution of impact reduces a country's leverage, which is essential to its vital interests on issues of trade, economy, environment and the general world order.

In recent years, quite a gap has formed between Canada's professed geopolitical commitments and its actions. Canada's current defence spending is 1.1 percent of GDP, ranking it 17th among 18 North Atlantic Treaty Organization (NATO) members,[8] with only tiny Luxembourg, nestled safely in the middle of Europe, spending less. Even France — for all its public opposition to American interventionism — spends proportionately more than twice what we do on defence. While Canada ranks 17th out of 18 (percentage-wise) in defence spending, it ranks 7th among the 30 Organization for Economic Co-Operation and Development (OECD) countries in terms of total GDP performance. Canada is also the largest country geographically in NATO and second largest in the world. Many defenders of the present level of military expenditure will argue that the absolute dollars Canada spends on defence puts it in the top 10 of NATO and top 25 in the world, and that Canada is one of the few countries that regularly dispatch large parts of their militaries abroad — but then fully neglect the pressures that being understrength puts on our troops.

Many Canadians take pride in our peacekeeping accomplishments, but Canada is now ranked 36th among UN peacekeeping contributors (and at the same time, with minimal public awareness, it has recently pushed the Canadian military to its limits outside UN auspices). In fact, Canada has only 239 participants on UN peacekeeping operations worldwide.

As Dalhousie's distinguished international affairs scholar Denis Stairs comments, "It is likely that most Canadians would be surprised if they knew how little their country is actually doing...in real terms, relative to how much noise it makes about its efforts in rhetorical terms."[9] Canada could do less as a country, or more as a country, but our rhetoric should be consistent with our actions: "There is no argument...for doing less while claiming more, for speaking loudly while carrying a bent twig."[10]

Various recent polls underscore other gaps in knowledge and complacent assumptions among Canadians. While 55 percent of Canadians "believe that there are international terrorists in Canada, waiting to attack Canadian civilians," only 44 percent "believe that Canadian security forces are capable of preventing terrorist attacks."[11] Similarly, where only one-third of Canadians "think that Canada should increase its level of defence cooperation with the United States,"[12] 79 percent of Canadians "believe Canada would need the United States to protect it in the case of an attack."[13]

So taken together, this all implies little faith in our own national security capacities, a clear belief that threats are real, and a genuine acquiescence to the notion that assistance from American security forces might well be required. Whatever this may add up to, it does not add up to geopolitical integrity.

These and other gaps indicate that we need to develop an adequate framework for promoting our own security alongside our broader democratic values. Canada needs a coherent set of policies around which we can advance our geopolitical interests, including our most vital trade, economic and development links, as well as our underlying humanitarian commitments that transcend our immediate interests.

We need to contemplate a "grand strategy for a modest country" that integrates military, diplomatic, civil society and foreign aid instruments in a thrust that preserves security and opportunity at home, advances leverage with our allies, and responds in an integrated way to the real threats from abroad.

In our work abroad, we need to shape a strategy that — as we learned from the experiences of East Timor, Bosnia, Kosovo, Haiti and Iraq — combines military, civil, private-sector, democratizing, and post-conflict-transition skills. These need to be built into real plans and models that maximize the ability of each to engage constructively on Canada's behalf and, where appropriate and helpful, enhance the leverage generated by combined application.

At home, from disaster relief to protecting the environment and the fisheries, to defending its sovereignty, Canada has a duty to build and train its military to match its growth as well as the complexities of Canadian and global society. We should be able to defend the sovereignty of Canada and the safety of Canadians both at home and abroad. We should have the capacity to move meaningful troop strength quickly — within days, not weeks — to global trouble spots to prevent humanitarian crises or to render civil assistance, as required. And we should be committed to partnering with alliances and coalitions of the moment where circumstances call for intervention, with or without the UN, as the case may require.

To address the roots of terrorism, our strategy should unify diplomacy, foreign aid, intelligence and military deployment. This is about more than intervention merely to attack the purveyors of terrorism; it is also about addressing some root causes — like the absence of democracy in most geographic points of origin for terrorists, and the prevalence of oppression, distorted ideology and intense poverty in many of the recruiting centres.

Finally, any framework of geopolitical integrity suggests that Canadian civil society needs more interactive engagement with our military; there should be meaningful forms of interaction and integration between the two. Ordinary Canadians should see our military as an expression of our collective aspirations and values, as one important embodiment of our identity. The civilian population has the right to benefit from, and more readily understand, the military culture and exigencies that are the requisite circumstances under which Canadians in uniform serve their country at home and abroad, and to understand the realities of sacrifice and discipline that define the military role and profession in our democracy. Geopolitical integrity is meant to draw all these elements together and unify them in a single vision and policy framework.

Whether combating ethnic cleansing from the air, patrolling our seas in defence of North American security, providing food and humanitarian relief in Africa, making a valued and strategic contribution as part of the International Security Assistance Force (ISAF) in Kabul or in combat with al-Qaeda in the Pakistan-Afghanistan border region, or patrolling the waters of the Persian Gulf as part of Operation Apollo, Canadian Forces reflect what we are for and what we are against as Canadians. This is as true of a military reservist helping civilian forces during forest fires or floods as it is of a deadly accurate Canadian Army sniper in the hills of Afghanistan. It is also true that a Canadian Forces intelligence major heading a coalition cyber-analysis task group in a hardened US facility in southern Asia reflects who we are and what we believe in, as does a young Royal Military College (RMC) cadet with aspirations for public service, or the commander of a Canadian Navy frigate on a NATO naval exercise. The values they reflect are the same that inspire our collective domestic commitments to individual freedom, democracy and equality of opportunity.

Why Geopolitical Integrity?

Canadians affirm democracy, freedom, tolerance, pluralism and community. We stand opposed to terrorists who seek to undermine our way of life, just as we oppose tyrannical regimes of the far right or far left. Yet there have been two compelling popular notions that have in the past prevented us — and continue in the present to prevent us from more active engagement in the world: an inordinate commitment to state sovereignty, and an almost theological belief in the UN as the only acceptable instrument of multilateral intervention.

Any genuine notion of geopolitical integrity that is value-based has to mean that the veil of state sovereignty cannot serve as perpetual insulation when core humanitarian norms are breached. There are good reasons for wanting to recognize state sovereignty under legitimate circumstances, but there are also good reasons for wanting to alleviate extreme human suffering on a universal basis, and there are some occasions in which the latter clearly supersedes the former. If the 1648 Treaty of Westphalia is the classical expression of state sovereignty as absolute, an even earlier text, Hugo Grotius's 1625 *De Jure Belli ac Pacis*, embodies the views that sovereignty is not absolute and that individuals ought to be helped across regimes in cases of extreme suffering and persecution.[14] Geopolitical integrity for Canada means siding with Grotius over Westphalia.

When Lieutenant-General Romeo Dallaire was in charge of UN peacekeepers in Rwanda, he "warned the UN about death squads and plans for mass killings" and sought to prevent them from happening. Despite the gravity of the threat, he was "overruled by UN headquarters in New York" and ignored in his appeals to several countries. The genocide killed more than 800,000 people, and General Dallaire has recently testified against the people who allegedly engineered the slaughter.[15]

With the publication of his book, *Shake Hands with the Devil: The Failure of Humanity in Rwanda*, Dallaire's experience has become part of the Canadian conscience, our collective sense that there are clear instances of right and wrong in the world. The lesson is clear: we need to be prepared to act on our convictions, before rather than after the fact.

General Dallaire's experience reminds us that where the threat of genocide is real and imminent, or, more clearly, where it is already taking place, democratic nations have a responsibility to intervene, either because we affirm that universal human rights trump the abstract claim of state sovereignty in certain instances, or because we believe that in such instances the government has *already* given up its legitimacy, its autonomous claim as a sovereign administration, because a responsible government by its very definition could never countenance genocide of any of its people. On either justification, state sovereignty should not be seen as an impregnable juridical barrier capable of preventing humanitarian engagement in the world.

This is not a radical departure from foreign relations practice and theory in our era and time. Prime Minister Martin enunciated a similar approach in his address to the United Nations General Assembly in the fall of 2004. And the NATO action in Kosovo very much reflects this kind of response to a genuinely perceived threat of ethnic cleansing.

Similarly, while the United Nations is an important instrument of humanitarian and security measures, it is not the final arbiter of such actions, nor is it the only acceptable mechanism for multilateral action. Ostensibly, Chapter VII of the UN Charter "provides a possible mechanism for dealing with democratization as a legitimate concern of the Security Council, through a liberal interpretation of the term 'a threat to peace.'" But as Reginald Austin points out in his *Democracy and Democratization* work in 2003, "the veto seems...to have frozen the possibility of this working," and even "the extreme circumstances of an ongoing genocide [in Rwanda] could not provoke a conviction...of a 'threat.'"[16]

In 2001, Canada supported a study by the International Commission on Intervention and State Sovereignty, *The Responsibility to Protect*, which essentially supports the Grotius position. Former Canadian foreign minister Lloyd Axworthy was a key part of that exercise. Although not quite yet "doctrine," the study's position is strongly supported as an extension of the Universal Declaration on Human Rights, which itself is regarded by many as an extension of the UN Charter.

Although the right to democratic governance has now matured to the point that it "has achieved at least 'soft' normative status"[17] in international law, the United Nations' decision-making processes and democracy-building capabilities have not kept pace with the maturing of this right. In the early 1990s, for example, the United Nations took on the task of establishing a liberal democracy in Cambodia by means of the UN Transitional Authority in Cambodia (UNTAC). Despite formally assuming this responsibility, however, the UN failed to plan for "vital democratizing activities" there, a failure that, for Austin, serves as "damning and dramatic evidence of the state of democratic myopia prevailing in the international community, and in particular the Security Council, at the time."[18]

There is also considerable evidence of the United Nations' inability to act in circumstances that clearly warrant action, including that cited in the UN Brahimi Report.[19] Moreover, the United Nations is less than democratic in its functioning, as it gives equal votes to both tyrannies and democracies. The United Nations would be a much more effective avenue of democratization and peace if it afforded greater influence to countries that are democratic, or demonstrably on the path to democracy. Canada needs to champion this kind of reform of the United Nations as part of the integrity of our overall view.

In fact, if our relationship with the United Nations, so long a pillar of our multilateral foreign policy, is itself not to be a violation of a legitimate commitment to geopolitical integrity, Canada has to refine that relationship in a fashion consistent with our values. In contemplating any possible international intervention, it would make sense for Canada to (1) communicate our desire for greater democratic reform within the UN; (2) signal that we are skeptical of the UN's capacity to act, or act comprehensively, in warranted cases; and (3) reserve our right to work outside it as warranted.

We should define, in advance, the types of circumstances in which we would consider crisis intervention or comprehensive democratization activities outside the UN to be more important than waiting for UN processes and plans to take effect. It is clear that some scenarios do warrant independent action. If the United Nations were to refuse to act in the face of another imminent threat of genocide, for example, the Dallaire experience tells us that Canada should choose to help avert that threat, rather than to affirm an abstract commitment to UN procedure and stand complicit in the face of wanton mass murder. In other words, Canada cannot on the one hand embrace the responsibility to protect and on the other hand hide behind UN multilateral gridlock in the face of clear humanitarian and state sanctioned, or permitted, cruelty aimed at captive and targeted populations. To be fair, the Eminent Persons Panel, structured by the UN secretary-general, is attempting to address these very challenges.

All this might sound like an ambitious agenda. But geopolitical integrity does not mean trying to do everything, and choices will have to be made based on our core capacities. At the same time, we should be clear at least on the principles animating our engagement in the world, and we should be able to make clear distinctions between legitimate and illegitimate objections to involvement. The framework of both geopolitical integrity and the implicit responsibility to protect is, of course, meaningless if Canada's support of "responsibility to protect" is not backed up by a serious "capacity to deploy."

Building Blocks of Geopolitical Integrity

Our capacity to act in the world depends first and foremost on a renewed policy commitment to the critical role of the Canadian Forces (CF) as an integral part of the core infrastructure of sovereignty and national security for Canada and of our legitimate international interests as served by our NATO, allied and unilateral decisions. Despite budgetary reductions, the Canadian military has been able to respond successfully to recent government requests, including the Gulf War, numerous peacekeeping missions, interventions in the former Yugoslavia and various engagements in Afghanistan. But the reality is that our armed forces are exhausted, both in terms of personnel and equipment, and in the last few years we have consistently pushed the same small number of dedicated members to the limit — with minimal public recognition of their achievements.

Soldiers from the First Battalion of the Princess Patricia's Canadian Light Infantry, for instance, have been to Afghanistan once and Bosnia three times in the last few years. The repeated six-month deployments abroad have been exhausting for the troops, and they have had a taxing, cumulative effect on the families left behind.[20] The concern for overtasking and overwork is often lost on the Canadians. Perhaps we all need reminding that the Second World War lasted a total of four-and-a-half years. Some Canadian soldiers, while overseas for four years, were in combat for three years (Far East), for two years (Italian campaign) or for less than one year (Northwest Europe); the Royal Canadian Navy (RCN) ships and Royal Canadian Air Force (RCAF) aircraft in the European and Far Eastern theatres were essentially in combat for five years; and then the war was over, and the survivors came home. While there is no comparison as to intensity, a very much smaller force of volunteers in the Canadian Army, navy and air force have been engaged rather consistently now for approaching 15 years! There should be no surprise at the impact that such operational momentum has had and will have on individuals and families. The CF does indeed need an operational pause to permit refurbishment, retraining, re-equipment and transformation.

Similarly, Operation Apollo, the Canadian Navy's two-year contribution to the War on Terror, involved 16 frigates, destroyers and supply ships, and "almost 97 percent of the navy's available sailors spent at least six months at sea during the operation." It is not just the people who are in need of renewal. As Dr. Richard Gimblett, a former naval officer and research fellow at Dalhousie University's Centre for Foreign Policy Studies comments, relative to our key supply ships *Preserver* and *Provider*, "We're holding some of these ships together with duct tape and binder wire. They're 35-year-old hulls."[21] None of which in any way diminishes the outstanding performance of Canadian infantry units in Afghanistan, the Coyote units in Bosnia and Afghanistan, or the high-tech patrol vessels and their crews in Operation Apollo.

There is also, again, the issue of public recognition and understanding of how well our troops performed in challenging theatres. Operation Apollo was the largest Canadian operation at sea since the Battle of the Atlantic in the Second World War, and yet it went largely unnoticed by the Canadian public. As one retired naval officer commented: "The Canadian Navy is like the proverbial tree falling in the forest — if you command a coalition [against terrorism] and nobody notices, did you really do it?"[22] In the pursuit of geopolitical integrity it is vital to give the forces the credit they truly merit while focusing on areas where more strength and investment are vital.

One of the key variables of a framework of geopolitical integrity relates to Canada's capacity to deploy in support of its own vital interests and core values. It is hard to look at geographically compact middle powers like Norway or the Netherlands and conclude that our needs are comparable. But a viable comparison can be made with Australia, a commonwealth partner with geography much larger than its population can occupy and with large parts of its geography uninhabitable. Like Canada it has low population density. And while not close to the United States, it comes under the American defence umbrella. Looking at Australia, its per-capita defence expenditure average, 1.6 times that of Canada's, is closer to the European average than is Canada's. And the proportion of the Australian population in their armed forces is also much higher than the proportion in Canada.

Australia's ratios suggest that we need to rethink total investment and complement. We should have a regular force level of 82,000 to 85,000 people (now under 59,000), and we should have an effective active reserve of 35,000 (roughly double present reserves). We need these numbers not simply to address present capacity needs, but so a robust strategy of geopolitical integrity is attainable. At present, the Canadian Forces can neither recruit to, nor maintain, approved levels. The regular forces and the reserve units need to be integrated more fully into society, and vice versa. And Canada's forces should be able to count on the best equipment and material the nation can afford.

Defence spending in Canada should be increased — substantially — closer to 2 percent of GDP, a commitment that could be phased in over several years. This level of defence spending could be secured and protected through a strong multipartisan call upon the government and minister of finance, a lesson we can take from one of our newest allies, the Poles. The Polish parliament, the Sejm, developed what is now a multiparty, nonpartisan position on defence spending. They have agreed, across all parties, that spending on defence will be no less than 1.95 percent of their gross domestic product.[23] This is a significant increase from present levels, which are already higher than the less than 1.05 percent of GDP Canada spends on its military.[24] The present minority parliament in Canada affords all parties a unique opportunity in this regard. And, even during majority government times, the Standing Committee on National Defence has often worked in an inspiringly nonpartisan fashion on the issue of investment. At current defence budget levels of C$12 billion, and a population of 32 million, Canada's per-capita cost for defence is approximately $380 and would over time (with this new benchmark) rise to approximately $760 in 2003-04 dollars. This should be acceptable and affordable in a society as wealthy as Canada's.

We need to increase defence spending so we can send our troops into the field with the best equipment we can afford, and so we have the capacity to act rapidly and flexibly in response to asymmetric threats. Our military should have enhanced capacity to deploy either independently or cooperatively with our allies, as the situation requires. We need initiatives on long-lift capacity, both air and sea, as well as increased recruitment, kit modernization and enhanced rapid deployment ability. We need initial investments in force-multiplying technology, material, kit and equipment that will allow staged incremental increases in complement to be real policy options and tactical choices for both our political leadership and our military command.

Interoperability — the capacity to join and combine operations with our allies, especially the United States — should be a key element of our strategy, not only because it is a way for us to obtain maximum leverage with more limited resources, but also because the doctrine reflects our national interest. And, we need to broaden the true meaning of interoperability to address present and foreseeable challenges. The traditional view of interoperability is narrowly focused on the technical and doctrine-based cooperation of naval, air, land and special forces of two or more countries. Since 2002, for example, Canadian troops have engaged "in combined combat operations with the United States in Afghanistan against al-Qaeda fighters."[25] American Colonel Frank Wiercinski, commenting on the soldiers of Princess Patricia's Canadian Light Infantry, captured the spirit of interoperability in his remarks: "We want to bring capability that we both can put together, and by doing that make ourselves stronger by using the best of each. And I think we've done that…I know [the Patricia's] are a great battalion."[26]

Building on such successes, we need to deepen our concept of interoperability so it embraces not only military components but also economic and humanitarian components of a more robust, capacity-building commitment in the world. Our scope of interoperability should incorporate stakeholders that specialize in the infrastructure of transition (physical, institutional and street level) to improve our capacity to help build healthy and stable postconflict societies.

We need a genuine and independent capacity to deploy special integrated forces in a limited way to protect or free Canadians in difficulty, or advance an explicit vital interest relative to terrorists or otherwise. After September 11, special forces — such as Canada's JTF2 — have taken on increasing importance in highly targeted operations. Accordingly, there is greater need for the air force to be able to provide support to such forces, including efficient transport and armed helicopter support.[27]

As another building block, Canada needs to develop strategic intelligence capacity. We need increased funding in both the Canadian Security Intelligence Service (CSIS) and military intelligence at home and abroad; we need an enhanced intelligence clearinghouse capacity in Ottawa that can better coordinate the actions of numerous actors. More effective intelligence would arise through a truly strategic and integrated security planning capacity, which the government of Canada's recent integrated *National Security Policy* clearly endorses. Geopolitical integrity requires Canada to recognize that strategic intelligence capacity is critical to identifying and tracking the trends — and, in particular, the mid-term domestic and international political, economic and cultural issues — that may affect Canada's interests. This is a serious weakness in Canada, and apparently one that is recognized only within the limited confines of the military intelligence community.

Ensuring an active ready reserve force of 35,000 people would also strengthen the connection between our military, security and civil society, and so help build our integrity. The naval, air and army reserves already significantly extend the reach and capacity of our regular forces. The reserves are also the key link between day-to-day society and the nation's military. They are the "face" of the CF in most communities across Canada, as the regular forces are concentrated in relatively remote areas (except for the navy). The reserves also represent an important community leadership resource in times of local need — natural disasters, major accidents, etc. In a country with a military as small as ours, real links with the general population are essential so the citizens can appreciate the important roles our military plays in Canada's way of life.

In pursuing these intelligence and reserve forces goals, it is vital that we do not adopt budgetary policies that force us into an either-or situation, a stark choice of either new equipment or an effective number of fairly paid regular forces.

The tasks for our armed forces should always be centred around a core combat capacity, with all the appropriate and necessary equipment, complement, technology, training and mobility exigencies implicit in that capacity. Without combat capacity, the ability to provide security for development work, peacekeeping and humanitarian intervention is diminished. Without security, other postconflict or prevention activities are simply not possible. So it is the duty of Canada or any country that aspires to intervene — through the UN or alliances — to combat aggression, ethnic cleansing or civil brutality, not only to have the military plans, capacity and kit necessary to take key enemy targets, but also to be able to contribute to the follow-on infrastructure to sustain postconflict transition for what is usually a largely innocent civilian population. Our interventions should be robust not for their own sake but to achieve a lasting difference.

Canada should therefore be able to assemble multidisciplinary task groups that can deploy with military leadership in support of such transitional processes. At a bare minimum, there must be coordination of critical components from the Canadian International Development Agency (CIDA), the RCMP, public and private infrastructure capacities, plus well-rooted nongovernmental organization (NGO) and local intelligence resources. Post-war reconstruction needs to draw on a range of regular soldier and reserve skill sets, especially in the critical security vacuum that opens immediately after the cessation of conflict.[28] This initial moment requires effective disarmament, demobilization and reintegration of ex-combatants, as well as police expertise to uphold the rule of law (the RCMP played this role in Kosovo). Second-tier experts and organizations, including NGOs, should then work to deepen capacity-building and the development of civil society, including reform in the justice sector, and to provide for elections and ongoing education, and for vocational training for ex-combatants.

In this multidisciplinary and capacity-building approach, however, we must not allow nonmilitary aspects to be so promoted at the expense of combat capacity, or we will lose the diplomatic coherence and international leverage that this more robust conception of interoperability seeks to develop.

Our domestic security is also a key building block of our geopolitical integrity, which is why this essay began with the potential for real terrorist attacks at home. Paradoxically, domestic security becomes easier if we are more outward-looking, engaged, proactive and collaborative — not only because of early detection of threats where they are first born, but also because of the greater likelihood of strong support and intelligence-sharing from our allies. And Canada has specific responsibilities on homeland security to our American neighbours not shared by any other ally.

Military leaders at various levels should be linked to their counterparts representing other components of public infrastructure and domestic security, including our public health network, transportation infrastructure and municipalities.[29] Public health networks should have contingency plans to cope with biological or chemical attacks, as well as potential attacks on our nuclear facilities. We should have terrorism prevention and response plans at federal, provincial and municipal levels, and these plans, in turn, must be reciprocally integrated with the national infrastructure and security capacity, and be drilled and rehearsed regularly.

At the same time, our lives as citizens of a free and open democratic society must continue. It is not only by Canadian constitutional convention that we understand emergencies as temporary events; we can plan for the worst while hoping for and assuming the best. But we need security experts and policy-makers to engage these issues so we can get on with our lives with greater confidence, knowing that we are doing everything we can to prevent a possible terrorist attack and to respond swiftly if one occurs. Canada's new *National Security Policy* (spring 2004) is an encouraging point of departure in this engagement.

The Department of National Defence (DND) has offered a vision for how it wants to evolve by the year 2020, a vision that connects well with some of the building blocks discussed here. The DND wants the capacity to "generate, employ, and sustain high-quality, combat-capable, interoperable and rapidly deployable task-tailored forces," and it wants to be recognized, "both at home and abroad, as an innovative, relevant, knowledge-based institution."[30] It seeks to be proactive and agile, and envisions "360-degree interconnectedness" with related organizations, allies and the general public.[31] It aims to draw on principles of interoperability and multinational collaboration "to conduct integrated combat and non-combat operations across the spectrum of conflict,"[32] a spectrum that includes threats as diverse as international terrorism, regional conflicts and failing states causing international instability, the spread of weapons, organized international crime, humanitarian disasters, epidemic disease and disruption of resource flows.

If we do not support the Canadian Forces in some version of this vision, we are abandoning them precisely at the point in our nation's history when we need them the most. Conversely, increased Canadian military and security engagement can also create welcome synergies. Increased attention to defending our borders, for instance, is in our immediate interest, but it is also a contribution to the American-led war on terror. Similarly, pulling our weight in international peace operations and humanitarian missions will generate greater goodwill among our allies, who will be much more likely to come to our assistance quickly — in Canada and abroad — when our citizens require it. For some Canadians, however, questions remain about the ways in which greater engagement will affect our sovereignty and our distinctive outlook on the world.

Often, among Canadian internationalists, scholars and diplomats, the question arises of how increased Canadian interoperability, particularly with the

United States, would diminish our sovereignty (a question that is also addressed elsewhere in this monograph).

Some critics advance the cynical view that, after September 11, "Canada is the most exposed of the Lilliputians [compelled to either] share responsibility for continental defence or be tossed aside as Gulliver [the United States] stirs."[33] In fact, however, we have collaborated on continental defence and other military initiatives for decades, and we have done so because it has been in our national interest to do so.

In the Second World War, the First Special Service Force (FSSF) was an elite Canadian-American force that fought with great success throughout Europe. But as a Canadian journalist of the day, Sholto Watt, commented, its significance went beyond its fighting achievements: "It was the first joint [multinational] force of its kind, drawn from two neighbouring democracies."[34]

Between Canada and the US, there are "over 80 treaty-level defence agreements, more than 250 memoranda of understanding between the two defence departments, and approximately 145 bilateral forums in which defence matters are discussed."[35] Key agreements between the two countries include the Permanent Joint Board on Defence, established in 1940, which provides for military and diplomatic contact at the highest levels; the Military Cooperation Committee, formed in 1946, which manages cooperation at the military planning level; and the North American Aerospace Defence Command (NORAD), established in 1958 to monitor and defend North American airspace — among numerous other significant agreements. From the Canadian side, the primacy of Canada-US defence cooperation was reaffirmed in the 1994 Defence White Paper, long before September 11, 2001.

In short, the Canada-US partnership is not an exigency occasioned simply by post 9-11 anxieties. Rather, it is a collaboration rooted in shared values and deepened through shared historical achievements. The Ogdensburg Agreement during the Second World War clearly committed both countries to preventing their territory from ever being used by anyone for an attack on the other.

There is no doubt, however, that the events of 9-11 served to heighten the importance of Canada's existing partnership with the United States. It was appropriate that Canada enhanced its security cooperation with the US following that day, as we did formally in a new agreement announced in December 2002.[36] It would be foolish for Canada to work *less* with the United States on continental security and other defence issues out of a naive assertion of sovereignty, since the two countries share so many of the same concerns and since the alternative would leave Canada more vulnerable.

Part of Canada's leverage in the world, part of what Prime Minister Martin recently called Canada's capacity to punch above its weight class, comes from its ability to influence the United States through our unique relationship with them. Tending to that relationship — pruning, weeding, seeding, watering and growing it — is never wrong: it neither implies lock-step agreement with America nor gives Canada the appearance of being free-riding neighbours. No matter who is in the White House or what the balance is in the US Congress, the relationship with America is always better when we have the capacity to pull our weight, deploy in alliance or joint operations, and maintain a diplomatic *amicus curiae* position wherever possible. We need to be frank with ourselves about what the loss of leverage with the US would cost us in terms of geopolitical leverage, broadly expressed.

So, far from diluting the responsibility of partnership with the United States, as some have suggested, Canada needs to further its commitment to mutual security on the continent, an original commitment that has been called "the Kingston Dispensation,"[37] after the exchange of assurances between President Franklin Roosevelt and Prime Minister Mackenzie King.

Building on a strong Canada-US relationship, we can also work to strengthen our security throughout the Americas. American defence expert Col. Joseph Nunez has recently called for a new "hemispheric security architecture" in the Americas, from Canada to Chile. He argues that we need "to transform our security architecture from an archaic collective defence framework to one that reinforces our incipient hemispheric community," and that this architecture "must be able to capitalize on the synergy resulting from shared values and interests." The architecture must be effective, organized, professional and "able to respond quickly to problems." To do so, it will require "standing multinational forces (SMFs) that can handle humanitarian assistance missions, peacekeeping, peace enforcement and other small-scale contingencies." SMFs in the Americas would represent a form of interoperability, multilateralism and global engagement all at the same time, and they could serve to reinforce the growing trade relationships throughout the Americas.[38]

Finally, it is helpful to think of the question of interoperability with the United States in terms of analogies. For some Canadians, the notion of Canada-US interoperability is ideologically charged, and questions of sovereignty understandably arise because we are partnering with the most powerful country in the world. But we must not lose sight of the fact that the partnership nevertheless makes sense for us, and if Canada were located elsewhere we would look for similar partnerships. Canada's relationship with the United States is sometimes compared with New Zealand's relationship with Australia. New Zealand is not afraid to say that its government "recognizes the need to work collaboratively with like-minded partners" and that "there is no strategic partnership closer than that with Australia."[39]

We also see New Zealand building on this primary relationship and expanding outward in a hemispheric pattern: it has formed "Five Power Defence Arrangements" (FPDA) with Australia, Malaysia, Singapore and (exceptionally) the United Kingdom to maintain security in the South Pacific and the Asia-Pacific region. New Zealand also endorses the concept of interoperability, both in bilateral operations with Australia and in multilateral operations through the FPDA.[40]

Europe is also moving more toward common security and interoperability. The European Union has been developing a common European Security and Defence Policy (ESDP) since at least 1993. Where NATO "retains sole responsibility for collective defence," the ESDP is meant to enable "the EU to respond to crises, coordinating its civilian and humanitarian assistance with, as necessary, a military element."[41] The ESDP recognizes that to respond to crises effectively, forces must be "readily and rapidly deployable to crisis areas."[42] The development of the ESDP and other collective institutions as an expression of common values in Europe is an extraordinary achievement of goodwill, reconciliation and trust when we recall that after the Second World War "the chasm that separated the Germans from the French, the British, the Danish and the Dutch was absolutely huge and conceivably unbridgeable."[43]

But work still remains to be done for the ESDP to be agile and effective. In early December 2003, the European Union approved its first common security doctrine, known as the Solana doctrine, but the approved doctrine was watered down from earlier drafts. Specifically, it backed away from endorsing "the possible use of first-strike military action,"[44] a dilution in the draft that occurred because of widespread European opposition to the US-led invasion of Iraq, and instead now speaks only of "preventive engagement." As experts point out, "there is a fundamental difference between 'pre-emption' and 'prevention,'"[45] and the watered-down version shows that "Britain, France and Germany have yet to overcome their fundamental differences over how Europe should respond to crises."[46]

Even so, through its policy of preventive engagement the EU has already succeeded in "persuading Iran to sign an additional protocol that would allow the International Atomic Energy Agency to carry out unannounced inspections of any nuclear site in the Islamic republic." But as one critic commented, what will happen if Iran does not comply with the protocol? Will Britain, France and Germany "be prepared to take tougher action? That's the acid test of the doctrine."[47] The European security doctrine is still limited, tentative and in its early stages, but it does signal a commitment toward interoperability and common military engagement in Europe as part of a broader trend among democratic nations.

Our military must have the flexibility to respond with a strategy that is global in scope. From 1980 to very recently, the Canadian Forces have been active to varying degrees in the Middle East, Africa (including Somalia, Rwanda and Eritrea), the Americas, Cambodia, East Timor, Kosovo and Afghanistan. Canada should be prepared to work independently or with the UN, the United States, NATO, others in the Americas, the ESDP or ad hoc coalitions of like-minded countries — and Canada-US interoperability does not diminish our capacity to do so.

Peace support operations constitute an important avenue for future global engagement, but we need to understand how peace support operations are distinct from the traditional Canadian idea of peacekeeping. The performance of our troops in countless difficult UN peacekeeping and peacemaking theatres has been outstanding. But these fields of action usually are less stable than the diplomats and politicians who committed the troops to begin with had hoped. The rules of engagement often create an unfair advantage for the sniper, the local militias, armed local combatants or the suicide bomber. Outdated peacekeeping approaches limit our capacity to intervene in unstable regions by requiring impartiality and restrictions on the use of force. The world does not always accommodate this level of idealism, as was shown recently by the bombing of the UN headquarters in Iraq, and it is clear that keeping and building the peace often requires more active forms of intervention.

A new model of peacekeeping has been developed bilaterally and through NATO, known as "peace support operations." This model recognizes that peacekeeping involves a spectrum of operations, including peace building, peacemaking, peace support and combat-based security enforcement. Some peacekeepers work in a tranquil environment, while others face very high security threats and might be forced to revert to fighting conditions. Afghanistan illustrates this view of peace operations as a spectrum: conditions in Kabul are characteristic of a classic peacekeeping operation, while in the northwest region of the country, the warlords are still supported by local society, and local fighting factions are kept apart by nothing but minefields; so international forces there would need to be more vigilant.[48]

How does Canada-US interoperability affect Canada's capacity to conduct peace operations abroad? First, as Desmond Morton argues, "If close links with the United States are incompatible with UN peacekeeping or peacemaking, the Canadian armed forces would have been equally unacceptable since the Ogdensburg Agreement of 1940."[49] Moreover, peace support operations now make use of rapidly advancing computer, digital and space technologies as these are being developed by the United States. Such advancements are most welcome from the standpoint of our soldiers: "Investments in sealift and airlift, better communications, precision-guided munitions and lighter but robust military vehicles are precisely what PSOs needed in the past and will need in the future." So Canada-US interoperability presents no contradiction to our traditional legacy of peace building; moreover, our access to American-led technology enhances our capacity to achieve results in such operations.

The final foundational building block of geopolitical integrity is the diplomatic element. As we have been reminded in Iraq, military victories by themselves do not guarantee peace and stability. International security requires attention to issues of governance, the rule of law, transparent management of resources, effective delivery of services and other forms of postconflict transition. And there is another critical dimension, often neglected, that our diplomatic channels ought to take into account: the intellectual dimension — the play of ideas and promises and dreams that influence the hearts and minds of potential peace builders and terrorists, allies and enemies.

In a recent perceptive essay, Captain Serge Bertrand of the Canadian Navy points out that in the Arab world, especially, "the adversary [of America and its allies] has successfully transformed his civil war into a regional struggle over Arab identity."[50] Osama bin Laden wins followers not primarily because there is poverty in the Arab world, though his converts may be poor, but rather by convincing his followers that there is a clash of civilizations, and that this clash ought to be accelerated and intensified. But it is crucial to keep in mind, as Captain Bertrand reminds us, that "the Islamic world contains a chorus of voices, both secular and religious, many of which are working toward a [radical] political and cultural revival," and that "the road to the hearts and minds of the *umma* [the universal Islamic community] is through the Arab body politic itself."[51]

Geopolitical attention to tangible security issues is not enough. As is clear from the struggle for Arab identity, the play of ideas and the forms of communities have the potential either to inflame conflict or to forge linkages, and this intangible dimension has very real consequences in the world. Canada's response to Turkey's application for membership in the European Union is an important example of how our diplomacy might take account of this intellectual dimension. As Thomas Friedman has argued recently, "If we want to help moderates win the war of ideas within the Muslim world, we must help strengthen Turkey as a model of democracy, modernism, moderation and Islam all working together."[52] Friedman points out that "Turkey has undertaken a huge number of reforms to get itself ready for EU membership," and if it is denied admission, extremists in the Muslim world will capitalize on the refusal and spin it rhetorically in the Arab media as evidence that there is no point in a Muslim country seeking admission to a Christian club. Canada, of course, is not an EU member, but we could well play a role in supporting (our NATO ally) Turkey's application through informal representations, as well as formal protocols such as the Canada-EU Joint

Statement on Defence and Security, which recognizes the common values and interests between the two entities and reaffirms "their commitment to reinforce their contributions to conflict prevention and crisis management."[53]

Through a robust foundation of geopolitical integrity in the world, built with interoperability and active partnerships worldwide as well as other flexible, yet principled tools of diplomacy, Canada has the potential to once again unite our professed commitments with our actions. If we choose to go it alone, with declining military resources, we will continue to lose diplomatic and political leverage on the world stage until we are regarded as an even more irrelevant minor player with an unnecessarily vulnerable domestic security infrastructure.

The Challenge Ahead

Clearly, if Canada is to pursue a reasonable measure of geopolitical integrity, there are priorities we must address. One of these is ensuring that a meaningful review of defence and foreign policy priorities reflects not only the core issues of geopolitical integrity for Canada but also the core organizational principles that can facilitate those principles being made real. Can Canada afford to have the Canadian International Development Agency (CIDA) separate from Foreign Affairs, and should Foreign Affairs be divided from International Trade? Would the prospect of genuine integrity in terms of Canada's presence abroad, not to mention its perceived competence and impact on the ground, not be enhanced if the same assistant deputy minister for a region or subregion had oversight for development, trade and foreign relations priorities? In terms of the organization of Canada's military, a plausible alternative to current structures might well be organization by global priority (the Americas, the Middle East and Africa, NATO territorial, civilian defence and support), with each arm given appropriate access to in-place mobile, lift, land, sea and air assets.

We need to be frank about the extent to which our domestic Canadian economic and social interests are put at risk if, say, a Caribbean country were to collapse into a drug-driven kleptocracy or a failed state. Our national interests are directly affected by events such as immigrants fleeing poverty, oppression and disorder; destabilization of our economic interests; or terrorists using failed states as recruiting centres and launching grounds for attacks.

We need to ask what we can learn from our Atlantic allies. The UK, for example, now pools resources between military, diplomatic and aid funds for explicit use to prevent the outbreak of hostilities in areas of the world where the UK sees it has critical interests. We need to ask what geopolitical integrity means in terms of joint forces training between allies, and to what extent should there be more focus on joint forces training and officer education with existing and potential allies in various key theatres?

National defence, unlike many contested federal programs, is clearly an uncontested federal responsibility, and the prime minister of the day goes a long way, in terms of his or her leadership, toward determining Canada's capacity for geopolitical integrity as a country. Where Prime Ministers Diefenbaker and Chrétien had little concept of how military, diplomatic and democratic capacity were jointly and severally essential to the advancement of our genuine national interests and shared values, Prime Ministers Pearson and Mulroney were more emotionally and intellectually engaged in the geopolitical context. We can only hope, therefore, that Mr. Martin and whoever may follow him fall more within the Pearson/Mulroney camp.

Initial signals from the newly sworn-in Prime Minister Martin were promising. His first official departmental visit was to the Department of National Defence, in contrast to Prime Minister Chrétien who (as prime minister) never once set foot in the building. Mr. Martin also recently announced new cabinet committees on global affairs; Canada-US relations; and security, public health and emergencies. He has proposed a national security standing committee in the House of Commons, whose members would be sworn in and then briefed on national security issues; he has appointed a new national security adviser in his own office; and he has called for a comprehensive review of Canada's place in the world, led by the minister of foreign affairs, to develop "an integrated and coherent international policy framework for diplomacy, defence, development and trade."

These were all welcome developments. In particular, the fact that the prime minister was planning to examine foreign affairs and defence in an integrated manner was a good sign. As Professor Douglas Bland argues (in another paper in this volume), the 1993-94 "defence inquiry was completed in detail almost before the foreign policy review began, an outcome many thought backwards in every sense of the word," and a process that "unintentionally illustrated the chronic schism between these essentially related policies and their parent departments."[54]

Yet the prime minister's spirited announcements and committee formations should not be cause for complacency, not least because no one can predict how a parliamentary committee will respond to the planned review of Canada's place in the world. In the light of the June 28, 2004, general election results, the decision by the new minister of defence, the Honourable Bill Graham, to forgo a full public review of defence policy, while lamentable, is a rare opportunity for the Standing Committee on National Defence (on which opposition members have a clear majority) to structure a stance of ongoing oversight on critical issues, such as the recent (Sikorsky) helicopter procurement and the proposed (and highly compelling) naval strategy relative to hybrid ships that can project Canadian presence (troops, supplies, light armoured vehicles and helicopters) to areas of vital or geopolitical Canadian interest. As well, they have an opportunity to review new strategy documents released by the army, navy and air force, and to make the ongoing public parliamentary discussion of Canada's military capacity in support of geopolitical integrity a permanent feature of national and parliamentary discourse. If internal reviews are referred to Senate or House committees, as Minister of Defence Graham has indicated, that would be a redeeming step toward parliamentary and public debate.

If Canada is to more competently embrace geopolitical integrity, then substantial linkages must now be forged between foreign, defence and aid priorities on the one hand, and our domestic economic and social agenda on the other. This is the only way we will see real and meaningful investment in Canada's geopolitical mission. What can minority party MPs do in a minority parliament? If they wish to send a signal on geopolitical integrity (as was suggested above), engaged parliamentarians from all parties could follow the nonpartisan example of Poland and take the lead in recommending that defence funding be raised until it is equal to 2 percent of GDP.

The world is a dangerous place, one in which Canada cannot afford to look the other way. We must be ready and able to defend our interests and what we believe in as a country, either through effective nonviolent measures or, when required, with force. We need to defend and affirm our values and be prepared to act on them. Canadians have shown leadership many times before on the world stage: from John Humphrey framing the UN declaration of human rights, to Prime Minister Pearson's peacekeeping initiatives, to Prime Minister Mulroney taking leadership against apartheid. We need to recall this proud legacy, and we need to reflect with some humility on General Dallaire's experience.

There was a core geopolitical integrity to Canada's role in World War II, in Korea, and in peacekeeping and pure combat engagements since. A lack of integrity diminishes Canada and Canadians — especially when there is a growing gap between Canada's moralizing and self-righteous postures and its capacity and will to do anything in their support.

This essay began with a discussion of Canadian uncertainties about our country's proper role in the world. If geopolitical integrity for Canada is grounded in solid affirmations of democracy, the rule of law, swift responses to the threat of suffering and the loss of human life, and decisive opposition to terror — if these are the principles animating our vigorous new commitment to engagement — we cannot be simultaneously championing the cause of insularity, retreat and passivity, or adhering too rigidly to Westphalia over Grotius. Medium-term history has yet to pronounce on whether Mr. Chrétien was right not to support our American and British allies in Iraq, but Canadians did clearly endorse that decision; and a short-term perspective on weapons of mass destruction and what may have been less than accurate intelligence on their existence, or the risk of their immediate deployment, would seem to sustain the Chrétien decision. But the goal of geopolitical integrity forces us to remember that at the time of the Canadian decision, it was not clear by any means that those weapons might not exist. So the challenge we face is one of contextualizing a decision made with the risk of those weapons having been perceived as very real, and the reality of mass murder against both Shiites and Kurds having been established facts. Why we acted in the face of less-established evidence in Kosovo with NATO allies and not in Iraq raises serious issues upon which Canadians need to reflect.

We are not Americans or Berliners, we are Canadians. If geopolitical integrity is to be in any way a real component of our domestic and international stance, being Canadian should imply a set of values we are prepared to promote and defend. And if their defence is required, Canada must be prepared with effective, deployable combat and humanitarian capacities of no small significance in the world at large, as well as the will to use them around the world.

Notes

1. *Britain and European Security* (Ottawa: British High Commission, 2003), p. 1.

2. Jeff Sallot, "Fraser Sees Big Holes in Canada's Security," *The Globe and Mail*, March 31, 2004.

3. Colin Freeze and Kim Lunman, "RCMP Lay Terrorist Charges," *The Globe and Mail*, March 31, 2004.

4. Michael Barkun, "Defending Against the Apocalypse: The Limits of Homeland Security," in *Governance and Public Security* (Campbell Public Affairs Institute, Maxwell School of Citizenship and Public Affairs, Syracuse University, 2002), p. 19.

5. Jack Pritchard, "What I Saw in North Korea," *New York Times*, January 21, 2004.

6. This paragraph draws on threat characterizations by Dr. Fitz-Gerald in a presentation to the IRPP board, December 2003.

7. Mark Landler, "Trafficking in Nuclear Arms Called Widespread," *International Herald Tribune*, January 24, 2004.

8. *The Military Balance, 2003-2004*, vol. 103, no. 1 (International Institute for Strategic Studies, University of Oxford Press, 2003).

9. Denis Stairs, "Canada in the 1990s: Speak Loudly and Carry a Bent Twig," *Policy Options*, Jan.-Feb. 2001, vol. 22, no. 01, 43-9, p. 49.

10. Denis Stairs, "Canada in the 1990s."

11. Ipsos-Reid poll, September 2002, cited by Jill Hawkins in "Summary of Canadians' Views on Security and Defence Issues, Various Polls, Fall 2002," IRPP Internal Research Document, p. 1.

12. Environics poll, June 2002; *ibid.*, p. 2.

13. Navigator poll, August 2002; *ibid.*, p. 2.

14. "The Peace of Westphalia (1648) recognized a society of independent, territorially sovereign states each having jural rights, which all states were bound to respect. In his seminal *De Jure Belli ac Pacis* (1625), Hugo Grotius had posited a right to intervene in the internal affairs of a sovereign if the crown were to maltreat its subjects." Christopher Doary, "*Pax Civitas Maxima*: Fighting the Good Fight," *Canadian Military Journal*, vol. 3, no. 2, summer 2002, pp. 25-31, p. 26.

15. "Former Rwandan Colonel Threatened to Kill Me: Dallaire," CBC online report, January 19, 2004.

16. Reginald Austin, "Democracy and Democratization," in *From Civil Strife to Civil Society*, (New York: United Nations University Press, 2003), 180-204, p. 182.

17. Reginald Austin, "Democracy and Democratization," p. 183.

18. Reginald Austin, "Democracy and Democratization," p. 196.

19. Dr. Douglas Bland, "Canada and Military Coalitions: Where, How and with Whom?" *Policy Matters*, IRPP, February 2002, vol. 3, no. 3, p. 22.

20. "Minister Says Canada's Soldiers Are 'Going to Take a Rest,'" CBC News online, January 16, 2004.

21. Chris Wattie, "After the War, Now What?" *National Post*, January 3, 2004, p. B3.

22. Chris Wattie, "After the War, Now What?"

23. Hugh Segal, "Defence in a One-Party Democracy: Polls Report a Consensus that Our Military Needs More — But Who'll Sway the PM?" *The Kingston Whig-Standard*, November 19, 2002, p. 5.

24. Major-general Doug Dempster, Director-General, Strategic Planning, Department of National Defence, "NATO 2003 Defence Expenditures as % of GDP," from NATO statistical memo, June 20, 2003, cited in *The CF in the Domestic and International Strategic Environment*, a presentation to the Conference of Defence Associations, November 22, 2003, p. 12.

25. Joseph R. Nunez, *A 21st Century Security Architecture for the Americas: Multilateral Cooperation, Liberal Peace, and Soft Power* (Carlisle, Pennsylvania: US Army War College, Strategic Studies Institute, 2002), p. 14.

26. Joseph R. Nunez, *A 21st Century Security Architecture for the Americas.*

27. This connection between the increased importance of special forces and air force support was made by the chief of air staff for New Zealand, as cited in "2000/01 Financial Review of the Ministry of Defence and the New Zealand Defence Force," online at Royal New Zealand Navy Web site: http://www.navy.mil.nz

28. Dr. Ann-Fitzgerald, IRPP presentation, 2003.

29. I am building on William T. Gormley Jr.'s observations in the American context. See "Reflections on Terrorism and Public Management," in *Governance and Public Security* (Campbell Public Affairs Institute, Maxwell School of Citizenship and Public Affairs, Syracuse University; Symposium on Governance and Public Security, January 18, 2002), p. 3.

30. Dempster, *The CF in the Domestic and International Strategic Environment*, p. 14.

31. Dempster, *The CF in the Domestic and International Strategic Environment*, p. 21.

32. Dempster, *The CF in the Domestic and International Strategic Environment*, p. 30.

33. Jim Travers, quoted in Nunez, p. 11.

34. Joseph R. Nunez, *A 21st Century Security Architecture for the Americas*, p. 31.

35. "Canada-United States Defence Relations," internal IRPP research paper.

36. Office of the Minister of National Defence, Office of the Minister of Foreign Affairs, and Department of Foreign Affairs and International Trade. Joint press release, "Canada and the U.S. Enhance Security Cooperation," December 9, 2002.

37. "In August 1938...President Franklin Roosevelt told an audience at Queen's University [in Kingston] that America 'would not stand idly by' were the physical security of Canada threatened...For his part, Prime Minister King, speaking a few days later (though not in Kingston), pledged that Canada would ensure that nothing it did would jeopardize the physical security of the United States." Dr. Michael Fortmann and Dr. David G. Haglund, "Canada and the Issue of Homeland Security: Does the 'Kingston Dispensation' Hold?" *Canadian Military Journal*, vol. 3, no. 1, spring 2002, pp. 17-22. More recently, George Haynal used the term "An Area of Mutual Confidence." See George Haynal, "Interdependence, Globalization, and North American Borders," in *Governance and Public Security* (Campbell Public Affairs Institute, Maxwell School of Citizenship and Public Affairs, Syracuse University; Symposium on Governance and Public Security, January 18, 2002), p. 54.

38. Joseph R. Nunez, *A 21st Century Security Architecture for the Americas: Multilateral Cooperation, Liberal Peace, and Soft Power* (Carlisle, PA: Strategic Studies Institute, US Army War College, August 2002), p. 29. Also available at http://www.irpp.org/miscpubs/archive/nunez.pdf

39. From "NZ Defence Policy Framework," online at Royal New Zealand Navy Web site: http://www.navy.mil.nz

40. From "NZ Defence Policy Framework."

41. *Britain and European Security* (Ottawa: British High Commission, 2003), p. 1.

42. *Britain and European Security* (Ottawa: British High Commission, 2003), p. 1.

43. Jean-Pierre Lehmann, "How to avert the risk of war with China," *Japan in the Global Era* column, *Japan Times*, Monday, June 17, 2002.

44. Judy Dempsey, "Words of War: Europe's First Security Doctrine Backs Away from American-style Pre-emptive Military Intervention," *Financial Times*, December 5, 2003, p. 11; "The possible use of first-strike military action was mentioned in the first draft, written principally by Robert Cooper, former foreign policy adviser to Tony Blair and now working with Javier Solana, EU high representative for common foreign and security policy."

45. George Schöpflin, Jean Monnet Professor of Politics, School of Slavonic and East European Studies, University College London. Quoted by Judy Dempsey, *ibid.*

46. Steven Everts of the Centre for European Reform, a London think-tank. Quoted by Judy Dempsey, *ibid.*

47. George Schöpflin, quoted by Judy Dempsey, *ibid.*

48. Dr. Ann Fitz-Gerald, IRPP presentation, 2003.

49. Desmond Morton, *Understanding Canadian Defence* (Toronto: Penguin Canada, 2003), p. 211.

50. Captain Serge Bertrand, "Fighting Islamist Terrorism: An Indirect Strategic Approach," *Canadian Military Journal*, vol. 3, no. 4, winter 2002-2003, p. 18.

51. Captain Serge Bertrand, "Fighting Islamist Terrorism," p. 19.

52. Thomas L. Friedman, "War of Ideas, Part 2," *New York Times*, January 11, 2004, op-ed, sec. 4, p. 15.

53. "Canada-EU Joint Statement on Defence and Security," *Britain and European Security* (Ottawa: British High Commission, 2003), p. 8.

54. Dr. Douglas Bland, "Canada and Military Coalitions," p. 14.

3

Force Structure or Forced Structure? The 1994 White Paper on Defence and the Canadian Forces in the 1990s

Sean M. Maloney

The *1994 White Paper on Defence,* Canada's only existing formal manifestation of defence policy, is a decade old. Designed as an interim policy during the first years of the post-Cold War world, the 1994 White Paper is generalized and tentative. We have entered a new era since September 11, 2001, the al-Qaeda War, leaving the stabilization campaigns of the 1990s behind. Canadian policy-makers, the punditocracy and professional military personnel are, however, still struggling with force structuring and equipment issues that were not adequately dealt with after the 1994 White Paper became policy. Indeed, current debates in the media over the adequacy or "burnout" of the army, the protective capabilities of the 20-year-old Iltis jeep, the infamous Sea King EH-101 helicopter debacle and the adoption of the Stryker gun-over-armour vehicle are all symptomatic of the problem.

The accession of Paul Martin as prime minister and the impending changes to Canada's national security policy apparatus will, some hope, be a catalyst for change in three areas: a new defence policy, a new foreign policy and a more rational approach to force structuring to implement these policies. Whether these three objectives are in fact achievable remains to be seen. That said, the 1994 White Paper continues to be formal policy, despite attempts to incrementally rationalize it with Defence Planning Guideline statements and other examples of structural "mid-course guidance." If a new defence policy is formulated, it will in all likelihood require some change in force structure. If a new defence policy is not formulated, then the 1994 White Paper and its fundamental premises that were established in the period 1992-93 will remain policy well into the twenty-first century. Equipment acquisition will continue to be a patchwork project and, without clear guidance, interservice rivalry will escalate and scarce resources will be wasted. Ad hoc responses to global crises that affect Canadian interests will continue, with increased prospects for unnecessary risk, catastrophic failure and national humiliation. It is critical that we discuss in frank terms the basis for future force structure. If we do not, future Canadian governments will be continuously saddled with geometrically expanding limitations on their ability to project Canadian military power.

Similarly, the recognition, by those commenting on national security issues, of the need for an interest-based national security policy and the means for its formulation has emerged in the past few years.[1] Such a process is dependent upon a clear understanding of past performance. This study is intended to contribute to that understanding.

A caveat: the projected force structure described in this study takes into account what was known about the world and about the existing Canadian Forces structure in the period 1993-94. It does not take into account technological, doctrinal or organizational developments that occurred in the late 1990s subsequent to the tabling of the 1994 White Paper or the change in the strategic environment after the events of September 11, 2001.

The 1994 White Paper on Defence

Background

The 1994 *White Paper on Defence* (hereafter "1994 White Paper") was conceived during a period when the world order was extremely unsettled. A variety of possible world models were proposed, mostly by American academic theorists — Samuel Huntington, Joseph Nye and Francis Fukayama were three, while the Israeli academic Martin Van Creveld took the lead in attempting to formulate controversial "post-Clauswitzean" military theory. Media commentators like Thomas Friedman explored the new economic terrain that appeared after the collapse of communism, while the Internet emerged to alter life as we had known it a decade before.[2] The withdrawal of Communist power and the retreat from the Cold War front lines left extremely violent regions in areas of Africa, Asia and even Europe. Theorists and commentators attempted to explain what they saw as bursts of ethnic conflict and determined that the nation-state was on its last legs. At the same time, utopian internationalists focused on the United Nations as the primary vehicle capable of driving through all of this mayhem and bringing stability. The apparent American acquiescence to UN legalities over the Gulf War of 1990-91 and its support of UN endeavours in Somalia were seen as indicators that the United Nations was perhaps the way of the future. The United Nations' authorship of *An Agenda for Peace* and conceptual development of a number of tools for peace (peacemaking, peace building, peace enforcement) also gave it a perceived authority in this area. In Canada, some saw this as a means to develop what they viewed as an independent policy by which to pursue their utopian aims and use the United Nations as a counterbalance to the American "hyperpower."[3]

The new defence policy debate in Canada was instantly politicized. In broad terms, the utopian internationalist faction asserted that the Canadian Forces should limit its capabilities to peacekeeping operations and eschew "warfighting." Warfighting, the argument went, was a Cold War construct that was expensive — a capability that should not be retained or improved upon, and that could be safely disposed of since "peacekeeping" and arms control could solve all of Canada's overseas security problems. Another faction, led by pragmatic realists, argued that the future was uncertain and that a spectrum of capabilities, including warfighting, had to be retained (Gray 1993).

The process that produced the 1994 White Paper was similar to that which influenced the 1964 White Paper. A broad array of public opinion was canvassed: hearings were held and specialists from both factions presented their best arguments. The newly formed Department of Foreign Affairs and International Trade (DFAIT) ensured that the views of a new faction of utopian internationalists were put forward. The Department of National Defence (DND) and DFAIT bureaucracies then set to work softening all the hard edges. During this process, detailed force structuring was eliminated from the draft documents. The result was the 1994 White Paper.

Premises

Cold War-era defence White Papers released in 1964, 1970 and 1987 (and the 1989 corollary of the 1987 White Paper) established that there were four areas through which Canadian defence activity was expressed:[4]

- continental defence
- NATO
- UN peacekeeping
- domestic operations

Essentially, each White Paper reprioritized these four items according to the circumstances of the day and the proclivities of the incumbent government. There was, over the years, a constant decline in funding for defence, starting in the mid-1960s. There was an implicit understanding within DND and the Canadian

Forces that the order of the four areas in defence White Papers was to be reflected in the allocation of resources. In some cases, lip service was paid and certain capabilities retained, but in all cases the decline in overall resources made the internal debate over how to implement the priorities more acute, particularly during the 1970s.

The 1964 White Paper emphasized the need to be able to intervene in areas peripheral to the North Atlantic Treaty Organization (NATO) area, while at the same time maintaining a strong deterrent force in North America and Europe. These objectives were generally achieved in the 1960s, though the forces' structure was curtailed in key areas due to poor policy coordination by the Pearson government. The 1970 White Paper shifted priority to domestic operations: drastic cuts were made to continental defence and overseas forces, and little emphasis was placed on peripheral intervention. The priorities established by the Trudeau government were not recognized by world events, and on several occasions Canada was forced to intervene overseas and deter attacks on the NATO area with a force structure that was suboptimal. The 1987 White Paper was delayed in its production (it was supposed to be released two or three years earlier), which put it out of phase with American and British defence policies established in the early 1980s. The 1987 White Paper was ultimately undermined by international events before it could be fully funded and implemented (Bland 1997).

Was the 1994 White Paper a departure from the Cold War-era periodic reordering of priorities? Its primary manifestations of Canadian defence activity were

- protection of Canada
- Canada-United States defence cooperation
- participation in international security

In "Cold War speak," this amounted to:

- domestic operations
- continental defence
- UN and NATO operations

Note that there is no explicit prioritization of these areas in the 1994 White Paper. However, it is easy to detect a defensive posture: we will deal first with problems at home, then with problems overseas. This is not surprising: the problems at the Mohawk reserves of Akwesasne, Kanesatake and Kahnawake throughout 1990[5] and the resurgence of nationalism in Quebec were dominant events of the day (as discussed in Armstrong 1995), and to ignore the possibility of "Bosnia in Canada" would have been sheer negligence on the part of the policy-makers and those influencing them. Indeed, the 1994 White Paper attempted to define, in general terms, the tentative direction and influences of the new world disorder on the future allocation of defence resources.

First, the document acknowledged that the Cold War was over and used an inordinate amount of space discussing the Organization for Security and Co-operation in Europe and various arms control efforts: this was obviously intended to mollify those who wanted to believe that Canadian arms control efforts during the Cold War had paid off. The same self-congratulatory tone was applied to "regional conflict resolution," with the discussion being dominated by anodyne statements like "The Middle East peace process has also yielded progress."

"International security concerns" were legion, according to the authors. "Global Pressures," including refugees, failed states, overpopulation, environmental degradation, resource depletion, the "resurgence of old hatreds" and the proliferation of weapons of mass destruction, were seen as the main drivers of instability. If these concerns were meant to take the place of an estimate of the Cold War threat, the 1994 White Paper was inadequate in this area. This is important when one is examining the document from a force-structuring perspective. In theory, some process by which interests are identified, threats to those interests evaluated and resources applied to deal with those threats should dominate. The 1994 White Paper provided the barest of bones and shied away from projecting into the future, its authors claiming that it was impossible to do so.

Indeed, after the cursory examination of the broadest possible range of "threats," several reasons were advanced to explain why Canadian money could not be spent on the armed forces. "Globalization" (never defined in the 1994 White Paper) was part of the problem, apparently, but the main constraint was the federal debt and the simultaneous need to maintain a good quality of life in Canada. Buried in this section of the document are the following lines:

> The Department and the Canadian Forces have absorbed past reductions in a variety of ways. Canadian defence commitments have been revised, personnel levels cut back, operations and maintenance budgets shrunk, defence infrastructure reduced, and capital programs cancelled or delayed...cuts will be deeper, and there will be more reductions, cancellations, and delays. In some areas, [DND and the CF] will do less [and] operate more efficiently to deliver elements of the policy outlined in this paper.[6]

This is all cunningly coded language: not all areas in which Canada used its military force could or would be funded. After identifying those areas, the authors offered no explicit prioritization. Flexibility was the main principle, which meant that there really could be no long-term policy derived from the 1994 White Paper, and that therefore no realistic and comprehensive force structuring could occur to adapt to this new era. In theory, there should have been another analysis and tabling of defence policy within five years, and some form of incremental annual evaluation should have been undertaken immediately in 1994. In time, Defence Planning Guidance documents were developed as a means of fulfilling this function, but they were no substitute for clearly expressed political guidance and leadership.[7]

We must confront the possibility that the government of the day did not value clearly expressed political guidance. In general, the political mindset seeks to retain as much flexibility as possible in order to stay in power. Making firm decisions on long-term policies in the face of possible opposition may be anathema to some personalities. A defence White Paper that says a variety of things to a variety of people is ideal — except to those who have to find a means to implement such "policies."

If a "force structurer" were to pick up the 1994 White Paper and scan it for information in carrying out his or her job, what would he or she find? By decoding the language, he or she might be able to derive the following information from the three "priority" sections.

Protection of Canada

Canada is big and rugged: we cannot afford to defend all of it. Resource protection is related to economic livelihood; therefore, military forces must play a role alongside other government departments in protecting poachers from grabbing our natural resources and damaging our economy. There are also internal threats to Canadian sovereignty amongst the Native peoples and in Quebec: Canada must be prepared to intervene with military force to ensure that the federal government is able to maintain control of the country in the face of any armed threat. Other government departments do not have the equipment, basing or training to handle the natural disasters that occur from time to time; therefore, the Canadian Forces will have to pick up the slack.

Canada-United States defence cooperation

Arms control is our primary means of ensuring a reduction in the nuclear threat to North America, not the allocation of resources to military functions like ballistic missile defence. At the same time, we have to find a way to demonstrate that we can participate in some fashion. If Canada does not collaborate with the United States on continental defence initiatives, our sovereignty will be lost, since the Americans will do what they want regardless. Interoperability with American forces is key and will provide Canada with benefits elsewhere. If we do not help defend North America we will be unable to rely on American space power, which we need for overseas operations.

Participation in international security

The United Nations will be Canada's predominant expression of international concern for increased global stability. The United Nations is not a perfect institution. We wish it could be as effective as NATO in military operations and will try to make it so. NATO is a necessary evil and should be harnessed to support the United Nations, so we will contribute to UN and NATO operations but only in some areas — deterrence, peacekeeping — and will not fight wars unless the United Nations gives its approval or unless there is an outright attack on a NATO member.

Our hypothetical force structurer would be able to conclude, with no ambiguity whatsoever, that the Canadian government of the day was unwilling to pay for a force structure to do all of this and had to make compromises without expressly stating so in order to avoid political embarrassment with our allies and the Canadian people. To be fair and balanced, however, we would do well to recognize that the post-Cold War economic order altered the very basis of Canadian financial security. Without a stable budgetary situation, the federal government would be hard pressed to spend any money at all. Indeed, there is an argument to be made that Canada's national debt was increasingly funded by potentially unreliable sources in the early 1990s. If enough of those creditors collapsed (perhaps due to the unreliable domestic situation vis-à-vis the Mohawks and/or Quebec), or otherwise generated turbulence in the economic structure that propped up Canadian governmental spending, Canada could conceivably have been plunged into an economic crisis rivalling the one it endured in the 1930s. Such a catastrophe would have made a globally deployable Canadian armed forces irrelevant.[8]

Force Structuring and the 1994 White Paper

Drawing on the generalized potential roles and missions for each section, a force structurer could easily have concluded that a number of core capabilities existed for each.

Protection of Canada

- demonstrate, on a regular basis, monitoring and controlling capabilities within Canada's territory, airspace and maritime areas of jurisdiction
- assist other government departments in protecting fisheries, counter-drug efforts and environmental protection

- conduct disaster relief
- conduct national search and rescue operations
- respond to terrorist incidents
- respond to aid to the civil power requests from the provinces

Naval and maritime air forces capable of operating on all three coasts (Pacific, Atlantic and Arctic) in all three dimensions (air, surface and subsurface), in addition to manned and unmanned sensor systems, would be necessary for the first two points. Search and rescue would require air transport capability over land and sea, while disaster relief and aid to the civil power missions would be best handled by land units, which have the command and control apparatus and manpower. Antiterrorism functions demand special operations forces.

Canada-United States defence cooperation

- maintain interoperability with American air, sea and land forces so that the Canadian Forces can contribute to the defence of North America
- cooperate in the surveillance and control of North American air space
- contribute to the acquisition and dissemination of ballistic missile warning

Forces for these endeavours would have included airborne, space-based, and surface-based sensors capable of monitoring missile and bomber attacks; antisubmarine forces to handle submarine-launched missile attacks; air forces capable of identifying and dispatching hostile aircraft; and air- or sea-portable land forces to confront any intruders attempting to establish a lodgment or otherwise challenge North American sovereignty.

Participation in international security

- assist DFAIT in conducting noncombatant evacuation operations
- participate in multilateral operations anywhere in the world under UN auspices or in the defence of a NATO member
- expand relationships with a variety of partners dealing with peacekeeping, confidence building and civil-military relations
- support the verification of arms control agreements

Unlike the other two sections, that on international security included a detailed list of force commitments. In essence, Canada committed to having the following forces ready for global operations: a joint task force headquarters, a naval task group of four combatants and operational support ship

(AOR) and maritime air support, three battle groups or a brigade group, a wing of fighter aircraft, and a squadron of transport aircraft. Portions of each of these were to be deployable within three weeks and full capability was to be available in three months. In addition, a standby battalion group for UN or NATO duty was to be immediately available. In "peacetime," Canada committed one ship to the Standing Naval Force Atlantic and another to Standing Naval Force Mediterranean, and aircrews were to serve with NATO's Airborne Warning and Control System (AWACS) component and NATO headquarters personnel.

The Canadian Forces 1990-94

What did the Canadian Forces look like in 1994?

The army

The Cold War-era brigade groups had dissimilar equipment and were slotted for different roles. The 4 Canadian Mechanized Brigade and 5e Brigade Mechanisé du Canada (BMC) formed 1st Canadian Division. In this role, these two brigades were equipped with Leopard C-1 main battle tanks, M-113A2 armoured personnel carriers and M-109 self-propelled guns, all fully tracked vehicles. Flyover sets of equipment were starting to be pre-positioned in West Germany for this role. In effect, each brigade group was to consist of a reconnaissance (recce) squadron, a tank regiment, a self-propelled (SP) gun regiment, two mechanized infantry battalions and a service battalion. The 1 Canadian Brigade Group, based in western Canada, was mostly equipped with light armoured vehicles of the Armoured Vehicle General Purpose (AVGP) series plus the M-109A2 SP gun. The AVGPs were wheeled training vehicles and were not designed to be used for combat in a mid- to high-intensity war. The 5 Brigade had the AVGP vehicles in Canada but was being retrained to handle mechanized equipment. A fourth brigade group, called the Special Service Force, or SSF, was based in Ontario. It consisted of a mixed brigade that included the Canadian Airborne Regiment (a parachute light infantry battalion), a mechanized infantry battalion with AVGP, an armoured regiment with AVGP and Lynx recce vehicles, a mixed artillery regiment with M-109A2 SP guns and a battery of air-droppable towed 105 mm guns, and a service battalion. The airborne artillery battery, plus an airborne recce troop and the Airborne Regiment, when combined, formed the Canadian Airborne Battle Group (Fox 1988; Last 1988; O'Connor 1988).

It should be noted that a series of equipment programs was initiated in the 1980s in response to the 1987 White Paper. The entire Canadian Army was suffering from "rust-out" since the Trudeau government had not provided enough money to replace aging equipment. The 1980s equipment projects were established for a new main battle tank, an armoured recce vehicle, a mechanized infantry combat vehicle and upgrades to self-propelled artillery. All of these projects were cancelled, delayed or "rationalized" after 1989. Some equipment projects, however, were implemented before 1989: these included the TOW Under Armour antitank vehicle (a tracked antitank missile vehicle based on an M-113A2 hull) and the ADATS (Air Defence Anti-Tank System) with the missile system mounted on a tracked M-113A2 hull (Last 1988).

In the early 1990s, the two brigades that made up 1st Canadian Division focused on mid- to high-intensity warfare for NATO operations in Europe. The SSF handled rapidly deployable commitments like NATO's Allied Command Europe (ACE) Mobile Force (Land) and the UN Standby Battalion role. No. 1 Canadian Brigade Group was in transition: some units were slated to become divisional troops for 1st Canadian Division. In 1990, Canada's overseas commitments involving combat troops outside of NATO Europe consisted of a battalion or regiment committed to UN duty in Cyprus every six months.

By 1994-95, the army had changed drastically. The 4 Canadian Mechanized Brigade Group was by this time disbanded, in the newly united Germany. The Canadian Airborne Battle Group was disbanded, and the SSF was converted to 2 Canadian Mechanized Brigade Group. The 1 CBG became 1 Canadian Mechanized Brigade Group, with 5e Groupement Brigade Mechanisé du Canada established from 5e BMC. The post-Cold War brigade groups were standardized as much as possible. The three armoured regiments each had a tank squadron plus two AVGP squadrons and a recce squadron. Some infantry companies were equipped with AVGP Grizzlies, while others had M-113A2 APCs. The artillery regiments all had tracked M-109A2 SP guns. There were two types of infantry battalion: mechanized and light. Each brigade had two mechanized infantry battalions and one light infantry battalion. One company of each light infantry battalion was a parachute-trained company.[9]

Maritime forces

By 1994, the transition from the Cold War-era *Mackenzie*-class antisubmarine warfare destroyers (DDEs) and *St. Laurent*-class helicopter-carrying destroyers (DDHs) to the *City*-class Canadian Patrol Frigates (FFHs) was well underway. By 1995, all 12 FFHs would be operational. The New Shipboard Helicopter program to replace the 30-year-old Sea Kings had been aborted by the Chrétien government in 1993. The four existing DDH-280-class destroyers were in the process of being altered to include superior anti-aircraft and command-and-control capabilities (the TRUMP upgrade). Three 30-year-old O-class hunter-killer submarines were still in service but were nearing the end of their service life. A new class of ship, the Maritime Coastal Defence Vessel, had been designed and was being built. Two mine countermeasures ships were in service. Eighteen CP-140 Aurora Maritime Patrol Aircraft (MPA) were also in service; three Arcturus MPAs, essentially emasculated Auroras, were also under procurement. By this time, the C2SF Tracker MPA fleet had been removed from the order of battle. Three operational support vessels, again each at least 30 years old, were still in service (numbers drawn from Jockel 1999, chap. 3).

Air forces

Air Command acquired 122 CF-18 Hornet fighter aircraft, or approximately six squadrons worth, in the early 1980s. Two of these squadrons were deployed to West Germany and four to North America as part of Canada's NORAD commitment. Two squadrons of CF-5 Freedom Fighter aircraft were committed to NATO's ACE Mobile Force (Air) during the Cold War, but this commitment ended sometime before 1994 and the CF-5s were put into storage or sold. No CF-18 squadrons replaced that commitment, and the Germany-based squadrons returned to Canada. By 1994-95, the number of operational CF-18 squadrons was four. In terms of airlift, five CC-150 Polaris jet airliners replaced the aging 707 fleet in 1992, while some 32 CC-130 Hercules, some of which were acquired in the early 1960s, remained in service. Twelve 30-year-old Labrador search and rescue helicopters continued on, accompanied by a dozen or so Buffalo turbo-prop tactical transports converted to search and rescue duty (Jockel 1999, chap. 4).

As for tactical aviation, the government chose to eliminate the Chinook heavy-lift helicopters, the Iroquois tactical transport squadrons and the Kiowa light recce helicopter units: all three machine types were acquired in the 1970s. In 1994-95, all were replaced with one type, the CH-146 Griffon, a helicopter based on a civilian executive transport. There were one Chinook squadron and five mixed Iroquois/Kiowa squadrons, plus a number of reserve squadrons equipped with Kiowas. In the new order, there were five Griffon squadrons supporting ground forces and four "Combat Support" squadrons that handled utility and local search-and-rescue tasks at Air Command bases in Canada.[10] Not all of these units were deployed in 1994-95: the transition to the new machine took time.

Reserve forces

By 1990, the intended expansion of the reserves generated by the 1987 White Paper was put on hold (Gervais 1992; Thompson 1992). At that time, militia (land force reserve) units were grouped into four militia areas that approximated (in numbers, not equipment or actual organization) five brigade groups. Naval reserve forces were geared to Cold War mobilization missions related to seaward defence and naval control of shipping, while the Air Reserve lost its

Kiowa recce helicopters and Air Reserve Augmentation Flight missions that supported NATO operations in Europe. By 1994, the future roles and missions of Canada's reserve forces were in question. Operationally, reservists were employed as individual augmentees to numerous domestic and global missions between 1990 and 1994:[11]

Domestic operations
- Operation FEATHER (Akwesasne) 1990
- Operation SALON (Kanesatake and Kahnawake) 1990

Global operations
- Operation CALUMET (MFO Sinai) entire period
- Operation MARQUIS (UNTAC Cambodia) 1991-93
- Operation SULTAN (ONUCA Nicaragua) 1989-91
- Operation MATCH (ONUSAL El Salvador) 1991-94
- Operation CAVALIER (UNPROFOR II Bosnia) 1992-95
- Operation HARMONY (UNPROFOR I Croatia) 1992-95
- Operation DELIVERANCE (UNITAF Somalia) 1993-94

Command and control

In 1994-95, the Canadian Forces command-and-control structure was in flux. Prior to the 1990s, Canadian Forces units were committed to alliance operations that had coalition headquarters for higher-command functions; these headquarters were single service rather than joint (two or more services). For example, 4 Canadian Mechanized Brigade was part of NATO's Central Army Group (CENTAG), which in turn reported to two other NATO headquarters responsible for operations in NATO's Central Region. At sea and in North America, naval and air forces had similar arrangements. It was only in the late 1980s that elements within the Canadian Forces leadership realized that some form of expeditionary capability was required. Since there was no coalition or alliance headquarters for noncombatant evacuation operations, for instance, Canada needed an operational-level joint headquarters to control operations involving more than one element. Operation BANDIT, a planned noncombatant evacuation operation conducted off Haiti in 1988; Operation VAGABOND, the United Nations Iran-Iraq Military Observer Group; and Operation MATADOR, the Canadian contingent for the United Nations Transition Assistance Group (UNTAG) in Namibia, demonstrated the need for joint operations control in expeditionary operations. In addition, on several occasions in the 1980s, National Defence headquarters was unable to control joint operations: it had become too reliant on coalition command structures overseas, structures to which Canadian officers were posted within a coalition framework but did not command. When National Defence Headquarters was confronted with two simultaneous crises in 1990, Operation SALON in Quebec (the Mohawk standoffs) and Operations FRICTION and SCIMITAR in the Persian Gulf, the need for a higher level of strategic joint command became evident. The resultant structure, J-Staff, was formed on an ad hoc basis in 1990 and then retained after both crises subsided (Maloney 2002-03). In 1994, J-Staff remained in existence and 1st Canadian Division functioned as an interim joint force headquarters while a more permanent solution was researched. There was no deployable joint force headquarters as such, though elements of the divisional headquarters supported by the 1st Canadian Division Headquarters and Signals Regiment and a Strategic Communications Regiment could act in this capacity on an interim basis (Garnett 2002-03).

Strategic sensor systems

During the Cold War, Canada deployed a number of strategic sensor systems geared to provide early warning in the event of a Soviet attack on North America. These included the Distant Early Warning (DEW) Line, Mid-Canada Line and Pine Tree Line radar stations; NORAD-committed Canada-US AWACS and their crews; and the Sound Surveillance System (SOSUS). A ring of signals intelligence (sigint) stations in British Columbia, the Canadian Arctic, Newfoundland and Bermuda were the first line of defence. On the whole, these sensor systems were not structured for sovereignty surveillance operations in support of other government departments. By 1994, the only systems left were the DEW Line, which was being replaced with the Northern Warning System, and flights by AWACS radar and control aircraft. Greater reliance on American space-based systems, which were developed in the 1970s and 1980s, was the norm. Some of the sigint activities were shut down, reducing the amount of high-grade intelligence that Canada brought to the Canada-United Kingdom-United States (CANUKUS) intelligence-sharing "table."

Basing

During the 1980s, the Canadian Forces had substantial infrastructure in North America and overseas to support its operations. There were pre-positioned depots in Norway to support the Canadian Air-Sea Transportable (CAST) Brigade commitment until this was withdrawn in 1987. In West Germany, there were two large air fields (Baden Soellingen and Lahr). These facilities supported Canadian global operations on the other side of the world: Baden and Lahr handled air traffic and other logistic matters for the Canadian UN contingents throughout the Middle East. Some of the forces deployed to fight in the 1990 Gulf War originated from units in West Germany and received substantial logistic support from Cold

War stocks. By 1994, all Canadian facilities in Europe were closed and infrastructure in Canada was significantly reduced. Army units were progressively centralized into "superbases," one for each brigade group. The Navy retained Halifax and Esquimalt, while Air Command retained its primary Cold War air bases with the exception of Summerside, Prince Edward Island.[12]

Special operations forces

The Canadian Forces took on the direct action counterterrorism role from the Royal Canadian Mounted Police (RCMP) Special Emergency Response Teams (SERT) organization in 1993. The new special operations unit, JTF-2, was in the process of being manned while the 1994 White Paper was being formulated. The role of special operations forces within Canadian national security policy was also under development at that time.

Canadian Forces' Overseas Commitments and Operations to 1994

While the 1994 White Paper was being developed, the Canadian Forces embarked on a series of new commitments whose pattern generally did not resemble that of its Cold War commitments. The collapse of the Soviet-supported Communist power projection infrastructure and the USSR's retreat from empire left power vacuums in areas that had been the site of front-line Cold War confrontation. These included Central America, southern Africa and Asia. Retreat from empire left tumultuous areas in the Caucasus, Central Asia and the former Yugoslavia. A number of UN observer missions were established to ease the transition in some of these areas. Canada usually provided a handful of military observers and sometimes transport capability. These missions were[13]

- Operation SULTAN (ONUCA, Nicaragua) 1989-92
- Operation PYTHON (MINURSO, West Sahara) 1991-94
- Operation MATCH (ONUSAL, El Salvador) 1992-95
- Operation CONSONANCE (ONUMOZ, Mozambique) 1993-94
- Operation MARQUIS, the UN Transitional Authority in Cambodia (UNTAC), was part of these post-Cold War disengagement missions but had a larger Canadian contribution in addition to military observers; it included a combat engineer troop and a logistics company. Canada's commitment to UNTAC lasted from 1992 to 1993.[14]

Humanitarian operations were not new to the Canadian Forces, which had conducted them since the 1950s. These missions usually employed one or two CC-130 transport aircraft or a 707 plus a ground crew support element. The Canadian Forces undertook the following humanitarian operations in the early 1990s:

- Montserrat and Nevis (volcano) 1991
- Ethiopia (famine) 1991
- Russia (famine) 1992-94
- Sudan (famine) 1993
- United States (hurricane) 1993

In the early 1990s, however, a new form of armed humanitarian intervention emerged. Conducted in nonpermissive environments, humanitarian aid efforts were accompanied by combat troops. The best known was Operation DELIVERANCE, the Canadian component of the American-led Unified Task Force (UNITAF), which conducted operations in Somalia in support of the UNOSOM I and UNOSOM II UN missions there. In this case, a joint Canadian expeditionary force based on the Canadian Airborne Regiment Battle Group was put together for operations in Somalia, the first time this had been done since Operation BANDIT off Haiti in 1988 (Oliviero 2001; Pupetz 1994). The other operations were LANCE, PASSAGE and SCOTCH (1994-95). These three missions were the Canadian components of the relief effort after the genocide in Rwanda. These operations included a medical unit, an engineer unit, a logistics unit and a protective element provided by the Airborne Regiment.[15]

Peace Observation, disengagement and humanitarian assistance were all relatively short-term commitments using mostly logistics and support units. The Canadian Forces, however, were committed to several longer-term missions in three primary theatres of operations in the early 1990s. Two of these theatres would remain active well into the late 1990s and even into the twenty-first century.

Canada's involvement in Haiti ran through the period when the 1994 White Paper was being formulated. A planned noncombatant evacuation operation, BANDIT was deployed in 1988. From 1990 to 1993, a Canadian-American attempt to pressure the Cedras regime into sharing power failed; soon after, a combined humanitarian effort involving Canadian and American troops (Operations CAULDRON and DIALOGUE) was peacefully rebuffed and a United Nations-sanctioned blockade imposed. Canada contributed substantial naval forces, including two destroyers and an operational support ship (Operation FORWARD ACTION) in 1993-94. After an American intervention in 1994-95 (Operation RESTORE DEMOCRACY), a UN stabilization force, the United Nations Mission in Haiti (UNMIH), was deployed. Canada contributed a 1,200-man infantry battalion group supported with helicopters. This force was annually reduced in strength between 1995 and 1997 (Operations PIVOT, STANDARD and STABLE) and then withdrawn (see Griffiths, Gimblett, and Haydon 1998).

Next to the Balkans, Canada's longest military commitment in the 1990s was in the Persian Gulf. Canada committed to the first two phases of a campaign designed to force the Hussein regime out of Kuwait in 1990-91 and comply with the demands of the UN Security Council. Operation FRICTION, the deployment of a naval task group (two destroyers and an AOR plus embarked helicopters) to join the American-led Multinational Interception Force (MIF) in 1990, was followed up with two squadrons of CF-18 fighters and an infantry company (Operation SCIMITAR). A field hospital and an infantry company to guard it were also deployed (Operation SCALPAL). After Iraqi forces were ejected from Kuwait and the Operation FRICTION, SCIMITAR and SCALPAL forces went home, a series of operations was implemented to stabilize the situation. One of these was the 1991 multinational Operation PROVIDE COMFORT in southern Turkey, which included a Canadian medical unit and transports (Operation ASSIST). A UN stabilization force, UNIKOM, was deployed to establish a buffer zone between Iraq and Kuwait; Canada contributed an engineer regiment on two rotations in 1991-92 and military observers from 1991 to 1999. Canada also contributed to coalition containment operations: Canadian Forces specialist personnel were part of the UNSCOM weapons of mass destruction search-and-destroy operation (Operation FORUM), and Canadian Forces AWACS personnel monitored the no-fly zones (Operations NORTHERN and SOUTHERN WATCH) (Maloney 2002a).

The largest series of Canadian Forces operations between 1991 and 1994 was in the Balkans. If the Persian Gulf was one primary theatre of Canadian operations, the Balkans was the other. In the period 1991-94, the Canadian Forces deployed military observers with the European Community Monitor Mission (Operation BOLSTER), led by the Conference on Security and Cooperation in Europe, to all of the former Yugoslavian republics. The inability of the European Community to develop a peacekeeping force in Croatia led to Operation HARMONY, the deployment of the United Nations Protection Force (UNPROFOR). In 1992, Canada deployed a mechanized infantry battalion and an armoured engineer regiment from its NATO forces based in Lahr, Germany. As the situation in Bosnia deteriorated, Canada offered a mechanized battalion group (replaced with a light armoured regiment) to serve with UNPROFOR II from late 1992 (Operation CAVALIER). A logistics battalion was also deployed to support both, while Canada also sent in a number of UN military observers for the duration. Rotations of all three units occurred every six months. As the humanitarian situation in Sarajevo worsened, Canada allocated one specially equipped CC-130 to the resupply effort (Operation AIR BRIDGE: from 1992 on). At sea, Canada joined NATO naval forces in enforcing sanctions: a Canadian destroyer or frigate rotated every six months from 1992 to 1995, while two CP-140 Aurora patrol aircraft joined that effort (Operation SHARP GUARD) (Maloney 1996, 1997a; Maloney and Llambias 2002).

Canadian Forces' Domestic Operations to 1994

Ongoing search and rescue and maritime surveillance operations continued throughout the period. Naval forces assisted with RCMP counter-drug operations on occasion (Operations HOGAN and JAGGY in 1993), while a hunter-killer submarine was involved in covert surveillance of illegal scallopers in 1993 in support of the Department of Fisheries and Oceans (Operation AMBUSCADE). The two largest domestic operations were both conducted in 1990. Operation FEATHER was the deployment of an infantry battalion to Akwesasne when armed violence broke out amongst Mohawk factions. Later that year, the entire 5e BMC was deployed south of Montreal to contain armed insurrectionists at Oka and Châteauguay (Operation SALON). A CP-140 Aurora MPA and a naval patrol vessel were involved in surveillance operations during Operation SALON. Compared to the overseas operational tempo, the domestic operational tempo was low from 1991 to 1994.[16]

Was There a Commitment-Capability Gap by 1993-94?

Overseas operations

The pattern of deployments and the types of forces deployed in a cumulative sense gives an indication of the high operational tempo in the period 1990 to 1994. From 1990 to 1992, a mechanized brigade group was deployed to Europe. During the 1990-91 Gulf War, the Canadian contribution consisted mostly of CF-18 squadrons and a naval task group. The period 1992 to 1994, however, saw the deployment, at any one time, of three infantry battalions, one light armoured regiment, two tactical helicopter units, an engineer regiment, a logistics battalion (plus), three destroyers, two patrol aircraft and two operational support ships. Other UN commitments, like the long-standing UN Disengagement Observer Force (UNDOF) commitment on the Golan Heights (1974 to the present), took the better part of a logistics battalion every six months. Compare this level of deployment with the available units in the downsizing Canadian Forces:

- 2/3 operational support ship
- 3/16 destroyers/frigates

- 2/18 maritime patrol aircraft
- 1/3 armoured regiments
- 3/7 infantry battalions[17]
- 2/3 service battalions
- 1.5/3 engineer regiments
- 1/3 field ambulances
- 2/3 tactical helicopter squadrons

The need for rest, refit and training usually means that at least three times the number of forces a nation wishes to deploy are needed to sustain commitments — assuming no casualties are being generated and personnel are not mentally breaking down with stress and post-traumatic stress problems. Indeed, perhaps four times the number are required for land forces outside of a mid- to high-intensity war scenario. Therefore, had the Canadian government wished to sustain the overseas operational tempo of the 1992 to 1994 period, the following forces would have been required:

- 12 infantry battalions
- 4 armoured regiments
- 4 service battalions
- 4 engineer regiments
- 4 field ambulances
- 4 tactical helicopter squadrons
- 16 destroyers/frigates
- 3 operational support ships
- 18 maritime patrol aircraft

For the land forces, these numbers correspond approximately to the Cold War force structure circa 1958 to 1970; that is, four complete, fully manned brigade groups (Maloney 1997b). Note that the maritime force structure had enough platforms and was capable of handling the strain of the operational tempo, though an additional operational support ship would have been an important addition. Fighter squadrons, obviously, were not deployed overseas during this period, nor were artillery regiments; they therefore do not figure in the calculation due to operation type. In other words, had the Canadian government wished to sustain the operational tempo it imposed on its armed forces, it would not have reduced the size of the armed forces from 1990 levels downwards, but instead would have increased them to pre-1970 levels or reduced the number of overseas commitments.

Neither was the thorny issue of reserve employment in operations addressed at this time. The cultural barriers between the reserve and regular components of the Canadian Forces remained as strong in the early 1990s as they had been in the preceding 30 years. Individual augmentees were occasionally employed, but subunits or units were not. The army, particularly, did not examine or develop creative means of burden sharing with its reserve component.

It must be pointed out that this analysis merely looks at numbers. It does not examine and cannot measure the human cost of such a high operational tempo. By failing to increase the size of the armed forces and at the same time imposing these commitments, the Canadian government essentially burned out large numbers of men and women. Having dumped service personnel into politically and geographically complex environments, usually with no clear justification in terms of the national interest or even a discernible UN mandate, this country is fortunate not to be confronted with a military disaster on a par with Hong Kong or Dieppe. Indeed, the long-term effects of these operations on Canadian Forces personnel constitute, in some cases, individual "Dieppes" or "Hong Kongs" of the mind.

There were other deficiencies. The principal one was the lack of strategic lift. In the 1950s and 1960s, Canada's forces could self-deploy using existing sealift and airlift (Maloney 2002c, chaps. 4, 8). By the 1970s this capability was drastically reduced as the lift vehicles aged and were not replaced. Canada's Cold War force structure was generally forward-based in West Germany. When Canadian UN missions were deployed after 1970, they were light interpositionary peacekeeping forces and the deployment was conducted in a permissive environment. The taking on of the CAST brigade group commitment to Norway in the 1970s should have generated the acquisition of Canadian sealift shipping. As a cost-saving measure, Norwegian roll on-roll off (RO-RO) ships were contracted for the only full-scale CAST deployment exercise, BRAVE LION, in 1986. No strategic airlift aircraft were acquired: Air Command was content to rely on CC-130 Hercules tactical transports pressed into a strategic role. The 707s that were on hand were incapable of operating from unimproved airstrips and incapable of transporting vehicles (Maloney 2002-03).

In the early 1990s, the strategic lift problem was not obvious. The peace observation missions were small in number and did not require anything beyond the Polaris airliners. UNPROFOR I and II relied initially on forces and equipment deployed from Lahr. When additional armoured vehicles were deployed, they were brought in using a chartered RO-RO ship that landed in a permissive environment. UNTAC's logistics vehicles could be flown in by CC-130, while the Griffon helicopters slated for UNMIH in Haiti could self-deploy through the United States and the infantry battalion could fly into a permissive environment. Forces assigned to operations in the Persian Gulf (UNIKOM, UNSCOM and the AWACS flight crews) could piggyback onto US Air Force airlifts.

The cracks started to show with Operation DELIVERANCE in Somalia. A Canadian AOR was used in addition to CC-130s, but the force's Grizzly, Cougar and Bison armoured vehicles had to be brought in by US Air Force C-5 Galaxy transports. What capability existed to evacuate the Canadian DELIVERANCE force if the situation became untenable? The United Nations Assistance Mission in Rwanda (UNAMIR) and its collapse should have been a wake-up call. What happens if the Canadian contingent of a UN mission in a remote, dangerous area requires immediate reinforcement to prevent a genocide, or requires extraction to prevent its demise? The Polaris had no defensive capability in the face of surface-to-air missiles or anti-aircraft artillery; it is essentially an airliner. A small number of CC-130 Hercules had defensive equipment (armour, flares, chaff), but they were assigned to Operation AIR BRIDGE in Sarajevo. In terms of heavy lift, the Polaris cannot carry vehicles, and the CC-130s can carry only one light armoured vehicle at a time and nothing else.

At sea, the three operational support vessels were multirole and could carry a limited number of troops and light vehicles. The operational support vessels, however, had other tasks such as refuelling the ships deployed off Haiti and the former Yugoslavia and serving with the MIF. If Canada had wanted to land or extract a force from a port in Africa, Europe, the Caribbean or the Middle East, it would have had to borrow or rent RO-RO shipping from another country. If an extraction were required under nonpermissive conditions, would the civilian crew of a rented ship (or aircraft) have been prepared to go in and pull Canadians out?

Continental defence and domestic operations

Measuring a commitment-capability gap for continental and domestic operations in the 1990-94 period is more difficult than an equivalent analysis for overseas operations. The types of domestic scenario envisioned by the 1994 White Paper did not demand unique forces: existing forces geared for overseas operations were adequate, as demonstrated during Operations FEATHER and SALON (on land) and Operation AMBUSCADE (at sea). There was always the possibility, however, that a situation requiring a brigade group to contain armed native unrest would occur while the bulk of the army was committed overseas. The army's response to the Front de Libération du Québec crisis in 1970 did not detract from the readiness of forward-deployed forces in West Germany. During Operation SALON, however, 5e BMC was committed to NATO in Europe as part of SACEUR's strategic reserve (Maloney 1997b, 442-5). Could the army have handled all of the 1993-94 commitments in the former Yugoslavia, Cambodia, Haiti, Rwanda and Somalia while simultaneously conducting a SALON-like operation in Canada? We are perhaps fortunate that such a situation did not present itself.

As for continental defence, the existing CF-18 force was more than capable of handling any NORAD commitments in addition to deploying an expeditionary fighter squadron or two (demonstrated during Operation SCIMITAR in 1990), while the CP-140 Aurora MPA squadrons were in a similar position when it came to maritime surveillance and control missions (again, demonstrated by the fact that CP-140s were deployed as part of Operation SHARP GUARD). Canada's naval forces were in a strong position to deal with continental defence threats while retaining forward-deployed forces for sanctions-enforcement missions off Haiti, in the Adriatic and in the Persian Gulf/Red Sea maritime interception operation (MIO). The main deficiency was the over-tasked operational support vessels and perhaps the aging *Oberon*-class submarines. The Sea King issue was, naturally, a problem. These aircraft should have been replaced with the EH-101 as scheduled, but political decisions by the Chrétien government prevented the acquisition of a New Shipboard Aircraft and needlessly put the lives of the crews at risk. The same applied to the ancient Labrador search-and-rescue helicopters and their crews.

An Alternative Force Structure

What would the force structure of the Canadian forces have looked like after 1994 had the government chosen to fully implement the premises of the 1994 White Paper? This is a trick question. Given that the document is tentative, vague and unspecific in terms of expressing Canadian interests and the threats to those interests, how could it possibly have been used as policy guidance for detailed force structuring?[18] One way would be to delve beneath the coded language and attempt to discern the bureaucracy's subtextual meaning and then base a force structure on that analysis; another would be to decide on a force structure and then argue that the policy guidance was vague enough to accommodate anything.

Given what knowledgeable analysts knew and had written in 1993-94 about the changed global situation (as opposed to what we now know happened between 1995 and 2001), a better contextual section could have been developed in the 1994 White Paper. Clearer identification of Canadian interests in this context was in fact possible.[19] Throwing one's hands in the air and declaring that we cannot predict the future was not really the best move: Royal Dutch Shell analytical departments perform similar exercises all the time with some degree of accuracy.[20] Consequently, developing a force structure to protect Canadian interests within the budgetary constraints of Canada's debt situation was in fact possible, but would have required clear strategic objectives to shape a coherent national security vision. Forcing conservative institutions like the bureaucracies of National Defence Headquarters and the three armed services to accept radical change in such a short time, however, would have been next to impossible. Incremental change would have been the order of the day.

Canadian interests in the early 1990s

In the most essential terms, Canadian interests have always been the physical and economic protection of Canada, its institutions and its people. The Canadian strategic tradition is based on Forward Security: stop security problems overseas from getting out of control before they have an adverse effect at home in North America. This has not fundamentally changed since the 1800s; nor will it change as long as Canada continues to exist and her people continue to occupy the territory we define as Canada. The 1994 White Paper is correct in stating that

> As a nation that throughout its history has done much within the context of international alliances to defend freedom and democracy, Canada continues to have a vital interest in doing its part to ensure global security, especially since Canada's economic future depends on its ability to trade freely with other nations.[21]

The 1994 White Paper, however, strays in its examination of threats. It labels these "concerns," a word that is not strong enough. Indeed, use of the terms "global pressures," "refugees," "failed states," "resurgence of old hatreds" and "proliferation" to replace specific threats was too broad and vague even in 1993-94. For example, there is no discussion of specific geography: the conflict areas threatening Canadian interests were in specific areas, and there were specific causes. By casting the concerns in the broadest of terms, the White Paper suggests that Canada had the ability to intervene anywhere globally at any time for a whole range of reasons. It could have chosen language that pointed to specific regions with caveats that Canada should keep an eye on other areas and be prepared for military involvement in these areas if necessary.

As we have seen, the main issues confronting the world system in the early 1990s related primarily to the withdrawal of the Soviet Communist empire and its affiliates in what used to be called the Third World. The Soviet withdrawal from empire was, in many ways, similar to the British, Belgian and French decolonization exercises of the 1950s and 1960s. Unlike the British, however, the Soviet leadership did not prepare many of its former colonies or client states for independence; the violence in the Caucasus is the best example. In the case of Yugoslavia, the Soviet decolonization drama was played out at the micro level: the Communist empire formally led by Tito collapsed into its constituent elements, which had been colonized in the 1920s. As for Cambodia, the Communist effort there could no longer sustain itself in the face of new realities.

Cold War proxy-war front lines were in the process of disengagement as early as 1989: examples include South Africa's withdrawal from Namibia and the associated withdrawal of support for UNITA (Union Nacional Por La Independence Totale Do Angola) forces in Angola; the acceptance of UN mediation in Nicaragua, El Salvador and Guatemala in Central America;[22] the reunification of Germany and the progressive withdrawal of the Group of Soviet Forces Germany; and the withdrawal of Soviet forces from Afghanistan.

Withdrawal from power produces, no matter how temporarily, a power vacuum. In many cases, the vacuum was filled by individual leaders and groups using ethnic animosity as a vehicle to achieve and retain power. Unfortunately, most analysts of the day ascribed violence in the Balkans, the Caucasus and elsewhere to "ethnic conflict" that had existed for decades/centuries/millennia and was unleashed when the Cold War system was suddenly removed. Instead of identifying these individuals and groups as totalitarian and antidemocratic, the authors of the 1994 White Paper concentrated on trying to understand them, identifying and addressing the "root causes." The key to this cipher should have been Saddam Hussein and his behaviour after the 1990-91 Gulf War. Here was a national socialist totalitarian repressing his population and generating instability in an economic area vital to the Western, now global, economic system.

The authors of the 1994 White Paper asserted that the four horsemen of the apocalypse were riding around the globe as if famine, pestilence, environmental damage and population growth were new to history. This litany can be found in the thinking of academics of the 1960s and 1970s. Why were these considered new problems? Why, exactly, did a Canadian defence White Paper need to address them?

It should have been obvious to analysts that there were two major threats to Canadian interests in the early 1990s. These threats should have been clearly stated and prioritized over all other global problems.

Eastern Europe and the former Yugoslavia

There is an assumption that the collapse of the Berlin Wall in 1989 meant that Communism was dead in 1989. But Russia, for example, did not withdraw her armed forces from East Germany until 1993. The end of the Soviet Union in 1991 generated a dangerous situation, particularly with regard to the security of its weapons of mass destruction, in addition to its arsenal of high-technology conventional weapons. The coup attempt in August 1991 indicated that, in the event of miscalculation, the situation in Eastern Europe could have deteriorated into a war (Cockburn and Cockburn 1997; Pry 1999). The collapse of Yugoslavia and the multiple wars that emerged there could have caused instability within adjacent former members of the Warsaw Pact. It is conceivable that if the situation in the former Yugoslavia was not contained, it could have spread into the newly freed captive nations. The emergence of a totalitarian leader in Belgrade, Slobodan Milosevic, and the continued interference by his regime in the affairs of adjacent states was recognized early on (see Ash 1999). The complicating factor was the existence of other totalitarians in the region, such as Franjo Tudjman in Croatia: everybody is bad, so who do the "good guys" back? Keep in mind the proximity of Yugoslavia to NATO's European members: Croatia is an eight-hour drive from Lahr in Germany. Having several wars going on in the NATO area with attendant spin-off problems like refugee flow into Italy and Greece and consequent economic and social pressures, in addition to the dramatic increase in organized crime that accompanies instability, was, in a word, destabilizing. This state of affairs was deliberately generated by men who made choices: it was not an act of fate (Glenny 1992; Silber and Little 1995). The 1994 White Paper would have us believe that only the spin-off issues should be addressed, not the source of the problem.

Middle East and Persian Gulf

The first post-Cold War crisis in which Canada was involved was the aggressive behaviour of the Hussein regime in Iraq. The brutal occupation of Kuwait, the pursuit of weapons of mass destruction programs and the invasion of Saudi Arabia at Khafji generated threats to Canadian interests on several levels. First, the oil-bearing region of the Persian Gulf supplied the economies of Japan and Europe, which were Canada's two largest trading areas outside of the United States. Damage to the oil delivery process in the Persian Gulf would send shock waves throughout the fragile post-Cold War global economy, which would ultimately affect the Canadian economy. Second, the acquisition of weapons of mass destruction by the Hussein regime would contribute to the increased vulnerability of the oil delivery process (as discussed in Pollack 2002); when delivery systems with enough range were acquired, the Hussein regime would have the ability to target Israel with nuclear weapons. The small territory of Israel would be unable to absorb even one nuclear weapon. Consequently, the probability was high that Israel would strike first if the Hussein regime appeared capable of building and deploying nuclear weapons (Cohen 1998). Canada has no interest in a regional nuclear war in the Middle East and, despite tensions between successive Canadian governments and Israel, it is not in Canada's interest to see Israel destroyed by a Baathist totalitarian state with many political views not dissimilar to those of Germany's national socialists in the 1930s and 1940s.

In both crises, totalitarian entities emerged immediately after the end of the Cold War to threaten Canadian interests in two specific geographical regions. Canada's strategic tradition lies in countering totalitarian states that threaten

Canadian interests: Nazi Germany, Fascist Italy, Imperial Japan, the Soviet Union, Communist China; there is little or no difference between these states and the Hussein and Milosevic regimes in terms of repressive methods and aggressive behaviour. Some may argue that the conflicts of the 1990s were of lesser magnitude than previous ones and that other nations should take the lead and let Canada decide whether to commit.[23] Such a path of least resistance is, or should be, anathema to independent-thinking Canadians and their representatives, as citizens of a sovereign state with its own values and interests. Contributors to the 1994 White Paper had three or four years to figure out that there were totalitarian threats to Canadian interests and values, yet they failed to do so — or at least they buried such analysis under so much coded language that it was unrecognizable.

Haiti was a special case. The Cedras regime was antidemocratic and odious, if not as efficiently totalitarian as, say, the Hussein regime or as outwardly aggressive as the Milosevic regime (Dupuy 1997; Trouillot 1990). The Haitian refugee flow affected American interests, which in turn could have been interpreted as affecting Canadian interests.

What else was happening around the globe at this time, and how did it affect Canadian interests? The turmoil in the former Soviet Union produced a host of opportunities as well as threats. The 1994 White Paper mentions "proliferation" as a "concern." It was in Canada's best interests that nuclear, biological and chemical weapons not be distributed from facilities no longer under central control. The potential threats to the new order in Eastern Europe and in the Persian Gulf by illicit, or even overt, transfers of technology were also threats to Canadian interests. As for opportunities, Canada has substantial populations of Polish and Ukrainian descent. Closer commercial and military ties would have facilitated information sharing on proliferation issues as well as analysis of potential problem areas in the Soviet decolonization process.

There is, of course, the matter of Africa. Numerous Canadian policy-makers have a soft spot for Africa, possibly engendered by their relief and assistance work on that continent when they were young and idealistic, while others may be motivated by the pressures brought to bear by interest groups wielding the "white racism" guilt-trip stick. In a cold, hard realist view, the combination of threats to Canadian interests that existed in the Balkans and in the Persian Gulf did not exist in Africa. What existed in Africa was the failure of national economies based on obsolete and inefficient socialistic models and absorbed by African leaders in the 1960s during decolonization, coupled with corruption at all levels of government, which itself was a holdover from corrupt French and Belgian colonial administrations of days gone by (Ayittey 1998).[24] This situation was exacerbated by environmentally unsuitable Western "democractization" and governance templates. In many cases, Rwanda particularly, the former colonial players remained heavily involved for their own obscure, possibly nefarious, purposes (Dallaire 2003). Was it in Canada's best interests to intervene and stop them? Was it in Canada's best interests to sweep up the mess caused by these powers? Was it in Canada's best interests to "fix" Africa?

Priorities

Given the 1994 White Paper's premise that there is not enough money for the armed forces to do everything we want it to do, and given the realities of the threats to Canadian interests if a proper analysis had been presented, some prioritization should have been done. This might have propelled the force structure best suited to implementing Canadian national security policy. Canada needed and continues to need both a force structure that can handle domestic and continental operations and a force structure to conduct expeditionary operations abroad. In some cases, forces for North American missions can also be used for global expeditionary missions and vice versa. However, Northern Warning System radars cannot deploy to Bosnia and armoured regiments should not police the Canadian Arctic. Therefore, the forces necessary for the surveillance and control of Canada constitute one priority "stream," while the deployable forces to carry out expeditionary and North American missions constitute another. Canada cannot have forces for one stream or the other. It must have forces for both. On the expeditionary side, prioritization should be dictated by the magnitude of threats to Canadian interests where those threats are manifest, and the types of forces needed to operate effectively in those geographical environments.

Based on the above analysis, a geographical prioritization of threats to Canadian interests outside of North America would have looked like this in 1994:

1. Eastern Europe
2. Persian Gulf/Middle East
3. Western Hemisphere
4. Asia
5. Africa

What operations types should Canada have been prepared to conduct in these regions, keeping in mind Canadian national security priorities and budgetary constraints? Again, this is a bit of a trick question and must be heavily qualified. The types of military operations conducted during the Cold War did not necessarily apply to the post-Cold War environment. During the early 1990s, the terminology to describe what Canada did, was supposed to be doing or had done was in flux: it should have been clarified in the 1994 White Paper. For example, let us examine the term "peacekeeping." Peacekeeping during the Cold War was not a static concept. What we refer to as "peace observation" and peacekeeping today was called peacekeeping in the 1950s. By the 1960s, the use of military forces to impose order in a nonlinear situation involving entities that are not necessarily nations (what we refer to as "stabilization" today) was also called peacekeeping (Maloney 2002c, chap. 1).

If one looks at the 1987 White Paper, one can extract several Cold War-era operations types:

- maintenance of strategic deterrence
- conventional defence in the NATO Area, which includes North America
- protection of Canadian sovereignty
- peaceful settlement of international disputes ("peacekeeping")
- arms control
- aid to the civil power (the use of armed force in Canada)
- foreign disaster and humanitarian relief
- assistance to civilian authorities (the use of military forces in Canada without resort to arms)

In 1992, tentative moves had been made to establish or differentiate new operations types for the new era. A provisional differentiation in DND consisted of (Maloney 2002a):

Cyprus. This model represents a situation in which a force is put in place to monitor an existing agreement.
Cambodia. This model is taken to represent a situation in which a major civic action-type program is required to create a semblance of order.
Yugoslavia/Somalia. This model represents a situation in which a force intervenes to facilitate the delivery of humanitarian assistance and/or impose a ceasefire or settlement. An agreement may or may not exist, and the intervention may require a degree of force or coercion.
Persian Gulf. At the high end of the spectrum is the Gulf War model in which a force is put in place to impose a settlement.

The terminology for these operations types would not emerge until the late 1990s. Even in the mid-1990s, there was a tendency to call them all "peacekeeping" but to differentiate between Chapter 6 and Chapter 7 missions (based on the UN Charter), where Chapter 6 was conducted with acquiescence of the belligerents and Chapter 7 was conducted in a less permissive environment.

Note that Canadian defence policy, as established during the Cold War and in the 1994 White Paper, does not formally recognize several other means by which Canada has used her military forces to attain national objectives. This is a result of the dogged determination of policy-makers to make all Canadian military activity fit a legalistic straitjacket provided by the United Nations. For example, Canada conducted what is traditionally called "gunboat diplomacy" on numerous occasions from the 1930s to well into the 1990s. Canada has conducted several noncombatant evacuation operations in permissive and nonpermissive environments since the 1960s. It also conducted global strategic intelligence gathering operations during the Cold War and in the 1990s.[25]

Perhaps it would not be wise to present the full spectrum of Canadian military operations in a defence policy White Paper. However, if one is to derive a force structure capable of implementing policy, the details are important. It is critical for us to understand that operations conducted by UNPROFOR I and II or UNITAF are not "peacekeeping," but something else.

The analysis conducted by the Operational Research Division in 1998, based in part on the language of the 1994 White Paper (Maloney 2002b), could easily have been conducted five years earlier. It is likely that the operations types developed after 1998, or variants thereof, could have been established in 1993-94. In any case, these types of operations would have been applicable to any force structure exercise:

- search and rescue
- domestic disaster relief
- international humanitarian assistance (permissive environment)
- surveillance and control of Canadian territory
- evacuation of Canadians overseas
- aid to the civil power
- national sovereignty enforcement
- defence of Canada-US territory
- collective defence in a NATO context
- peace observation and arms control verification
- interpositional peacekeeping
- stabilization operations

What does not appear on the list is outright intervention with combat forces. Note that Canada has intervened with combat forces three times since 1990: the Persian Gulf (1990-91), Kosovo (1999) and Afghanistan (2001). At the time, however, those formulating the 1994 White Paper refused to conceive of such operations even though the mission in Somalia with UNITAF was an intervention, for example. Why the authors did not consider military intervention as a possibility for Canada's forces is unfathomable but probably relates to the misguided belief that there was no legal basis for such action in the new world order, in which the United Nations supposedly had primacy.

As with the Cold War defence policies, a force structure to carry out these operations would have certain basic characteristics. First, Canadian Forces would have to operate across the spectrum of conflict. Concentrating Canadian Forces in one area to the exclusion of all others is pointless; it does not serve Canadian interests and does not protect Canadian sovereignty. Second, we must recognize that some force types can be used for several roles. We must, however, also recognize that too much focus on multirole platforms can in fact be uneconomical when there are small numbers of an available capability; the operational support ships are one example of this problem. Third, forces for all three dimensions (air, sea, land) are necessary. No one dimension can serve all of Canada's national security needs.

Note that Canadian strategic tradition includes the concepts of forward security, alliance and coalition warfare; operational influence; and saliency. The Canadian Forces employed will never be as sizeable as, say, those provided by the Americans. In general terms, Canada will operate as part of an alliance or coalition and will not undertake independent operations outside the Western Hemisphere. There are exceptions: non-combatant evacuation operations and humanitarian aid delivery, for example. To ensure that Canadian Forces are not misused by larger alliance or coalition members, Canada usually provides salient and effective forces to ensure access to alliance or coalition command structures (Operational Influence). In theory, this translates into some form of strategic influence in the coalition if handled effectively by diplomatic personnel. It is not enough to provide symbolic contributions to alliance or coalition efforts (Maloney 2002d).

We should recognize that the concepts of war and peace in the twentieth-century sense were in many ways obsolete by the early 1990s. "War" is usually a declared legal state between nation-states. In a globalized world in which a plethora of influencers that are not recognized nation-states are operating and can affect Canadian interests with varying levels of violence, we must recognize that Canada is essentially "at war" all the time. This type of conflict does not resemble the First or Second World War in intensity and may be akin to the Cold War in some ways, but it is a state of continual tension where the threat of violence is real.

Coming back to our regional priority list, what operations types would have fitted into the regions based on the types of problem affecting Canadian interests?

Eastern Europe:

- collective defence in a NATO context
- international humanitarian assistance

- peace observation and arms control verification
- stabilization operations
- intervention

Persian Gulf/Middle East:
- evacuation of Canadians overseas
- stabilization operations
- peace observation and arms control verification
- intervention

Western Hemisphere:
- stabilization operations
- defence of Canada-US territory
- international humanitarian assistance
- evacuation of Canadians overseas
- peace observation and arms control verification
- intervention

Asia:
- international humanitarian assistance (permissive environment)
- evacuation of Canadians overseas
- peace observation and arms control verification

Africa:
- international humanitarian assistance (permissive environment)
- evacuation of Canadians overseas
- peace observation and arms control verification

What if the operations types for the priority geographical areas (Eastern Europe, Persian Gulf/Middle East, Western Hemisphere) are combined? These missions cover a broad band on the spectrum of conflict, from lower to higher intensity, and involve all three dimensions:

- international humanitarian assistance (permissive environment)
- peace observation and arms control verification
- evacuation of Canadians overseas
- stabilization operations
- intervention
- collective defence in a NATO context
- defence of Canada-US territory

There are essentially two groupings of missions: the first three on the list are not manpower intensive and are generally temporary or of short duration. These are short-notice contingency operations that can be mounted in all five geographical regions without the same level of preparation and duration as in the second group.

International humanitarian assistance (permissive environment). The forces necessary for the delivery of humanitarian aid in a permissive environment include an Air Logistics Control Element and between one and three transport aircraft like the CC-130 Hercules. Canada has conducted these operations types since the 1960s, and this level of effort has remained constant. In any one year, there has been at least one such operation. In terms of duration, the humanitarian aid delivery element has deployed for weeks, as opposed to months or years. Consequently, the effort is usually temporary and uses nationally controlled forces diverted from other missions (Maloney 2002b).

Peace observation and arms control verification. The average number of Canadian military observers deployed on a peace observation or arms control verification mission is 15. The observers are usually deployed in a permissive environment. At any time, Canada has had between one and five arms control and peace observation missions ongoing, with personnel "subcontracted" and under the command of coalitions or alliances established for the particular task. The effort in these cases can be protracted, sometimes over years, but the numbers of personnel so employed are low and virtually no major equipment is used (Maloney 2002b).

Evacuation of Canadians overseas. Noncombatant evacuation operations (NEOs) are short-duration rescue operations. They usually involve a small number of transport aircraft, a warship, or both, plus special operations forces and intelligence personnel. NEO assets are under national control. Canadian-led NEOs are not frequent, but over-reliance on our allies in their area calls into question Canadian sovereignty when embassies and diplomatic independence are involved (Maloney 2002b).

Stabilization, intervention and conventional defence operations are a breed apart in all areas, particularly in intensity, duration and risk. The forces required for them are more complex, as are the lift and command-and-control requirements. There is substantial scope when selecting the types of forces to deploy, and that will be related to alliance or coalition objectives, Canadian national objectives, the threat, and the level of force necessary to achieve the objective. In all three of the priority geographical areas, a pattern emerged in the 1990s. Stabilization forces were introduced, sometimes with the consent of the belligerents; economic pressure using maritime forces to implement sanctions was applied to coerce a belligerent to accept international efforts; ground and air intervention forces were readied in case the situation deteriorated, which augmented the pressure tactics; and then the intervention force was employed, if necessary, to compel the belligerent forces to comply with international demands.

In the early 1990s, however, there was some disagreement on the weight of the stabilization forces. Some suggested that stabilization forces be lightly armed and not equipped for combat operations, which in turn would generate confidence amongst the belligerent factions. Unfortunately, this permitted better-armed belligerent forces to coerce the stabilization forces, which rendered their efforts ineffective. Others suggested that forces equipped, trained and logistically supported to conduct combat operations were more capable of performing stabilization missions. As for conventional defence missions, the same types of forces used for intervention and heavy stabilization missions could handle these other tasks.

There is no generic force structure for stabilization, intervention or conventional defence operations. The exact makeup of each mission will depend on the terrain, how the belligerent forces are equipped and the level of force deemed necessary to deal with the situation. For example, armoured units and mechanized infantry supported with self-propelled artillery were part of an alliance operation to deter an enemy mechanized attack in NATO's Central Region with its open and rolling terrain and rivers; and light infantry and light armoured forces were deployed to Somalia to provide armed protection to humanitarian delivery organizations and to disarm belligerent factions interfering with these activities. These operations were conducted in a desert environment. In Croatia, a mechanized infantry battalion equipped with TOW antitank systems stabilized the lines in mountainous terrain in an environment where the belligerent forces used mainly battle tanks and multiple rocket-launch systems. In Haiti, light infantry supported with helicopters conducted quick reaction force airmobile operations against the remnants of the collapsed Cedras regime in mountainous jungle areas as well as urban terrain.

If a prioritization exercise based on national interests had been conducted for the 1994 White Paper, the emergent force structure should have had several components:

- Balkans Force
- Persian Gulf Force
- Contingency Deployment Force
- Continental and Hemispheric Defence Force

The three services would act as force generators of these four commands. The Balkans Force and the Persian Gulf Force would have forces assigned to the alliance efforts in these two regions. These forces would be salient and militarily effective within the alliance efforts. The scale of the Canadian deployment would depend on the level of effort being exerted by the alliance at the time, with preparations made to augment that effort when required. If a stabilization force was protecting Kuwait and pressure was being applied by sanctions enforcement, then a mechanized battle group and a helicopter-carrying frigate would be assigned. If an intervention force was being readied to augment a stabilization force in Bosnia, than a brigade group and air support would be positioned and readied for action, with the strategic sealift and strategic air transport forces earmarked to move the combat equipment needed for the effort.

In essence, the Balkans Force and the Persian Gulf Force would each have a battle group forward-deployed either on operations in a stabilization role or pre-positioned for intervention or defence operations. Flyover training in each region would be mandatory. One brigade group would be allocated to each force, with the balance of each brigade held in Canada or with its equipment pre-positioned in Europe or South West Asia. A CF-18 squadron equipped for precision-guided munition (PGM) delivery would also be assigned to each force: they could be forward-based in Germany together and deploy to the active theatre when required. A special operations force increment would also be assigned to each force. Note that each force would be able to contribute to alliance or coalition efforts in areas immediately adjacent to the primary operating area. For example, if forces were required for conventional defence in the NATO area, then the Balkans Force would be responsible for intelligence collection and contingency planning to use its assigned forces to contribute to the alliance effort.

The Contingency Deployment Force (CDF) would prepare for noncombatant evacuation operations, humanitarian aid delivery, and peace observation missions, with its strategic transport aircraft, special operations forces, light infantry brigade group, and culturally aware pool of experienced and mature observer personnel. The CDF would also have the capability to conduct limited duration intervention missions in nonpriority geographical areas with light infantry and special operations forces. A deployable joint force headquarters would command the effort: it would have its own lift and strategic communications capability. Naval forces would be assigned by Maritime Command on a case-by-case basis.

The Continental and Hemispheric Defence Force (CHDF) would include all of the strategic sensor systems and interceptor, antisubmarine, and surveillance assets necessary to observe and protect Canadian sovereignty. Land forces based in Canada not deployed as part of the Balkans Force or the Persian Gulf Force and not assigned to the CDF would be assigned to a brigade group headquarters reporting to the CHDF for aid to the civil power and assistance with civil authority tasks. The CHDF would also be the repository of corporate knowledge related to interoperability matters and continental defence.

The primary equipment deficiencies to implement such a force structure after release of the 1994 White Paper would have been strategic airlift and strategic sealift. Structurally, the CDF would have needed a deployable headquarters. An expanded special operations force and intelligence-gathering capability would also have been required. The maritime forces as they existed in 1994, with the exception of the Sea King helicopters, would have been adequate in number and quality to implement this force structure plan, though some rationalization vis-à-vis the AORs and the strategic sealift would have been required. As for air power, four CF-18 squadrons, two of them fully equipped and trained for PGM delivery, would have sufficed, with additional aircraft held in reserve in case battle losses required their use.

The land forces certainly would have required expansion. The three light infantry battalions would have needed to be brought up to strength, and some creative means to employ reserve units, at the subunit level at least, would have been necessary. The raising of three other infantry battalions would have been called for, as would the return to the order of battle of the fourth armoured regiment. The M-113A2 armoured personnel carrier would have needed replacement with a more capable mechanized infantry combat vehicle. The Lynx recce vehicle would have needed to be replaced. The Leopard C-1 would have required upgrading or replacement with cascaded (and therefore cheap and more effective) Dutch Leopard 2 or even American M-1 Abrams tanks. The M-109A2 could have received an upgrade. Expanded use of unmanned aerial vehicles would have been useful, building on existing Canadian technology.

The cumulative effect of these upgrades would have been to position Canada for the next generational evolution of fighting technology instead of now having to make the leap from the 1970s Cold War kit, which in any case would have limited Canada's military effectiveness in any coalition or alliance effort. It is unfortunate for the Canadian Forces and for Canadian taxpayers that this was not deemed affordable by the government of the day.

Back to the Real World: What Happened after 1994

The premises of the 1994 White Paper did not generally hold after 1995. The optimism over the place of the United Nations in global security collapsed after the setbacks in Rwanda, Croatia, Bosnia and Somalia. NATO- and ABCA-led[26] coalitions took on the stabilization tasks, as they were more efficient in their command and control relationships, could bring coercive firepower to bear and were comparatively unencumbered by the legalistic demands of the unwieldy

UN bureaucracy. Warfighting, in fact, was necessary, as the events in Kosovo and Afghanistan demonstrated. Peacekeeping was held up to derision in an increasingly lawless world where nonstate entities violently made hash of formal treaties and diplomatic arrangements. Weapons of mass destruction remained a serious issue, but, with the exception of its participation in UNSCOM, the Canadian Forces had no real role to play in countering proliferation. Humanitarian assistance missions as anticipated by the 1994 White Paper were not conducted on a sustained basis by the Canadian Forces: it is difficult to conclude that Canadian efforts in this area were able to stem the tide of refugee flows generated by the unrest gripping Africa, though the stabilization of Haiti was a useful exercise. The Canadian Forces shone, however, when they were prepared to go to war, and they used salient and effective warfighting forces in stabilization and combat operations: Bosnia during the Implementation Force (IFOR) and Stabilization Force (SFOR) periods, Kosovo and Afghanistan are the best examples.

The burnout of the Canadian Forces was eventually, by 1997, recognized as a problem within DND (Young 1997). Instead of increasing the size of the forces, particularly the army, the government continued to commit military forces overseas. The four new peace observation missions, in the Central African republic, Sierra Leone, Guatemala and the Congo, were not manpower or equipment intensive. On average, however, there were two infantry battalions, an armoured recce squadron, and a tactical helicopter squadron deployed at any one time to Bosnia and Haiti from 1995 to 1997, both commitments rotating every six months. From 1998 to 2000-01, there were two infantry battalion groups, two recce squadrons, and a tactical helicopter squadron deployed to Bosnia and Kosovo, both commitments rotating every six months, plus two independent infantry company groups deployed to locations as disparate as East Timor and Ethiopia-Eritrea (Maloney 2002b). The exact Canadian interests at stake in these last two locations are unclear and the scale of effort in both locations was virtually token, especially in East Timor.

Air operations conducted during Operation ALLIED FORCE over Serbia included a Canadian component. A CF-18 squadron was incrementally deployed, but its operations were adversely affected in that insufficient PGM delivery equipment, munitions or trained pilots were immediately available. Indeed, the force had to be cannibalized from all four existing CF-18 squadrons, in terms of both equipment and personnel.[27]

In terms of domestic operations, however, a policy was implemented whereby the Canadian Forces would be the "force of last resort" in any aid to the civil power operation. This placed the onus on the Royal Canadian Mounted Police (RCMP) and regional policing organizations to deal with civil unrest. The

Canadian Forces supported two police deployments against armed native groups, Gustavsen Lake (Operation WALLABY) and Ipperwash (Operation MAPLE), but did not deploy units as it had during the events of 1990. Deployments related to resource protection increased: the so-called Turbot War of 1995 produced a multiship and patrol aircraft maritime force deployment to confront Spanish overfishing, while a submarine was deployed to covertly observe illegal scalloping operations (see Maloney 2002b; "Gunboat Diplomacy" 1995).

The Canadian Forces got involved in the international disaster relief business at the insistence of DFAIT. A Disaster Assistance Response Team (DART) was formed in 1997 (Young 1997). A deliberately modest organization, the DART was a grouping of secondarily tasked medical and engineer subunits with a small protective force, all using the existing CC-130 airlift. It deployed briefly to Honduras in 1998 and Turkey in 1999.

In terms of domestic disaster relief, the Canadian Forces were used more and more by other government departments and the provinces, once they realized they could do so. Two prime examples are the Winnipeg flood of 1997 and the ice storm of 1998. The SwissAir MAJAID (Major Aircraft Disaster) operation in 1998 is another example (Maloney 2002b).

A Canadian presence was maintained in the Persian Gulf. One frigate was deployed on a regular basis with an American carrier battle group, while Canadian AWACS personnel continued to monitor the no-fly zones alongside their American comrades and Canadian UNSCOM personnel hunted down production capability regarding weapons of mass destruction. The cat-and-mouse game with the Hussein regime reached dangerous proportions when Iraqi forces started moving in the directions of Kuwait and Jordan in 1998. With the Hussein regime's continued lack of compliance on the weapons of mass destruction issue, a coalition was established to coerce Iraq. The Canadian component, Operation DETERMINATION, included a frigate and two tanker aircraft (Maloney 2002a).

Most of these operations were manageable within the Canadian Forces' existing resources. Serious deficiencies, however, emerged after 1994. In 1996, Canada attempted to lead a multinational force into Zaire when it appeared as though a genocide of similar proportions to the events of 1994 in Rwanda was in the offing. This effort, Operation ASSURANCE, lacked the strategic lift, strategic communications and intelligence capabilities necessary to conduct effective operations in this remote region of Africa. It was not enough to borrow everything from the United States or other allies. Unlike Australia, France or the United Kingdom, Canada was demonstrably incapable of conducting a self-sustaining expeditionary operation, even in pursuit of a humanitarian objective (Hennessy 2001).

The second event was the GTS *Katie* affair. To save money, DND embarked on a program called Alternate Service Delivery (ASD), whereby civilian corporations would be contracted to handle some defence tasks. In this case, a strategic lift RO-RO vessel was contracted to deliver the Canadian Kosovo Force (KFOR) contingent's vehicles and equipment from Canada to Greece during the Kosovo War, and then to retrieve the equipment when the operation was completed. On the return trip, a dispute over the contract resulted in the ship owners holding Canada hostage: the ship drifted around the North Atlantic with 20 percent of Canada's armoured vehicles aboard. A naval task group, Operation MEGAPHONE, was deployed to retrieve the ship and recover the equipment.[28]

Canada continues to globally deploy, on average, two infantry battalions and two armoured recce squadrons at any given time. The ongoing SFOR operation in Bosnia, rotating every six months, is one of these.[29] Operation APOLLO, Canada's commitment to Operation ENDURING FREEDOM in Afghanistan, brought a light infantry battalion and armoured recce troops to fight in the al-Qaeda War (though most of the force was delivered using American strategic airlift). When this commitment was withdrawn in the summer of 2002, another infantry battalion, this one a mixed light-medium battalion, plus an armoured recce squadron, were deployed to Kabul to serve with the NATO-led International Security Assistance Force in 2003. An ASD strategic sea- and airlift was employed to get the Operation ATHENA force to Kabul. The ASD airlift, however, remains a dangerous undertaking: the entire Spanish International Security Assistance Force (ISAF) contingent was killed when their ASD aircraft crashed in Turkey in 2003. In another case an ASD (not Canadian) transport conducting humanitarian relief in the Congo lost its rear door and 140 people were sucked out while the aircraft was in flight (CNN.com/world 2003; International Security Assistance Force 2003). Canadian maritime forces, including two CP-140 patrol aircraft and at least one ship (down from six in the fall of 2001), continue to support Operation ENDURING FREEDOM.

On a more positive note, it is now recognized, after the successful deployment of JTF-2 to fight in Afghanistan with Operation ENDURING FREEDOM, that special operations forces are, and will be, integral to Canadian expeditionary operations. The serendipitous development of the Coyote surveillance vehicle and the acceptance of ISTAR (Intelligence, Surveillance, Target Acquisition and Reconnaissance) as an integral part of operations are significant. No developments in either area were anticipated in the 1994 White Paper process, or, apparently, by those involved in force structure exercises early in the 1990s. The close-out of the old divisional headquarters structure and its replacement with a proper Joint Force Headquarters, Joint Operations Group and Joint Signals Regiment by the late 1990s (Young 1997) are other significant developments on the path toward achieving an efficient and effective deployable command and control structure. Canada continues, however, to be deficient in its use of unmanned aerial vehicles.

Conclusion

This exercise was conducted to answer a question posed by the Institute for Research on Public Policy: In order to implement the 1994 White Paper, what force structure should have been developed? Some may argue that the answer is a moot point — an exercise in alternative history — or that such an exercise is superficial. An examination of the 1994 White Paper in light of subsequent events, however, illuminates several areas that should apply in any future defence policy process:

- Future defence policy must be based on a realistic vision of Canadian interests, how they may be threatened and what military forces are needed to counter such threats. Use of vague language may be acceptable politically, but it smacks of deception, indecision and lack of leadership.
- The Canadian Forces must be able to function across the spectrum of conflict in order to protect Canadian interests. Selecting one band in the spectrum and confining the force structure to that band does not serve national interests. Canada can and does fight wars; it is not merely a peacekeeping nation. Obsolete concepts of peace and war must be jettisoned. They are products of the nineteenth century, not the twenty-first.
- Policy-makers should be mindful of Canadian strategic traditions and how they affect national security policy and the development of the force structure to carry out that policy. The principles of saliency, operational influence and forward security will always be at work in the background, no matter what terminology is employed to describe Canada's military operations.
- The human factor is critical when the balance of commitments, rotations and deployment capability is considered. An over-reliance on quantitative analysis overlooks this, which in turn will have negative effects on recruitment, retention and morale. To ignore force burnout is not only morally wrong but criminally negligent. Such behaviour will, over time, increase the costs of defence beyond the apparent short-term "savings" accrued through the introduction of too much "efficiency" into the system.
- Policy-makers must educate themselves on the problems of weapons and platform procurement and resist the temptation to inject political expediency into those processes for short-term gain. The Sea King-New Shipboard Aircraft program is a case in point. At the same time, the Canadian Forces

must stop second-guessing what the policy-makers want and instead present them with a professional view as to what equipment is required. The onus should be on the policy-makers, not on the Canadian Forces. The problems with Main Battle Tank acquisition since 1970 are a case in point. The equipment procurement process needs to be expedited, perhaps with the elimination of assistant deputy minister (ADM [Material]) and the re-formation, outside the DND, of the Department of Defence Production.
- Policy-makers are being encouraged by recent research to think more strategically about their national security interests and objectives. They must ensure that the same attention and care are directed toward operational and force structure considerations.

In the end, any examination of alternative force structuring and the 1994 White Paper tells a cautionary tale. Many will argue that there was no miscalculation, that Canada "got it just right," that the Canadian government pulled off a balancing act in the face of severe fiscal constraints. This is not the case, as many soldiers in the field would attest. In pursuing efficiency above all else, some of the nation's representatives may well have sacrificed the country's self-respect. Why is it, for example, that Canada, a G-8 power, can sustain only one or two battle groups, of fewer than a thousand personnel each, overseas, while other NATO nations such as France, Germany, Italy, Poland and the United Kingdom are able to deploy self-contained brigades, and even divisions, to stabilize critical areas? Why is it that Bangladesh and Nigeria deploy larger forces than Canada, for all of its peacekeeping rhetoric, to UN operations? Why is it that Uruguay has more overseas deployments than Canada? Have we become so used to small, nearly tactical, contingents, because of cultural factors within the armed forces, that we do not envisage other possibilities? Is our army an army of battalion commanders? Are we perhaps too smug in our estimation of the "worth" of a Canadian soldier versus a Dutch, Nigerian or American one? Quality does matter, but, as Stalin correctly put it, "quantity has a quality all of its own." The inability of the regular and reserve forces to deal with systemic cultural problems is an example of the failure of creative problem solving within the Canadian Forces. At the same time, the government's continual bowing to the efficiency god at the expense of the mental and physical health of its soldiers is abhorrent. At the Canadian Forces Officer Candidate School, junior officers are taught that there is "no excuse" when it comes to their failings. These junior leaders are expected to take responsibility for, and learn from, their mistakes. Why should those responsible for national security policy be held to a standard inferior to the one they hold up for their subordinates?

The lessons of the 1994 White Paper should resonate in any future evolution of Canadian national security policy.

Notes

1. The interest-based approach, and the need for an established national security policy framework based on it, was advocated in parallel by Macnamara and Fitz-Gerald (2002) and Maloney (2001a,b). Subsequent national security commentators have picked up the mantle, including Delvoie (2001-02) and the authors of the CDFAI report *In the National Interest: Canadian Foreign Policy in an Insecure World* found at *http://www.cdfai.org/index2.htm*

2. For an entertaining and interesting take on the 1990s' attempts to project into the future, see Directorate of Land Strategic Concept, *The Future Security Environment* (annex A), "Long Term Environment," Department of National Defence.

3. For example, Lloyd Axworthy was a proponent of these views, which are represented, in a somewhat modified form, in Axworthy (2003).

4. The various White Papers have been collected in Bland (1997).

5. The media event that Canadians saw on television and that the CBC called "Oka" was actually the continuation of what was essentially a Mohawk civil war begun in the spring of 1990. See Hornung (1991).

6. *1994 White Paper on Defence*, p. 10.

7. Confidential interview.

8. See Friedman (1999) for a primer on post-Cold War economic developments, crises and changes; see also Tremblay (1996).

9. The drawdown of the army and the disbanding of the Airborne Regiment after the Somalia affair combined to produce three light infantry battalions that each had two infantry companies without mechanized vehicles and a parachute company. The original three battalions, 3 RCR, 3 PPCLI and 3 R22eR, were reduced on the order of battle after the close out of the Germany-based units in 1992-93. Initially, the light infantry battalions were "10/90" battalions, where the bulk of the battalion was supposed to be made up of militia personnel; then the para companies were added and slowly they worked their way back into the order of battle as regular battalions.

10. These data are derived from a wall chart circa 1999, supplied by Air Command, listing bases, squadrons and aircraft types.

11. Sean M. Maloney, lecture to 70 Communications Group, "The Operational Employment of the Reserves, 1945-2003," November 21, 2003.

12. DND backgrounder, February 2004, "Facilities Closures and Reductions."

13. The data in this section are drawn from Maloney (2002b).

14. Access to Information, DND (August 24, 1993); briefing note to DM/CDS, "UNTAC."

15. Access to Information, DND (October 12, 1993); memo to DM/CDS, "Canadian Forces Participation in the United Nations Assistance Mission for Rwanda (UNAMIR)"; (February 22, 1995); briefing note, "Situation Update: Rwanda."

16. The data in this section are drawn from Maloney (2002b).

17. At this point, three other infantry battalions were 10/90 battalions and/or mixed regular/reserve light battalions, and not considered deployable by FMC HQ.

18. This study recognizes that detailed force structure planning did occur in the 1994 White Paper process and that the original drafts were less vague. The fact that these were discarded or not published means that we have to work with what was made available to the public. The exact process that produced the 1994 White Paper requires more thorough historical study than is provided here.

19. Like that provided by Gray (1993).

20. DND eventually paid for Dr. Mike Hennessy at the Royal Military College to attend GBN seminars in scenario analysis. Dr. Scot Robertson and the author worked with Dr. Hennessy in the creation of several scenarios for the government of Canada after the 9-11 attacks. The author was also involved in scenario/force structure development, a process initiated after the 1998 auditor general's report stated that there was "a poor link between Defence Policy and the planning and management of the Canadian Forces capabilities." Consequently, a Strategic Capability Planning Process was established within DGSP, which used operations and scenarios to illuminate certain critical areas related to force structuring.

21. *1994 White Paper on Defence*, p. 3.

22. These operations are covered in two volumes edited by William J. Durch (1993, 1996).

23. Indeed, one reviewer of this paper suggests that Canada should only have followed the American or European lead and policy in both areas and goes on to state that since neither the Milosevic nor the Hussein regimes were out for global domination, they were not really Canadian problems. This is manifest Cold War-era thinking and does not take into account the globalized economy that Canada was a part of in the 1990s and is today. The two regimes in question were odious, dangerous and abhorrent. They threatened Canadian interests in Europe and in the Persian Gulf basin. "Strategic" and "regional" threats are one and the same in the post-Cold War era.

24. For some personal observations overlooked by the African apologist crowd, see Klitgaard (1990); Richburg (1998).

25. These cases are examined in Griffiths, Gimblett, and Haydon (1998).

26. ABCA: American-British-Canadian-Australian and sometimes New Zealand. The armed forces of these countries have extensive connections at the military level related to interoperability, command and control, and doctrine. An analysis of post-1995 stabilization operations indicates that most are led by ABCA nations, and almost all four ABCA nations contribute forces to those operations, whether they are conducted under the auspices of NATO, the United Nations or other coalitions.

27. Bob Bergan brought these deficiencies to light in his paper "Balkan Rats and Balkan Bats" presented at the CDAI Students Conference held at the RMC in Kingston, Ontario, October 25, 2003.

28. Department of National Defence (August 3, 2000), "HMCS *Athabaskan* Carries Out Boarding of GTS *Katie*"; fact sheet, "Operation MEGAPHONE: Retrieval of CF Equipment onboard GTS *Katie*."

29. Department of National Defence (January 1999), "Canada's Participation in the NATO-Led Stabilization Force in the Balkans."

References

Armstrong, Joe C.W. 1995. *Farewell the Peaceful Kingdom: The Seduction and Rape of Canada 1963 to 1994.* Toronto: Stoddart.

Ash, Timothy Garton. 1999. *History of the Present.* New York: Random House.

Axworthy, Lloyd. 2003. *Navigating A New World: Canada's Global Future.* Toronto: Alfred A. Knopf.

Ayittey, George B.N. 1998. *Africa in Chaos.* New York: St. Martin's Press.

Bland, Douglas L., ed. 1997. *Defence Policy.* Vol. I of *Canada's National Defence.* Kingston: Queen's University School for Policy Studies.

CNN.com/world. 2003. "Survivors Relive Jet Door Horror." May 11. *http://www.cnn.com/2003/WORLD/africa/05/11/congo.jet/*

Cockburn, Andrew, and Leslie Cockburn. 1997. *One Point Safe.* New York: Doubleday.

Cohen, Avner. 1998. *Israel and the Bomb.* New York: Columbia University Press.

Dallaire, Romeo. 2003. *Shake Hands with the Devil.* Toronto: Random House.

Delvoie, Louis. 2001-02. "A Not So Benign New Century: Conventional Security Challenges to Canadian Interests." *International Journal* 57, no. 1 (winter).

Dupuy, Alex. 1997. *Haiti and the New World Order: The Limits of Democratic Revolution.* New York: Westview.

Durch, William J. 1993. *The Evolution of UN Peacekeeping: Case Studies and Comparative Analysis.* New York: St. Martin's Press.

———. 1996. *UN Peacekeeping, American Policy, and the Uncivil Wars of the 1990s.* New York: St. Martin's Press.

Fox, J. 1988. "The Restructuring of the Canadian Army." *NATO's Sixteen Nations* (October): 19-22.

Friedman, Thomas L. 1999. *The Lexus and the Olive Tree.* New York: Farrar, Straus and Giroux.

Garnett, G.L. 2002-03. "The Evolution of the Canadian Approach to Joint and Combined Operations at the Strategic and Operational Level." *Canadian Military Journal* (winter): 3-8.

Gervais, J.C. 1992. "Moving SLOWLY Towards Total Force." *Forum* (spring).

Glenny, Misha. 1992. *The Fall of Yugoslavia: The Third Balkan War.* London: Penguin.

Gray, Colin S. 1993. *Canadians in a Dangerous World.* Toronto: Atlantic Council of Canada.

Griffiths, Ann L., Richard Gimblett, and Peter Haydon, eds. 1998. *Canadian Gunboat Diplomacy: The Canadian Navy and Foreign Policy.* Halifax: Centre for Foreign Policy Studies.

"Gunboat Diplomacy." 1995. *Maclean's*, March 20, 10-12.

Hennessy, Michael A. 2001. "Operation ASSURANCE: Planning a Multi-National Force for Rwanda-Zaire." *Canadian Military Journal* 2, no. 1 (spring): 11-20.

Hornung, Rick. 1991. *One Nation Under the Gun: Inside the Mohawk Civil War.* Toronto: Stoddart.

International Security Assistance Force. 2003. *ISAF Soldiers Killed in Plane Crash.* Press release. Brunssum, Netherlands: Author, May 26. *http://www.isafkabul.org/press/pr26mayplanecrash.htm*

Jockel, Joseph T. 1999. *The Canadian Forces: Hard Choices, Soft Power.* Toronto: Canadian Security Intelligence Service.

Klitgaard, Robert. 1990. *Tropical Gangsters: One Man's Experience With Development and Decadence in Deepest Africa.* New York: Basic Books.

Last, David M. 1988. "Equipping the Total Force Army." *Canadian Defence Quarterly* (fall): 9-31.

Macnamara, W.D., and Ann Fitz-Gerald. 2002. "A National Security Framework for Canada." *Policy Matters* 3, no. 10 (October).

Maloney, Sean M. 1996. *The Hindrance of Military Operations Ashore: Canadian Participation in Operation SHARP GUARD 1993-1996.* Halifax: Centre for Foreign Policy Studies.

———. 1997a. *Operation BOLSTER: Canada and the European Community Monitor Mission 1991-1994.* Toronto: Canadian Security Intelligence Service.

———. 1997b. *War Without Battles: Canada's NATO Brigade in Germany 1951-1993.* Toronto: McGraw-Hill Ryerson.

———. 2001a. "Canadian Values and National Security: Who Decides?" *Policy Options* 22, no. 10 (December).

———. 2001b. "Helpful Fixer or Hired Gun? Why Canada Goes Overseas." *Policy Options* 22, no 11 (January-February).

———. 2002a. *War With Iraq: Canada's Strategy in the Persian Gulf 1990-2002.* Kingston: QCIR Martello Paper 24.

———. 2002b. *ORD Project Report PR 2002/01: Canadian Forces Operations 1970-2000.* Ottawa: Operational Research Division, Department of National Defence, March.

———. 2002c. *Canada and UN Peacekeeping: Cold War by Other Means 1945-1970.* St. Catharines, ON: Vanwell.

———. 2002d. "The Canadian Tao of Conflict." In *Forging a Nation: Perspectives on the Canadian Military Experience,* edited by Bernd Horn. St. Catharines, ON: Vanwell, 271-286.

———. 2002-03. "Purple Haze: Joint Planning in the Canadian Forces from Mobile Command to J-Staff (Part 1)." *The Army Doctrine and Training Bulletin* 5, no. 4 (winter): 57-72.

Maloney, Sean M., and John Llambias. 2002. *Chances for Peace: Canadian Soldiers in the Balkans, 1992-1995.* St. Catharines, ON: Vanwell.

O'Connor, Gordon J. 1988. "Developing a Total Force Army." *Canadian Defence Quarterly* (fall).

Oliviero, Charles S. 2001. "Operation DELIVERANCE: International Success or Domestic Failure?" *Canadian Military Journal* (summer): 51-58.

Pollack, Kenneth M. 2002. *The Threatening Storm: The Case for Invading Iraq.* New York: Random House.

Pry, Peter Vincent. 1999. *War Scare: Russia and America on the Nuclear Brink.* Westport, CT: Praeger.

Pupetz, Ron. 1994. *Canadian Joint Forces in Somalia: In the Line of Duty.* Ottawa: Department of National Defence.

Richburg, Keith B. 1998. *Out of America: A Black Man Confronts Africa.* New York: Basic Books.

Silber, Laura, and Allan Little. 1995. *Yugoslavia: Death of a Nation.* New York: Penguin.

Thompson, Roger. 1992. "Total Force Army or Total Farce: A Critical View of Canada's New Military Structure." *Forum* (spring): 15-16.

Tremblay, Miville. 1996. *A Country Held Hostage: How the World Finances Canada's Debt.* Toronto: Stoddart.

Trouillot, Michel-Rolph. 1990. *Haiti: State Against Nation.* New York: Monthly Review Press.

Young, M. Douglas. 1997. *Report to the Prime Minister: Compendium of Changes in the Canadian Forces and National Defence.* Ottawa: Department of National Defence. March 25.

4

A National Security Framework for Canada

W. D. Macnamara and Ann Fitz-Gerald

Introduction

The September 11 terrorist attack on the World Trade Centre in New York sent a security chill across the world. It was viewed as a terrorist attack not just on New York but also on the core values and interests of the American people. It was an attack against a North Atlantic Treaty Organization (NATO) partner. It was something that could easily repeat itself elsewhere.

Since that day, world leaders have been forced to review their own national security at the highest strategic level, and citizens are asking searching questions about security policy. As Air Chief Marshal Sir John Slessor reminded us in 1953,

> It is customary in the democratic countries to deplore expenditures on armaments as conflicting with the requirements of social services. There is a tendency to forget that the most important social service a government can do for its people is to keep them alive and free.[1]

Many people and nations suddenly came to realize that without national security, nothing else matters, a mantra-like statement the media has echoed across the world.

In Canada, quick to support and join the coalition response, public debate has arisen concerning the relevance and adequacy of our national response and the capability and readiness of the Canadian Armed Forces. Questions include:

- Does Canada devote an appropriate level of effort toward security and developing new security policies?
- Do we know enough about how other countries tackle similar problems?
- Are governments really capable of thinking strategically and in the long-term?

A key strategic policy issue is Canada's perceived capability to "fight against the best alongside the best," not only in terms of equipment, but also in terms of availability, deployability and sustainability of our forces outside of Canada — issues of interoperability with our allies. The events leading to the tragic loss of four Canadian soldiers in a training accident have served to reinforce these questions.

Since 1945, when Canada and her allies celebrated the end of the Second World War, Canada's national security and defence identity has undergone numerous transformations, while experiencing varying degrees of public and political support. Since the 1989 end of the Cold War, many in Canada have failed to see the defence portfolio as significant to Canada's overall security concerns and have underestimated the strong links and interdependencies it has with the country's democratic value system and its economic well-being. The shock of September 11, however, has caused some reaction and raised questions regarding Canada's defence and security posture. It is now widely acknowledged, both inside and outside of government in Canada, that there is no national security framework or clear and distinct national security policy as such.

The lack of a national security framework or "policy" reflects a general lack of strategic vision and meaningful strategic objectives — and the associated resources — required to implement such a vision across the whole of government in general, but in particular related to both the Department of Foreign Affairs and International Trade and the Department of National Defence. This has been highlighted in the most recent report from the Canadian Senate[2] and confirms similar and long-standing observations by others, including Byers (1986), Dewitt and Leyton Brown (1995), Bland (1987), Boulden (2000) and Delvoie (2001).

This paper will therefore examine the national security strategy and policy framework needs of Canada, starting with clarifying definitions of common terminology. This will be followed by a discussion of the way in which Canada defines its national interests and core values, which should serve as the foundation for a Canadian national security strategy. It will then examine how these strategic interests shape defence policy and military power, which, together, represent one of the main instruments for implementing a national strategy. Lastly, this paper will discuss the approaches to formulating national strategy and the linkages between defence policy and national strategy within the British and American governments. Conclusions will propose a national security policy process framework that not only reflects Canada's interests but also is comparable in terminology and scope with traditional coalition partners, the US and UK.

What Is Meant by National Security and National Strategy?

The term "national security" is used frequently to refer to matters ranging from domestic or internal security through to international security, but is seldom defined. Such imprecision can lead to difficulties, unless a Humpty-Dumpty-like[3] flexibility in meaning is intentional and, therefore, intentionally confusing. Clarity in the meaning and understanding of the scope of national security is essential at the outset.

Canada's National Defence College in 1980 adopted the following definition to form the foundation of its curriculum:

> National Security is the preservation of a way of life acceptable to the Canadian people and compatible with the needs and legitimate aspirations of others. It includes freedom from military attack or coercion, freedom from internal subversion, and freedom from the erosion of the political, economic, and social values which are essential to the quality of life in Canada.[4]

For the United States

> National security is a collective term encompassing both national defence and foreign relations of the United States. Specifically, the condition provided by:
> a) military or defence advantage over any foreign nation or group of nations;
> b) favourable foreign relations position; or
> c) defence posture capable of successfully resisting hostile or destructive action from within or without, overt or covert.[5]

For Canadian policy-planners, in keeping with both definitions, it should be understood, and be equally clear, that the concept goes beyond internal security or military issues, and takes into account the whole range of government activities. National security, then, as pointed out by Slessor, truly is the first and most important social service for a government to deliver to its people.

Throughout security discussions, there are other words and terms used which, to many both inside and outside the discussion, have different meanings in different contexts. How a national security policy may be developed and implemented leads directly to the notion of a national or grand strategy.

Strategy itself is a word that carries many meanings and interpretations and is used regularly in private and public sector, and nonprofit organizational, activity. Although there is a common misperception that strategy is primarily concerned with military matters, it really reflects goals or objectives, an understanding of the environment (including resources available), and the means by which the resources will be organized or applied to achieve the objectives. Thus, strategy can mean different things depending on the context and the level at which it is being considered. For example, at the lowest level, it is possible for an individual to have a personal strategy, perhaps relating to his or her career aspirations. At the highest level, a government may have a strategy (or set of strategies) in order to meet its obligations to the country, across critical policy areas.

Over 60 years ago, Edward Meade Earle said

> Strategy is not merely a conception of wartime, it is an inherent element of statecraft at all times. In the present-day world, strategy is the art of controlling and utilizing the resources of a nation — or a coalition of nations — including its armed forces, to the end that its vital interests should be effectively supported and secured against enemies, actual, potential or merely presumed. The highest form of strategy — sometimes called grand strategy — is that which so integrates the policies and armaments of a nation that resort to war is either rendered unnecessary or is undertaken with the maximum chance of victory.[6]

In his recent article entitled "Grand Strategy in the 20th Century," Sir Michael Howard states that grand strategy refers to the state of a country in wartime and peace, and involves warfighting and war avoidance (or deterrence).[7] He acknowledges that most states maintain their independence, extend their influence and, at times, extend their dominion using national tools like armed force, wealth, allies and public opinion. In other words, certain strategic tools and techniques exist for a nation-state to use to protect its core values and national interests, which, for most countries, would include their political integrity and territorial sovereignty.

In the United States, the Department of Defence currently defines national strategy or grand strategy as "the art and science of developing and using the diplomatic, economic and informational powers of a nation, together with its armed forces, during peace and war to secure national objectives."[8]

Two other definitions of grand strategy draw on the work of Liddell Hart and Bernard Brodie, the first of which explains that it involves the coordination and direction of all the resources of a nation, or a band of nations, toward the attainment of political objectives sought.[9] Bernard Brodie, who sought a more rigorous approach to strategy and a systematic form of analysis rather than a narrow superficial approach to security problems adopted by the military, referred to grand strategy as "an instrumental science for solving practical problems."[10]

Edward N. Luttwak, in his 1987 book, *Strategy: The Logic of War and Peace*, describes grand strategy as something that is used to explain the totality of what happens between states and other participants in international politics, the conclusive level of strategy as a whole.[11] Thus, grand strategy is the pinnacle of the strategic edifice which offers the most general explanation of conflict.

For some states, military power is the leading instrument for pursuing grand strategic objectives. But it also presumes that the nation-state is a cohesive unit, which is why a national strategy seeks to protect the values and interests which ensure that cohesion. Professor Lawrence Freedman goes beyond the military context, stating recently that grand strategy is concerned with how far, and for what purposes states concern themselves with the international system. He suggests that the new challenge is protecting a state's international position, not vis-à-vis other more radical states, but vis-à-vis the more fundamental shifts in the system that contest the whole idea of a state.

> Governments are not carried forward by the tide of technology and the remorseless logic of markets but have real and difficult choices to make about how they look after their people and act beyond their borders. That is, they need a strategy.[12]

These discussions are reflected in the concise US definition of a National Security Strategy: "the art and science of developing, applying and coordinating the instruments of national power (diplomatic, economic, military and informational) to achieve objectives that contribute to national security."[13]

National security strategy — grand strategy — therefore is something far-reaching, all-encompassing, visionary and long-term, yet dynamic, in nature. So the art of strategic planning should first define particular objectives, based on the national values, to be pursued as the foundation of a national or grand strategy. At a private-sector corporate level, this involves the board of directors and the operating executives. At the nation-state level, this should involve the major government portfolios and the whole Cabinet.

National Strategy and National Interests

Commonly, strategic planning models and processes begin with the articulation of national or corporate values and goals, which collectively represent the fundamental interests to be pursued and preserved. The expected operating environments — events, issues and trends — are then projected into the future and their potential impact on the fundamental interests assessed. A strategy is then developed using the available resources (human, technical, financial) to promote national or corporate interests and achieve national or corporate goals in the perceived future environment.

A modification of the Ascher-Overholt Strategic Planning Model[14] for a nation might then be useful to clarify and simplify the concept and process.

The first step, identification of national interests, brings us to another matter, the concept of "the national interest" itself. The concept of national interest is sometimes criticized as being potentially too subjective or too specific. But a particular problem, like other terminology discussed above, is the misunderstanding and misuse of the words. Plischke, following an exhaustive review of the concept, states

> Much of the difficulty of comprehension and objection to its application in the conduct flows from its careless and equivocal use by political leaders and the biased and parochial interpretations of some of its analysts. But, these are scarcely valid reasons for justifying the invalidation of a concept as prevailing and important as national interests in managing and understanding the foreign policy process.[15]

Joseph Nye also has noted that "in a democracy, the national interest is what a majority, after discussion and debate, decides are its legitimate long-run shared interests in relation to the outside world."[16]

A clear and unequivocal statement of national interests, however difficult to develop, can be a valuable tool in policy development and in assessing reaction to domestic and international events. In addition, it can be used to encourage public discussion and involvement in how national interests may be protected or pursued. National interests are inextricably linked to national security and a national security policy.

In the US, national security interests are "the foundation for the development of valid national objectives that define US goals or purposes. National security interests include preserving US political identity, framework and institutions; fostering economic well-being; and bolstering international order supporting the vital interests of the United States and its allies."[17]

Figure 1

A Strategic Planning Model

Interests	+	Environment	Strategy
National values and national goals		Domestic/int'l • Political • Economic • Technological • Socio-cultural • Military/defence	National policies • Foreign/diplomatic • Economic • Technological • Economic • Military/defence

Source: Ascher and Overholt (1983), with modifications by author.

In Canada, there is no clear definition or overt statement of national interests as such, although the expression of something being or not being "in the national interest" is commonly used by politicians and commentators. However, a construct of Canada's national interests can be achieved by examining statements of foreign and defence policy goals, which incorporate statements of fundamental values. Such a construct could not only provide clarity of purpose, but also foster informed public discussion, as well as serving as an element of accountability for public policy makers.

In the 1995 foreign policy White Paper, *Canada in the World*,[18] Canadian values are clearly stated to be

- respect for the rule of law
- democracy
- human rights
- the environment

The foreign policy goals in that document, which may be seen to reflect national objectives, even if not overtly stated, are

- the promotion of prosperity and employment
- the protection of our security, within a stable global framework, and
- the projection of Canadian values and culture

The Department of National Defence document *Shaping the Future of Canadian Defence: A Strategy for 2020* states: "The Canadian values to be defended include"

- democracy and the rule of law
- individual rights and freedoms as articulated in the Charter
- peace, order and good government as defined in the Canadian constitution
- sustainable economic well-being[19]

In Canada, the development of foreign and defence policy is a somewhat ad hoc process, lacking in detailed analysis from first principles, but driven more by certain concepts enunciated by deputy minister-level officials in the Department of Foreign Affairs, the Privy Council Office or, indeed, the Prime Minister's Office. This process, as it concerns the *Canada in the World* document and many of its predecessors, is frankly discussed in David Malone's recent article, "Foreign Policy Reviews Reconsidered." Notwithstanding expectations to the contrary, Canada's foreign policy reviews have been essentially more of the same thing, but reflecting the personal interests of politicians and officials rather than fundamental analysis. Even so, Malone claims that *Canada in the World* still produced "a fluent, sophisticated, acute analysis of the world in 1995, with fairly insightful projection into the mid-term future."[20]

Notwithstanding this criticism, the combination of *Canada in the World* and *Shaping the Future of Canadian Defence* provides a foundation for at least some government perception of Canada's national interests based in economic well-being, security, and protection and projection of Canadian values: democracy (including the rule of law), individual freedom and human rights, and social justice. However, there is no consistent national security policy-planning framework within the Privy Council Office, the Department of Foreign Affairs and International Trade, or the Department of National Defence. Officers and officials in the departments and agencies involved admit there is no such framework and support the need for some coherent and consistent process that could be used to integrate various departmental interests. Some also indicated that there are opposing opinions on the need for such a framework, as "resisters" wish to avoid being constrained by a particular process or even an overt statement of interests.

This reinforces the assessment of Jane Boulden who, in her study, *A National Security Council for Canada*, stated, "there is a national security policy gap that needs to be addressed in Canada." The gap in question is a functional one related to the national security policy-making process and "the machinery of government." The process has continued to be ad hoc, based on Cabinet decisions made on advice received from ministers' individual departments, which may indeed be in conflict. The Auditor General observed the decision-making process for peacekeeping deployments in 1996. As Boulden reported, "The Auditor General found that a clear statement of Canadian objectives was not always provided" and the implications for Canada or for other government departments were given "insufficient" analysis.

The absence of a formal process, then, contributes to limited or poor analysis and less than coherent and consistent policy development or decision-making.

Systematic Approaches to National Security Policy Planning — United States

The development of process models or frameworks can be useful tools in explaining the sequence and content required in a dynamic system, as well as the impact that changes may have on other elements of the system. Richmond Lloyd has developed one framework to describe the strategy and force planning process in the United States (figure 2). Given that there can be many activities and elements involved in any one of the components, a detailed description of the components and inherent activities is included in his text and will not be repeated here.[21] However, it is clear from the model that the starting place is the statement of national interests and national objectives leading to a national security strategy, encompassing political, economic and military components and the external environments including resource, technology and international analyses. These feed into the national military strategy which — following resource, risk and deficiencies assessment — leads ultimately to the force posture.

In the United States, such a framework is additionally useful because of the formal statements made periodically by each administration, as well as a glossary of terms and definitions, some of which have been used above. For example, in the Clinton administration document entitled "US National Security Strategy for the New Century," the United States' national objectives are clearly stated to be

- enhancing security at home and abroad
- promoting prosperity
- promoting democracy and human rights

It is interesting to note the similarity with the foreign policy goals of Canada in *Canada and the World*.

The same document details the statement of the United States' national interests by category:

- vital interests ("those of broad overriding importance to the survival, safety and vitality of our nation")
- important interests ("do not affect national survival but they do affect our national well-being and the character of the world in which we live, e.g., NATO in Bosnia")
- humanitarian interests ("in the event of natural or man-made disasters, or gross violations of human rights, our nation may act because our values demand it").[22]

Figure 2

Strategy and Force Planning Framework for the United States

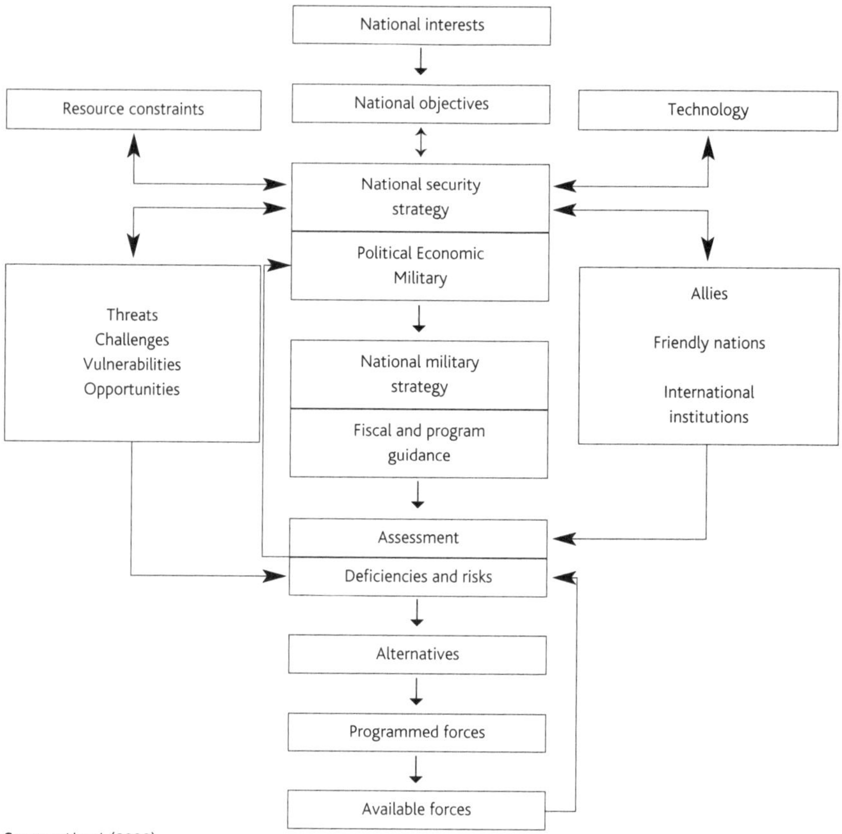

Source: Lloyd (2000).

More recently, the *Quadrennial Defense Review,* "America's Security in the 21st Century," articulated US interests and objectives as

Ensuring US security and freedom of action, including

- US sovereignty, territorial integrity and freedom
- safety of US citizens at home and abroad
- protection of critical US infrastructure

Honouring international commitments, including

- security and well-being of allies and friends
- preclusion of hostile domination of critical areas including Europe, Northeast Asia, the East Asian littoral, and the Middle East and Southwest Asia
- peace and stability in the Western Hemisphere

Contributing to economic well-being, including

- vitality and productivity of the global economy
- security of international sea, air and space, and information lines of communication
- access to key markets and strategic resources[23]

These statements all provide the starting point for the top-down analysis and policy development process and reflect the basis upon which national security — in its widest context — and military strategy planning are undertaken.

Given such statements of goals and interests, another very useful tool, termed the "National Interests Matrix," was developed by Donald Nuechterlein, in its original form in 1976 but most recently published in his 2001 book, *America Recommitted — A Superpower Assesses Its Role in a Turbulent World.*[24] This matrix (figure 3) starts by listing the basic national interests in the left-hand column and, across the top, the levels of interest: survival, vital, major, peripheral. A given world situation can be quickly assessed in "interest terms" by asking the question, "Which countries' interests are affected by this situation, and at what level?" Considering that survival or vital interests usually call for a military commitment, this matrix tool can quickly indicate which countries are likely to respond, and using which means. Equally, it serves to clarify which United States' interests may also be affected.

As useful as the policy framework or national interest matrix may be, they still require a fundamental first policy step: a clear articulation of national goals and values; i.e., the national interests.

Figure 3

Nuechterlein's (2001) National Interest Matrix

Intensity of interest / Basic national interest	**Survival level (critical)**	**Vital level (dangerous)**	**Major level (serious)**	**Peripheral level (bothersome)**
Defense of homeland				
Economic well-being				
Favourable world order				
Promotion of values				

The United Kingdom Approach

It should be remembered that the "national interest" approach to policy planning originated with British Prime Minister Lord Palmerston in 1848, when he stated, "We have no eternal allies and no perpetual allies. Our interests are eternal and perpetual and those interests it is our duty to follow."

The United Kingdom's approach to linking a national strategy with defence planning has evolved considerably since Palmerston and even the application of Winston Churchill's model during the early 1940s. Today, Prime Minister Blair's inner office continues to lead proactively on defence and foreign affairs issues, similarly to the Thatcher government. So although national interests should be arrived at by consensus in the Cabinet and Office of the Prime Minister, Mr. Blair has slimmed down the Cabinet and personally has a great sense of national interests. This has become known as the "presidentialization" of the British system of government: the contention is that the British prime minister increasingly resembles the president of the United States in his or her ability to take policy decisions and drive the government machinery, free (or relatively free) from the constraints of collective ministerial responsibility and other constitutional conventions.[25]

There is also a recognition that the Ministry of Defence (MoD) is a leading authority on national strategy and longer-term thinking, underlined by the way in which the department handles equipment procurement and recruitment policies. Despite the fact that the Cabinet is served by the Defence and Overseas Secretariat, the work of which feeds into the Defence and Overseas Policy Committee, there is a tendency for the current government to bypass the Cabinet committees and take advice directly from the director of policy within the MoD. Thus, the MoD is more political than it has ever been and is perceived to be more of a medium-term thinker than the Foreign Office, which is more pragmatic and short-term in its view. The Foreign Office has also been described as an office that has become overburdened with the development and application of management techniques, which has resulted in time lost for policy thinking, policy planning, and for the formulation and orderly pursuit of policy objectives.[26]

Whereas a defence policy flows from a country's foreign policy, a country's foreign policy should be based on national interests and values that serve as national policy at the grand strategic level. Although the term "national interests" can at times be deceptive and unclear, in a democracy, voters have a right to a foreign policy that promotes their immediate interests. However, the raft of real issues that confront governments and voters on a daily basis can dilute this, as they create conflicting priorities and, therefore, make difficult a consolidated view of what national interests may be perceived to be.

Similar to the way successful multinational corporations define their mission statement in fairly general terms (for example, Sony's commitment to provide "personal and portable sound to every household"), national interests do not need to be narrowly defined. From a British perspective, London School of Economics' professor Mary Kaldor stresses

> National interests include national security as well as jobs, environmental sustainability, as well as trade, the fulfillment of properly chosen humanitarian goals, as well as business interests of British firms.[27]

The next step in the British approach to grand strategic policy formulation is to define the assets that are available to pursue the national interest as defined. In this context, nation-states have four general categories of "tools" or "instruments" at their disposal: those that come under the control of the government, such as political skills, diplomacy, a country's armed forces, and resources that come under the control of other government departments; those that depend on the government but which operate outside its control, such as the national media and state councils; institutions and organizations such as think-tanks, academic institutions and nongovernmental organizations; and a country's foreign investments and assets.

The last step is to create a vision that describes objectives and activities in a meaningful way, not only to policy practitioners but also to the public and foreign audiences. Unless a country's voters believe that public expenditure is devoted to causes that relate to their interests, they cannot be expected to provide support. As John Coles has observed

> National interests should not be suppressed in order to construct an artificial consensus or a bogus unity. Influence was a means and not an end in itself. Occasionally, it might be appropriate to accept a loss of influence if that was the only way to protect our interests.[28]

For the United Kingdom's MoD, the implementation strategy for the process described above unfolds in a reasonably systematic way, bearing in mind the approach of the current government. An obvious and integral element of the national strategic planning equation is the Treasury which, every two or three years, calls for spending reviews, something which triggers the MoD to produce a Strategic Defence Review (SDR). The last SDR was conducted in 1998 and was sufficiently strategic in vision to still hold contemporary relevance after the events of September 11. The strategic analysis process involves defining the future strategic context and invites the participation of the other Cabinet members (usually led by the secretary for defence). If disagreement exists in terms of how this strategic context is defined, the work then goes to the relevant Cabinet committees. However, since the arrival of the Blair government and its commitment to "joined-up" government deliberations, this contingency has not been used.

The main framework of the strategic environment includes the analysis of political; economic; military; physical; scientific and technical; social and cultural; and legal, ethical and moral issues — discussing each in turn and then drawing them altogether in an overall description of the world as seen from a British perspective.[29]

Once an exhaustive analysis has been completed, UK national interests are defined within the strategic environment and categorized into essential, vital and marginal interests. While these definitions are still perceived as a function of the state, the UK military's strategic doctrine must be sufficiently flexible to cope with shifts in perceptions of national interests reflected in policy. From the analysis a national grand strategy is developed, which drives the application of national power within the broader international political system in support of national and collective interests, usually in conjunction with other allies and partners.[30]

The senior planners then revisit the instruments of national grand strategy, which include diplomacy, economic power and military power. The future strategic context is then reviewed and broader implications for defence are drawn from the analysis. This may include statements such as "to continue to structure forces so they are capable of operating as a coalition in a large-scale warfighting operation" or "to recognize a growing emphasis on multinational approaches and filling capability gaps." Policy priorities are then taken and grouped under their relevant headings, such as National and European Security, Force Capability and Development, People, Wider Security Interest and Conflict Prevention, etc. Aims and objectives for each of these categories are then articulated and serve as the executive text of the SDR. While the SDR is only published every few years, the strategic analysis and planning processes are perpetual and dynamic.

In the mid-1990s, the structure of the most senior UK MoD departments was altered slightly; the restructuring resulted in the creation of the Defence Management Board (DMB) that would sit directly under the Defence Council, the highest national authority responsible for defence. The Policy Directorate is responsible for producing the Defence Strategic Plan (DSP) through the same analytical process described above. The short-term version of the DSP is the Corporate Plan, which has a shorter-term vision of four years and is produced by the Defence Performance and Analysis Department under "Finance." The methodology which serves as the foundation for the Corporate Plan is the

Balanced Scorecard (BSC), which targets the deliverables and underlines the agreement the department has with the Treasury.

Thus the approach to defence planning and formulating defence policy in the UK has a definite strategic orientation and, moreover, there is a direct link between strategic planning surrounding the national security of the UK and the grand strategy of the country, based on core values and national interests. These tight linkages have resulted in coherent policy, which the UK MoD has been able to communicate clearly to the British public. This strategic orientation of the MoD has gained it an influential overall advisory position in the highest offices of the Blair government. Its continuous commitment to strategic analysis and influence in the Cabinet has triggered current work on the development of new levers of power, to be used when military power is not the best way to penetrate the centre of gravity of a threat to the UK's national interests.[31]

Lastly, the UK's dynamic approach to strategic analysis for defence decision-making led the secretary of state for defence to request that the MoD embark on a "new chapter to the 1998 SDR."[32] This decision was precipitated by the fact that the threat of terrorism has potentially been raised, which has significant strategic impact for the UK; the MoD is also keen to know if it is dealing with something strategically different in light of the fact that the UK itself may now be strategically different. In a recent House of Commons debate, the secretary of state for defence stated:

> The threat to the UK goes beyond the UK's borders and thus has implications for the UK's armed forces on deployment. The more they are used for overseas missions, for which changes under the SDR are designed to equip them, the more they risk presenting themselves as targets...Professor Lawrence Freedman has also said that, where conflicts in far-off places involve British troops, we cannot always assume that they will necessarily be fought out only where they arise.[33]

A framework model of the UK process is presented in figure 4.

A National Security Framework for Canada

It is clear from examining Canadian experience and both the US and UK processes that Canada lacks any process even remotely comparable in analytical rigour, multidepartment involvement, coherence and consistency. The absence of a clear process is possibly a major factor in the irregularity of foreign and defence policy reviews: the methodology is not clear and has to be reinvented each time, leading to the less than relevant results as reported by Malone.

Figure 4

UK Model

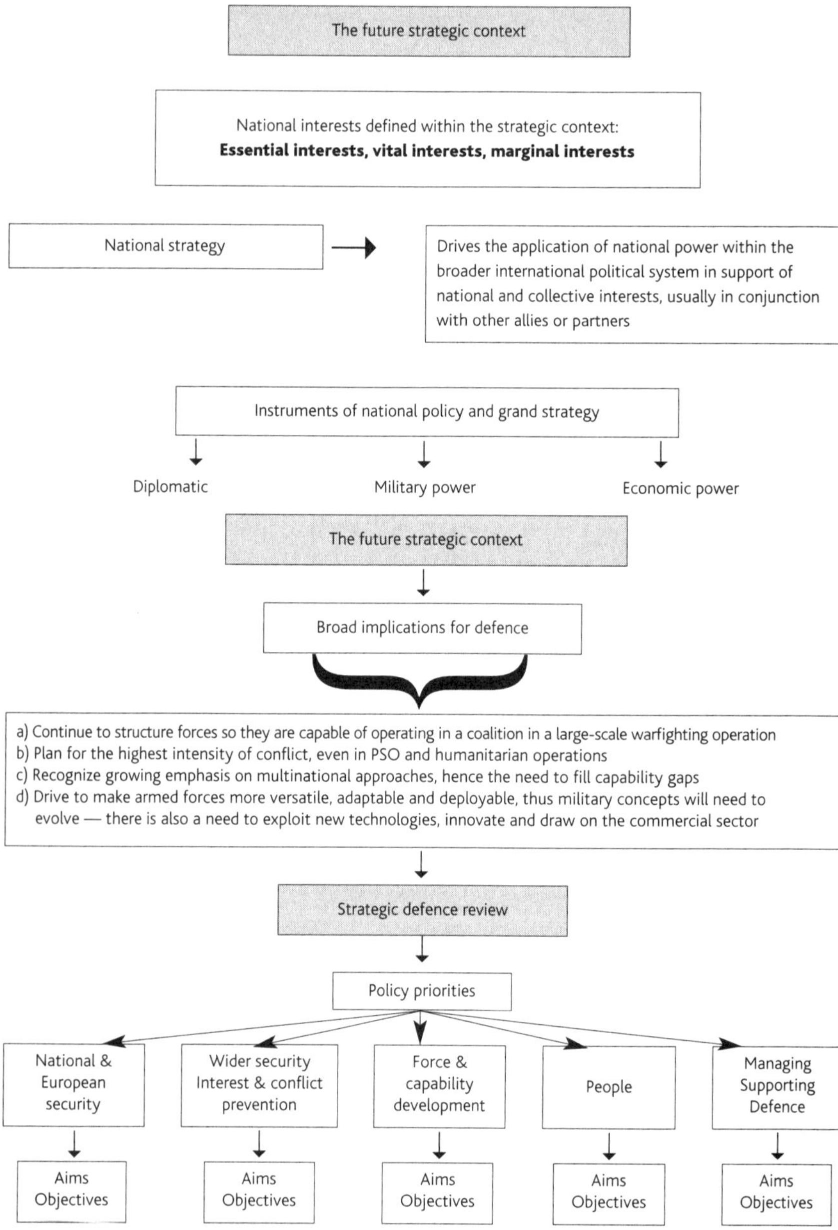

Further, as already observed by Boulden and the auditor general, the absence of stated relevance of deployments of Canadian Forces to Canada's interests — goals and values — indeed reflects inadequacies in the policy and decision-making process.

A national security policy framework for Canada has therefore been proposed, taking into account the processes in the US and UK, but recognizing certain special concerns for Canada. The first of these is the need for clear articulation of Canada's national interests, based on a clear statement of fundamental values and national goals. The second is to highlight both domestic and international interests against which an environmental assessment will be undertaken to identify the events, issues and trends which may have an impact on Canada's interests. Thirdly, a clear link between a national security strategy, national security policy and defence policy should emphasize that national security is about more than defence, but that defence is importantly integrated into national security; hence the important involvement of the other government departments concerned with foreign, economic and social policy. Finally, defence policy that articulates the roles and capability requirements consistent with the national security strategic situation ultimately leads to the location and role of individual service strategies in the overall scheme to ensure a comprehensive approach to the issue. This framework is presented in figure 5.

The first step, the statement of national interests — values and goals — can be derived from existing policy and constitutional documents. The Canadian values used here (democracy, the rule of law, freedom and social justice) are the core values or intrinsic values of virtually any Western democracy, and the shared values which are so often referred to in political speeches and policy statements. These are not, however, the Canadian values often expressed in terms of social or "instrumental" values, as described by Maloney[34] ("soft power," tolerance, education, medicare and the like), which are, in reality, derivatives of the core values. The core values should be clearly stated, as these are the values we expect our troops to defend and, if necessary, die for.

The national goals (sovereignty, regional cohesion, economic well-being, security, international stability and promotion of values) already articulated in both foreign policy and defence policy documents are consistent and appropriate.

The next step is a formal strategic analysis, an environmental assessment and stakeholder analysis, to identify domestic and international events, issues and trends that may affect Canadians and Canada's interests. This step requires a detailed analysis conducted by a team of analysts schooled in the analytical techniques and knowledgeable in the areas of politics, economics, technology, and military and international affairs. This analysis — strategic intelligence — should

Figure 5

A National Security Policy Framework for Canada

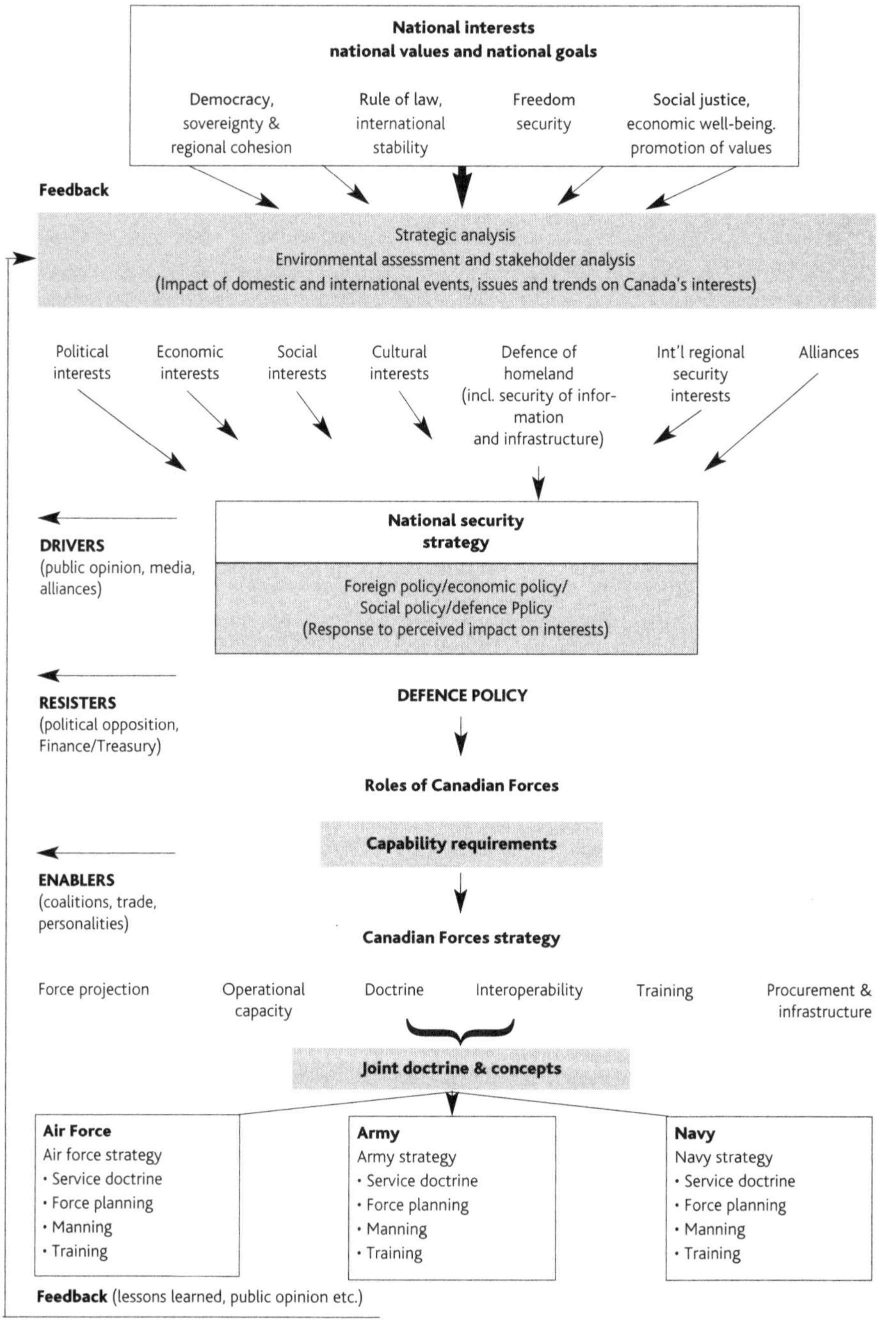

identify the major risks and threats to Canada's interests and, therefore, overall national security, in the broad sense defined at the beginning of this paper.

From the strategic analysis, a policy response is developed — a national security strategy. First, the impacts of domestic and global events on the various areas of concern to Canadian's national security are detailed: political, economic, social, cultural, as well as those pertaining to defence, international regional security, and our allies. This leads to the development of a set of major policies (foreign, economic, social and defence) and a national security strategy to react to those impacts and protect or promote Canadian interests. The coordination of the development and execution of the various policies is properly the role and responsibility of the Privy Council Office, which in the context of this national security framework and process will thereby be effectively a national security agency.

Defence policy, formulated in concert with the other security policy components, can then concentrate on the essentials as detailed in figure 5. The roles of the Canadian Forces are dictated by the policy that then leads to an analysis of the capability requirements to meet the perceived risks and threats identified in the strategic analysis. From this analysis, a Canadian Forces strategy can be formulated — how to organize and deploy the Canadian Forces given the threat environment, capabilities required and resources made available by the government. This will lead to decisions in a number of areas concerning activities; for example, the need for force projection; operational capacity; operational doctrine, including opportunities and needs for interoperability with allies; training; procurement; and infrastructure. Joint doctrine and concepts developed from this phase can then provide the foundation for the individual services to develop their own strategies, doctrine and planning.

On the left side of the chart are indicators of elements external to the process that can have an influence on the policy directions. "Drivers" such as public opinion, the media and alliance members will provide essential views of whether or not an open policy process includes public consultation. "Resisters" to certain policies may arise from political or bureaucratic opposition, such as the Department of Finance, the Treasury Board or even Foreign Affairs or National Defence. "Enablers" such as the trade environment, interest group coalitions or even public personalities may be less directly involved in the policy development but can be important factors supporting the intended policy direction.

Finally, the proposed framework provides for feedback that can and should be formal and public to ensure that the policies, in fact, address the concerns identified in the strategic analysis, or even identify changes in risks or threats.

An additional tool, a modification of the very useful Nuechterlein National Interest Matrix is offered (figure 6) to improve the utility of the matrix for Canadian conditions. This matrix can be used for analyses not only to clarify interests at risk

Figure 6

Nuechterlein's (2001) National Interest Matrix (modified for Canada)

Basic national interest / Intensity of interest	**Survival level (critical)**	**Vital level (dangerous)**	**Human level (serious)**	**Periheral level (bothersome)**
Defense of Canada/NA				
Economic well-being				
Stable world order/int'l security				
Promotion of Cdn. values				

in a given situation, but also to assess the level of response and to communicate more clearly with both the various relevant departments and the Canadian public.

Determining whether the proposed systemic process for national security policy development will lead to better sets of coordinated policies which may improve Canada's governments' domestic and international performance would perhaps attribute too much simply to process. On the other hand, with clear statements of national interests at the highest level and a coherent set of policies aimed at protecting or enhancing those interests, the Canadian public could become more directly involved in the discussions, and hence legitimize the assumptions in the foundations and the eventual policy outcomes, as befits an involved democratic society.

Conclusion

The need to have a systematic and consistent process for the formulation of Canada's national security policy must be recognized to be in the interest of all of Canada — the government, principal policy departments and the Canadian public, but especially the Canadian Forces. To have a process that uses terminology and elements similar to those of our allies and major coalition partners can permit the comparison of conclusions, needs and capabilities to the better benefit of sound decisions affecting Canada's security interests in the broadest sense.

It is hoped that the proposed framework and the modified Nuechterlein matrix in their proposed (or some modified) form will find their way into the responsible departments and agencies. Canada may then have, for perhaps the first time, a set of policies that truly represents a comprehensive national security strategy, from which will flow a national security policy that will be understood and discussed by all Canadians, and to which appropriate public resources will be dedicated.

Notes

1. Slessor (1995).

2. The Standing Senate Committee on National Security and Defence (2002).

3. "When I use a word," Humpty Dumpty said, in a rather scornful tone, "it means just what I choose it to mean — neither more nor less." "The question is," said Alice, "whether you can make words mean so many different things." (Caroll, 1865).

4. This definition, formed in 1980 by Major-General L.V. Johnson CF (Retired) when Commandant of the National Defence College, was derived from the discussion of national security in the opening chapter of *The Soldier and the State* by Huntington (2002).

5. *Department of Defence Dictionary of Military and Associated Terms*, Joint Publication 1-02 (as amended to December 19, 2001).

6. Earle (1943). Quoted in Howard, see note 7.

7. Howard (2001), pp. 2-3.

8. Department of Defence (2001).

9. Hart (1949).

10. Brodie (1949), p. 177.

11. Luttwak (1987).

12. Freedman (2001), p. 11.

13. Department of Defence (2001)

14. Ascher et al. (1983).

15. Plischke (1988).

16. Nye (1991), pp. 54-6.

17. Department of Defence (2001)

18. Department of Foreign Affairs and International Trade (1995).

19. Department of National Defence (1999).

20. Malone (2001), pp. 555-78.

21. Lloyd (2000).

22. Clinton (1999).

23. United States, Department of Defense (2001).

24. Nuechterlein (2000).

25. Coles (2000).

26. Coles (2000).

27. Kaldor (1997).

28. Coles (2000).

29. United Kingdom Ministry of Defence (2001).

30. United Kingdom Ministry of Defence (2001).

31. Based on discussions with Group Captain Mike Colley, UK MoD's Joint Doctrine and Concept Centre, January 23, 2002.

32. House of Commons (2001).

33. Hoon (2001).

34. Maloney (2001).

References

Ascher, William and William H. Overholt. 1983. *Strategic Planning and Forecasting.* Toronto: John Wiley and Sons.

Brodie, Bernard. 1949. "Strategy as Science." *World Politics* 1 (July): 177.

Clinton, William J. 1999. *A National Security for the New Century.* Washington, DC: The White House.

Coles, John. 2000. *Making Foreign Policy.* London: John Murray.

Department of Defence. 2001. *Department of Defence Dictionary of Military and Associated Terms.* Joint Publication 1-02 (as amended to December 19).

Department of Foreign Affairs and International Trade. 1995. *Canada in the World: Canadian Foreign Policy Review.* Ottawa: Department of Foreign Affairs and International Trade.

Department of National Defence. 1999. *Shaping the Future of Canadian Defence: A Strategy for 2020.* Ottawa: Department of National Defence.

Earle, Edward Meade. 1943. *Makers of Modern Strategy.* Princeton, NJ: Princeton University Press.

Freedman, Lawrence. 2001. "Grand Strategy in the 21st Century." In *Defence Studies* 1(1) (spring): 11.

Hart, Liddell. 1949. "Strategy as Science," *World Politics* 1 (July): 486.

Hoon, Geoff. 2001. Minutes of Evidence. House of Commons, Defence, Question #300, November 28.

House of Commons. 2001. "Threat of Terrorism." House of Commons Select Committee Report, #3481-3482, December 12.

Howard, Michael. 2001. "Grand Strategy in the 20th Century." In *Defence Studies* 1(1) (spring): 2-3

Huntington, Samuel. 1957. *The Soldier and the State: The Theory and Politics of Civil-Military Relations.* Cambridge: Belknap Press of Harvard University Press.

Kaldor, Mary. 1997. "Foreign Policy in the 21st Century." In *Report of the Seminar.* London: One World Action.

Lloyd, Richmond M. 2000. "Strategy and Force Planning Framework." *Strategy and Force Planning*, 3rd ed. Newport, RI: Naval War College.

Luttwak, Edward N. 1987. *Strategy: The Logic of War and Peace.* Cambridge, MA: Harvard University Press.

Malone, David. 2001. "Foreign Policy Reviews Reconsidered." *International Journal* 56 (fall): 555-78.

Maloney, Sean M. 2001. "Canadian Values and National Security Policy: Who Decides?" *Policy Options* 22(10) (December): 45-9.

Nuechterlein, Donald. 2001. *America Recommitted: A Superpower Assesses Its Role in a Turbulent World*, 2nd ed. Lexington: University Press of Kentucky.

Nye, Joseph. 1991. "Why the Gulf War Served the National Interest." *The Atlantic* 267-268 (July), 54-6.

Plischke, Elmer. 1988. *Foreign Relations — Analysis of Its Anatomy*. New York: Greenwood Press, 1988.

Slessor, John. 1954. *Strategy for the West*. New York: William Morrow.

The Standing Senate Committee on National Security and Defence. 2002. *Canadian Security and Military Preparedness*. Ottawa: Report of the Standing Senate Committee on National Security and Defence.

United Kingdom Ministry of Defence. 2001. *British Defence Doctrine*. MoD Joint Doctrine & Concepts Centre: RMCS Shrivenham.

United States, Department of Defense. 2001. *America's Security in the 21st Century*. Quadrennial defense review report. Washington, DC.

5

Canada and Military Coalitions: Where, How and with Whom?

Douglas Bland

If we will the ends, we had better will the right means. For the means we select may betray our ends.

Michael Ignatieff, *Virtual War: Kosovo and Beyond*

Reinforcing Tradition

Acting through coalitions is a defining and traditional characteristic of Canadian foreign policy. This tradition is rooted in Canada's political and cultural history, its relative power among states and in the *modus operandi* of the international community. Foreign policy by coalition is also, however, a pragmatic strategic choice, for Canada would be essentially isolated from the major events and decisions in the international community in the absence of coalitions, or through a reluctance to join them. The fundamental question, therefore, is not whether acting through coalitions ought to remain central to Canada's foreign policy, but how Canada can influence the shape and operating expectations of established and emerging coalitions to best benefit Canada's national interests

The question is particularly relevant today given the challenges facing traditional alliances like the North Atlantic Treaty Organization (NATO), the frequency with which states seek to form and act through coalitions — if sometimes only briefly and for specific and narrow purposes — and the emphasis in Canadian foreign policy over the last few years on building nontraditional "coalitions of the willing" to address humanitarian and other global issues. Indeed, there is general agreement among scholars and practitioners that even the major Western powers "must treat multinational action as a central organizing principle for defence [that will] affect every facet of their preparations, from equipment acquisition to operational planning and concept development."[1] In these circumstances, Canada ought to have appropriate policies and an agile bureaucratic machinery to allow governments to assess when, where and with whom Canada will act in its own interests in both traditional alliances and in coalitions of the moment.

A comprehensive paper on Canada and coalitions would necessarily address a long history, rancorous debates in foreign policy, the efficacy or lack thereof in more recent military operations around the world and the place of non-military coalitions in foreign policy. This essay, however, is mainly concerned with Canada's present and future military capabilities and what the realities of those capabilities mean to policies based on acting through coalitions formed inside and outside traditional alliances established for conventional military operations and "operations other than war"(OOTW).

"What's the use?" is a question that overhangs most discussions of Canadian foreign policy and armed forces. Does it matter whether Canada has a large, capable military force or mere token forces sufficient to buy a seat at international tables? Some argue that Canada, rich as it is, could not afford, for domestic political reasons, a force large enough to influence international events which involve the major powers, so why try? But this influence argument is misconstrued if one assumes (as this author does) that Canada requires armed forces not to influence others' decisions about their interests and actions, but to influence decisions others may take about Canada's interests and policies. In other words, Canadian history tells us that without adequate military forces (and the political will to use them appropriately), others will take decisions about vital Canadian interests in North America and internationally. Canadian politicians have usually been willing to buy a seat at the table — at times a very cheap seat to be sure — to have say in decisions that affect Canada and Canadians.

Canada was once a successful and influential peacekeeper because the Canadian Armed Forces had capabilities that elicited respect and confidence from the international community. Canada was once able to "protect itself from help" offered by the United States because Canada could assure Americans that their northern flank would be protected. During most of the Cold War, American leaders believed the claim because they could see the forces deployed for that purpose. One clear message from the 9-11 crisis is that this trust is greatly diminished, and now Canadians are exposed to intrusive American demands for changes to Canadian domestic policies. Pierre Trudeau once derisively remarked that Canada had armed forces "more to impress our friends than to frighten our enemies," although some might suggest that he never understood the enduring importance of this observation.[2] But the reality underlying the remark and its importance to Canadian foreign policy may become increasingly obvious if Canada, as many credible observers report, continues to run down its armed forces such that we can impress in our own interest neither friend nor foe.

For Canada, coalitions formed outside traditional alliances have posed, and will continue to pose, several challenges to policy coordination, doctrine, defence program and capabilities development, leadership, sustainment and national command. The underlying questions, however, concern assessments of the most likely configurations of future coalitions, the political requirements for building them, and what policies and procedures Canada might adopt in these modern circumstances to give greatest effect to its foreign policies and national interest while continuing to act through coalitions.

Coalitions have a few distinct characteristics: they are more or less formal undertakings of two or more states encompassed by a promise to act within some definite area, time or circumstance; and they supply a mechanism for consultation between partners from time to time. Security-related coalitions usually provide for a commitment "to some future action." This action "could entail almost anything — detailed military planning, consultations during a crisis, or a promise by one state to abstain from an upcoming war."[3] In Canadian foreign policy, coalitions tend to be positive undertakings between Canada and other states in order to aggregate their political, economic and military powers to accomplish national goals that none can effectively achieve alone. The promise to act in concert, however, may be more or less ephemeral depending on the circumstances.

The temporal nature of most agreements that underpin coalitions highlights another major feature. Coalitions exist in an environment of competing interests, attitudes and perceptions, which create what Michael Ward termed an "alliance dynamic" characterized by "contradictory tendencies operating within bureaucratic meshes."[4] In other words, coalitions are political creatures subject to changing international events and domestic attitudes, and they thus require continual maintenance and management within and between states. In reality, the more likely it is that "the promise" will be called (or the more critical the events then and afterwards), the greater is the need for reliable mechanisms to shape the concerted efforts of coalitions.

Long-standing alliances like NATO have developed elaborate mechanisms for consultation and military command, but even NATO suffers great pains in sustaining a united goal in a crisis — witness, for instance, the allied campaign in Kosovo. Coalitions of the moment cobbled together even with the best intentions in the face of a crisis are confronted with major difficulties in holding together and conducting dangerous operations. History is replete with examples of these dynamics. But today, the effect of media reporting on domestic and international audiences increasingly seems to overwhelm decision-makers and commanders, such that one must assume that coalitions agreed in peacetime will be greatly stressed in crisis and conflict.

At issue, especially in active military coalitions — including those not ostensibly aimed at combat operations — is the fact that coalition leaders are certain to make decisions about national "blood and treasure." There are few states in which citizens will comfortably assent to sacrifices seemingly imposed by outsiders, no matter their indifference to the terms of a coalition agreed in peacetime. These general tendencies ought to be a warning to Canadian political leaders and senior defence and foreign policy planners to beware of coalitions. At least, leaders ought to carefully consider decisions related to Canada and coalitions in the context of the national interest, and insist on a firm national voice in any coalition decisions that directly affect Canada and Canadians.

The Bookends of Canadian Military Interventions Abroad

In 1956, at the behest of allies and under the United Nations flag, Canada took the lead in organizing and directing the UN Emergency Force in the Middle East. Canada was able to lead this critical international endeavour because it possessed the commensurate soft and hard assets to allow Canadian diplomats to convince allies, foes, prospective partners and the belligerents in the region that an effective force could be deployed. Furthermore, because Canada had in its own armed forces the hard assets — the troops, transportation ships and aircraft, communications units, and commanders and staffs — it could conduct and lead the operation as though it were a neutral state acting solely in the interests of the United Nations. Lester Pearson might have won the Noble Peace Prize for his initiative, but he knew beforehand that Canada had the wherewithal to accomplish what he had proposed.

In 1996, at the behest of the United Nations and the desire of the government, Canada again took the lead in an attempt to deploy an emergency force to Zaire and the Great Lakes region of Africa in order to help alleviate the enormous humanitarian crisis there. Unfortunately, Canada simply did not have the requisite assets to make this commitment effective in any sense. Indeed, some contend that if it had not been for the sudden end to the crisis, Canada might have overtly suffered a major foreign policy embarrassment. The designated commander of the proposed multinational mission, Lieutenant General Baril, was dispatched without adequate staff or communications support. Moreover, the government expected Baril to command a multinational force even though the

Canadian Forces did not have anywhere near enough transportation units to deliver even its own small contingent to the area. Neither did it have any logistical units to support the Canadian Forces detachment, to say nothing of the demands of the proposed international force.

In the most understated way, General Baril concluded in his final report that Operation Assurance was hampered; "due to such factors as the active posting season, leave, equipment [un]availability, and other tasking, there exists a real life gap between our real readiness levels and those derived from Defence Planning guidance tasks."[5] This statement not only illustrates the weakness of the Canadian Forces, but also that policy-makers ought to be wary of defence documents that tend to describe the intended capabilities of the Canadian Forces but not the actual capabilities or the readiness of forces. Michael Hennessy observed that, for the Canadian Forces, "moving from notional capability to actual capability was problematic" during this period, mainly because "real operational readiness was impossible to gauge."[6] The operational readiness system, such as it is, has long been criticized as being unreliable. There is little evidence to suggest that this serious defect in national security planning has been corrected.[7]

This brief tale of two operations — the bookends of Canada's intervention history — highlights the inescapable relationship between ends and means, between foreign policy aspirations and policy outcomes. Although no governments complained publicly in 1996 about the Canadian attempt to organize a multinational coalition, few trusted Canadian competence enough to actually commit forces to the so-called Multinational Force. Combatants in the region showed little respect for the force commander or members of the Canadian Forces when they arrived in the field. The inability of the Canadian Forces even to get to Africa in any useful military state did not go unobserved in London and Washington.

But those failings and the consequences that followed did not begin or end in National Defence Headquarters (NDHQ). After all, General Baril and other dedicated people were trying to react to an incomprehensible national policy. In Baril's opinion, "ultimately, during Operation Assurance the only apparent elements of a national strategy were the objectives inferred from UNSC [UN Security Council] Resolution 1080." It was a telling assessment of the long 40-year slide from the coherence of the Pearson era to the "bungle in the jungle."[8]

The Canadian Forces throughout the post-Cold War period reacted to confounding demands from governments, the public and a radically changed international system. They, like most other Western armed forces, found themselves in 1992 with internal policies, force structures, force levels, major capabilities and doctrines built to meet national defence policies that were suddenly out of step with the international situation. In a few months senior officers were asked to look away from 50 years of experience and toward emerging threats and conflicts in the

Middle East, Africa and the Balkans. Moreover, they were asked to assume significant responsibilities to redress instability in governments, chaos in some states, humanitarian crises and famine relief, and to intercede in near-war situations where no one was sure who was fighting whom or why.

Any fair assessment would conclude that the rank and file members of the Canadian Forces overcame many of these challenges courageously and well in all the varied missions to which governments committed them. And they did so when governments inexplicably cut away many of the resources they needed and imposed radical infrastructure, organizational, force development and policy changes on the Canadian Forces and the Department of National Defence (DND). In meeting their duty in these circumstances, officers and members of the Canadian Forces suffered numerous physical and mental wounds in Somalia, Rwanda, Zaire, Bosnia and Kosovo, among other places, and while serving in the air and at sea with the equipment at hand.

Not all the problems that beset the Canadian Forces during the 1990s sprang from "externalities." The officer corps was found wanting. Indeed, the Inquiry Into the Deployment of the Canadian Forces to Somalia concluded that several senior officers "failed in their duty as commanders" and that others, including some high-ranking defence officials, may have "covered up" wrongdoings in Somalia and in Ottawa.[9] Some units while on duty in the Balkans behaved in ways that shocked Canadians. The confusion at National Defence Headquarters, along with the public sense that some leaders were untrustworthy and apparently disregarded the common soldier, led to the early retirement of one chief of defence staff. The Canadian Forces have recovered from many of these failings, but the reforms are still underway and incomplete.[10]

The hard experiences of the Canadian Forces in the 1990s exposed a number of recurring difficulties in modern coalition and multinational operations. Generally, they fall into six categories: weak mandates and directions; uncertain international command; confused civilian and military relationships, especially between international commanders and international officials; overtasking of individuals and some types of units; incompatible communications and logistics systems; and contradictory force protection orders and rules of engagement. These problems occur within national contingents and between contingents as well as between force commanders and international authorities, particularly during UN-mandated operations.[11]

Some elements of these difficulties can be resolved for particular operations and in some coalitions before anyone is deployed. Others can be addressed once the units are gathered in the theatre of operations, although allowing forces to deploy with the hope that "things can be sorted out on the ground" is a precarious and unmilitary way to do business.

This observation is particularly telling when people in the midst of a crisis expect arriving forces to immediately swing into action to remedy their sometimes desperate circumstances. However, many problems and the frustrations caused by recurring difficulties can be addressed only through the development of international regimes or codes and policies that set in place principles, norms and procedures for multinational coalitions. But before Canada enters into any negotiations, Canadian political leaders, officers and officials must decide what principles, norms and procedures best serve Canada's laws, foreign policies, military capabilities and interests.

In Canada, fundamental inconsistencies between foreign policy ends and Canadian Forces means plagued foreign policy and defence planning throughout the 1990s, creating, according to Louis Delvoie, "a policy vacuum" with dangerous consequences for Canada.[12] In 1993-94, the government convened two separate and distinct committees of the Senate and the House of Commons to study Canada's foreign and defence policy options. Each study followed similar procedures, welcomed wide public participation and appeared largely free from institutional and partisan biases. The resulting reports were well received, yet in time their procedural weaknesses undermined their value as guides for policy and administration. The committees' reports seemed uncoordinated, if not contradictory. The defence inquiry was completed in detail almost before the foreign policy review began, an outcome many thought backwards in every sense of the word. The process unintentionally illustrated the chronic schism between these essentially related policies and their parent departments.

The ideological differences between various individuals and organizations in the Canadian defence and foreign policy communities, which emerged during the review process, continued throughout most of the 1990s. Both proponents and opponents of "soft power" and the humanitarian agenda argued among themselves, while real operations were launched from Bosnia to East Timor. Critical questions concerning force development and capital acquisition for the Canadian Forces were left hanging or, more usually, officers and officials in NDHQ moved along as they saw fit. In time, as the effectiveness of the "human security agenda" lost momentum, the recovering Canadian Forces set off on an ambitious — if underfunded — Strategy 2020 of its own making.[13] Where once the divergence between ends and means in Canadian foreign policy may have hung on ideological nuances, some of the decisions that have been taken in NDHQ (unless they are reversed or disavowed by political leaders) may have already outpaced the relevance of a future policy review of coalitions and Canadian foreign policy.

Consider, for example, the defence minister's recent policy announcement that henceforth "early in/early out" would be the basis for Canadian Forces deployments in any future coalition operation. This decision was taken and put into effect in the context of a small Canadian Forces deployment to Eritrea. The mission framework closely mirrored a policy preference favoured in NDHQ. Officials argue that if Canada wished to avoid most "coalition traps" and to exert maximum control over its policies, choices, armed forces and other resources in coalitions, then it should strive to find balanced coalitions of lesser powers. These coalitions, committed to low-risk, short-term objectives, in specific areas and for specific durations, may be more compatible with the capabilities of the Canadian Forces and the DND's budget. However, it is unclear at this time whether the strategy was decided to meet Canada's foreign policy objectives or dictated by the continuing decline in Canadian Forces' capabilities.

There are apparent contradictions in the strategy even within the DND's own policies. For example, the 1994 *White Paper on Defence* and current defence plans commit Canada to provide forces for NATO without preconditions, except for the size of the force. The early in/early out policy certainly does not apply to the NATO Bosnia expedition where, in fact, the Canadian Forces are building near-permanent facilities and arranging for an indefinite stay.

Even though the defense minister's strategy may satisfy the needs of the Canadian Forces, it is unclear whether it satisfies Canada's foreign policy objectives. For instance, how will Canada, which has a major global economy, limit itself to this strategy and the criteria of limited engagement without being labelled a "dodger." The early in/early out strategy presupposes that other states (assumed in the halls of NDHQ as less capable states) will fill in behind departing Canadian troops. Is this a reasonable assumption, and is it reasonable to expect these lesser states to carry duties that are presumably too onerous for Canada? What message does this policy send to the international community concerning Canada's reliability as an international actor in global security and humanitarian issues?

If acting through coalitions is central to Canada's foreign policy and the conduct of that policy overseas, then what effect will a limited engagement strategy have on Canada's policies and diplomacy in the future? For instance, one must assume that Canada would commit itself to a multinational operation to have a say in how the issue is settled. But if Canada withdrew necessary forces from the coalition before the problem is resolved, then what role could Canada expect to play in determining final outcomes? In effect, early in/early out could in some circumstances undercut Canada's foreign policy and national interests. Thus if Canada is to adopt a strategy of early in/early out, then that policy decision and its application in particular cases should fall to the foreign minister and the Cabinet as a whole, but not to the defence minister alone. One would hope that these types of questions were debated in detail in Cabinet and in the departments before the defense minister made his announcement of a strategy that may profoundly affect the conduct of foreign policy and the image that important friends and allies of Canada hold.

Escaping the worst effects of coalition politics does not necessarily require abandoning the idea and the foreign policy of acting through coalitions. Nor is it unreasonable to consider how to get the most benefit for Canada from such policies at the least cost and risk. But it seems inappropriate for a rich country to plead "lack of resources" to a world of poverty, strife and conflict, or to the United Nations or to traditional allies who are accustomed to Canada's active participation in international affairs. Moreover, it may seem distinctly un-Canadian to many citizens and could eventually prove to be a political liability for politicians who might try to defend the notion.

The hard pounding of the last decade, the inconsistency of the government's policies and preferences, and the withering away of military capabilities — especially of well-trained, experienced people — have had a profoundly negative effect on the Canadian Forces. This result will doubtless limit Canada's foreign policy choices insofar as the latter depend on military assets. By some accounts foreign policy is already suffering and constrained; witness the unprecedented number of times Canada has had to refuse peacekeeping service with the United Nations.

A balance between foreign policy ends and national means was met generally, if not ideally, during most of the Cold War era and for UN operations conducted during the same period. Since 1989, however, the usual basis for foreign policy and defence coordination and planning has been upset and, arguably, no comprehensive national strategy has replaced the old "strategy of commitments."[14] Plans are in place in NDHQ to build a new "light and lethal" Canadian Armed Forces, and an outline of these plans and criticisms of them will be discussed later in this paper. However, the chief worry outside NDHQ is that these plans are being pushed forward without the benefit of any national security strategy, and they may be incompatible with Canada's foreign policy and security needs. Building such a strategy as a foundation for national security planning and operations for the twenty-first century ought to be of paramount importance to Canada's political leaders if they intend Canada to recover its international place and to act effectively through coalitions in the future.

A national security strategy

Plans and intentions. Military planning in the absence of a national security strategy has been complicated by a significant reduction in national defence budgets (some 30 percent over the past 10 years), the so-called revolution in military affairs (RMA) and the fact that old age has rendered much of Canada's defence capabilities obsolete. Chiefs of defence and other military leaders and defence officials have been forced to take decisions concerning what capabilities to produce in the short and long terms without much guidance from governments or coordination with foreign policy goals. For example, should planners prepare the future force according to the directive of *Defence 1994*, "to fight alongside the best against the best," a significant and expensive objective by any estimation; alternatively, should they support "soft power" humanitarian interventions worldwide in "coalitions of the willing," where the region and "the willing" may be unfamiliar to the Canadian Forces? Even if the choices were not as stark as these (and they are not always so), there are few clear beacons to guide military planners when making choices about where to direct Canada's long-term defence program.

Defence planners, however, are not completely innocent in these circumstances. Quite naturally, military officers and other authorities in the force-development process have their own notions of what kind of armed force Canada needs. They also have their own ideas about why, where and with whom Canada should make coalitions. These ideas and attitudes shape the decisions these individuals take regarding defence capabilities, the distribution of resources between capabilities and missions, and the military arrangements and procedures they make with allies.

The lack of a national strategy can create two types of problems. First, in the absence of a national strategy, military officers have no way of knowing whether their decisions will affect foreign policy goals for better or worse. But one thing is sure: their decisions will condition foreign policy insofar as hard assets drive foreign policy choices. Second, changing foreign policy goals and (more often) the aspirations of politicians in crisis situations may tempt political leaders to demand the deployment of the Canadian Forces in circumstances for which they are ill-equipped or otherwise unprepared. Such decisions, as in the Operation Assurance (Zaire) decision in 1996, may result in an international embarrassment for Canada or place members of the Canadian Forces in great danger, or both.

In an era of standing coalitions and coalitions of the moment, which might involve the Canadian Forces and other Canadians in anything from combat operations to humanitarian actions in insecure regions of the world, the government, and especially the armed forces, must be appropriately prepared for a wide-ranging operational environment. If Canadian governments believe that Canada ought to continue its alliance with traditional friends and be prepared to join coalitions of the moment under United Nations or other leadership, then they must support a full range of intervention capabilities. On the other hand, governments could choose in advance a smaller range of coalition possibilities and develop a national security and foreign policy strategy of ends and means appropriate to that choice.

A policy of "go small or stay at home," however, might carry penalties or sideline Canada in a world where multilateralism is the organizing principle of international relations and where influence accrues to those willing and able to share the burdens that flow from this principle. In the absence of a national strategy that both spells out with reasonable precision the objectives for foreign and defence policy and matches these goals to appropriate means, no one can predict with certainty where the diverse planning in departments and the Canadian Forces will take Canadian foreign policy.[15]

Ends and means

A Canadian national security strategy for acting through coalitions, by definition, cannot escape the joining of ends and means. Nor can foreign policy planners leave the fundamental decisions about what means are appropriate to policy ends entirely to others, lest the means drive the ends. Ends and means may become discordant through weak analysis, misunderstandings, incoherent direction and planning, or inattention over time. However, major difficulties in execution can arise from failure to maintain the resources necessary for coalition operations at reasonable levels of readiness.

Means in this context fall into two broad designations, soft and hard assets, each of several capabilities, some of which are controlled by domestic decisions and some by international agreement. Soft assets are highly flexible, readily assembled, unobtrusive and process oriented. Hard assets are physical, technical, obvious in deployment, normally requiring substantial and continuous preparation for employment, and costly; and they need to be deployed in large numbers to give much effect to events.

Soft and hard assets are (or ought to be) complementary and compatible. However, whereas soft assets can be developed quickly from national and international sources, hard assets are difficult and expensive to develop, something that has important consequences for foreign policy and defence planners. Generally, soft assets — money, people skilled in diplomacy and the technical functions of particular coalitions, and the mechanisms for intracoalition policy-making — are most important to diplomatic coalitions meant to unite declarations of intent and to confirm commitments before humanitarian and security coalitions are brought into action. Hard assets, on the other hand, are predominant whenever a coalition wishes to display its unity through a show of force and whenever a coalition resorts to overt operations.

Hard assets are the stuff of security coalitions in action. Foremost, hard assets are people in governments, international organizations, nongovernmental organizations (NGOs) and armed forces who, through their skills and initiative, move declared policies into actual policies. These people must be recruited, trained, deployed and cared for by governments and international organizations if coalitions are to have effect anywhere. Hard assets include, as well, equipment of various sorts; stockpiles of expendable resources; land, sea and air transportation means; communication devices and networks; and other tangible items that allow people to construct working coalitions in the field.

National armed forces are the most obvious hard asset in any coalition. Yet, developing and maintaining appropriate hard military capabilities to service coalitions is a considerable difficulty for governments and military leaders. Ideally, governments would define their foreign policy related to coalitions in sufficient detail to allow diplomatic and military planners to develop appropriate capabilities for these purposes. Afterwards, governments would modify commitments to coalitions carefully to maintain a reasonable balance between foreign policy ends and available military means. This logic has not always been followed in Canada.

Coalitions usually demand, if they are important, a commitment of both soft and hard assets. While some might believe that a commitment to coalition partners is a sometime thing — promissory notes, not cash, as was perhaps the case through much of the Cold War — coalitions today are based on cash, not credit. As coalitions are a central part of Canadian foreign policy and national defence planning, there is little room for naive and incomplete policies that may ultimately commit significant amounts of national treasure and prestige in coalition politics. Similarly, there is no room at all for strategic and institutional disorder when such commitments place Canadians at risk.

Commitments to coalitions that affect the well-being of other states can embarrass governments and Canadians when coalition partners point to the inadequacies of Canadian preparations to uphold their side of the bargain. Recently, the United States ambassador to Canada criticized Canadian defence efforts when he remarked: "I must note that many of our friends in Canada have expressed concern...that many on the US side of the border share. That concern is over resources for Canadian Forces...it has now reached the point where without significant increases the Canadian Forces could lose much of their effectiveness."[16] This official statement from the Bush administration, made before September 11, should warn foreign policy planners that Canada's incapacities are sidelining Canadian influence not only in Washington, but also in the international community generally. Allies may listen to Canadian rhetoric, but they can count our ranks for themselves and draw conclusions appropriate to their own interests.

Where in the World, How and With Whom?

Choosing where in the world Canada is willing and able to act in multinational operations, including humanitarian operations, is a difficult political decision, complicated by public and political perceptions that Canada is a leader in international peacekeeping. This national myth encourages Canadians to expect the government to participate in, if not lead, significant international coalitions. The nature of international crises and Canada's apparent enthusiasm for intervention gives an erratic shape to many aspects of foreign policy planning. However, coherent, coordinated diplomatic and military policy and plans suffer without prior basic consideration and choices about where Canada can and ought to act in international affairs.

Choosing among three obvious coalition leaders seems appropriate. Canada could continue its traditional emphasis in North America and Europe, to include, perhaps, a broad definition of where Europe begins and ends, and emphasize operations with NATO. On the other hand, Canada might emphasize coalitions of the moment, usually formed under the direct auspices of the United Nations. Third, Canada might more closely identify its defence and foreign policy with American aims and programs and ally itself mainly with American-led coalitions. Each of these general options carries its own costs and benefits, and any decision on where to go should be made in that context. Therefore, no matter the choice, governments ought to make an explicit and inseparable decision to build a defence and foreign affairs establishment capable of deploying and maintaining over a long period the requisite forces such operations demand.

Although Canada's coalition policies since 1950 have been shaped largely by the United Nations and NATO, and more recently by coalitions of the moment conceived and constructed at the edges of these institutions, are these the only options in Canada's foreign policy future? Might not other international organizations, the Organization of American States (OAS) or the Association of South-East Asian Nations (ASEAN), for example, sponsor some types of international actions that could not conveniently be arranged through the United Nations or NATO?[17] Could not Canada prepare itself to lead a common cause security and defence coalition constructed around a particular issue or regional matter, perhaps in the Caribbean or in Africa, as it has done recently on issues like land mines and international courts? These are questions for later in this paper. For the moment, it is safe and conventional to assume that Canada will always be a partner in NATO, the United Nations, coalitions of the moment and with the United States, and that the characteristics of each partnership will condition Canadians' particular responses to international affairs.

The North Atlantic Alliance

The North Atlantic Alliance is the principal coalition that oversees security matters in Europe west of the Russia-Eastern Baltic-Ukraine-Romania boundary and in the Mediterranean Sea. Its regime is well known to Canadians, and its political/military mechanisms are sound and practised. The alliance has overwhelming military power in the region and further afield in some circumstances. Although NATO functions by consensus, the major powers, and unquestionably the United States, play a dominant role in policy decisions and in any operation the alliance might undertake. In the past, allies have expected each member to join, as their capabilities permitted, every NATO mission and certainly those anticipated in "general war" circumstances. More recently, in the Balkan missions, states' operational commitments to NATO have become more discretionary. That is to say, national interests and sensitivities have become more significant in decisions about who will join operations and what they might bring to each mission.

During the Kosovo war, for example, Greece downplayed its involvement for important domestic reasons. The NATO deployment to Macedonia was approved by "silent voting" — an example of "the rule of the most affected" in action — and involved a discrete number of national armed forces under British leadership. These examples suggest that NATO is once again adapting to emerging circumstances, much as it has done in the past. But what is most striking is the degree to which the alliance is able to form coalitions within the coalition.

Once cohesion and unbreakable consensus were the dominant characteristics of the alliance standing before the Soviet Union. Today, carefully controlled "flexible responses" to particular situations by various combinations of states and military organizations allow the alliance to act together in name, but without unduly stressing states' interests. From another perspective, the Macedonia operation hints at a greater willingness of the European powers to lead alliance actions without significant contributions by the United States. Although this opinion might be overdrawn given the small scale of the Macedonian operation, the largely European International Stabilization and Assistance Force deploying to Afghanistan (without any Canadian Forces) may support this notion.

Recent developments in NATO and the evolving European defence entity may change Canada's response to coalition operations within the alliance. For instance, it may no longer be necessary for Canada to join every NATO initiative simply to show its solidarity with the allies. Canadian Forces deployments, especially, might be more discrete and concentrated (as they have arguably been recently) without fear that they will be criticized in Brussels or at home. On the other hand, once made, commitments to actual operations are difficult to undo — witness 25 years in Cyprus and 10 years in the Balkans. If the only prospect for withdrawal from the Balkans is "peace in our time," then Canada and the Canadian Forces might be committed to the region for a very long time. In this event, foreign policy officials might consider drafting a specific policy aimed at Canada's participation in NATO coalitions, and that policy could not avoid questions and recommendations on force development and capabilities planning in Canada that might be at odds with extant planning in NDHQ.

The United Nations

Arguably, as the United Nations takes a more prominent role as the legitimizing authority for international security operations — and especially for interventions by the international community in the affairs of sovereign states — coalitions that it has blessed may become the main mechanism through which most states act in their own interests and, presumably, in the interests of the global community. The United Nations is, therefore, the second foundation institution — and Canada's preferred institution — for coalition building.

In fact, building coalitions within a coalition has become a principal business of the United Nations Security Council and the secretary-general. Over the years, however, the United Nations has experienced considerable difficulty building effective coalitions to maintain stable conditions in contested areas; for instance, in the Middle East, Africa and the Balkans. Various ideas and reports, such as the UN Brahimi Report, point to the continuing need to reform the United Nations if it is to better anticipate, plan for, organize and support all but simple, small, less risky coalition operations. Unfortunately, there is no reason to expect that any of these ideas or recommendations will be put into effect soon.

Nevertheless, the United Nations will probably remain the "legitimizer" for most international interventions that involve member states, and it will continue to promote interventions whenever the secretary-general can build momentum for them within the Security Council and among public opinion. Canada, because of its support for the United Nations and because it is a rich country with limited advanced capabilities, can expect to be called upon by the secretary-general to commit soft and hard assets to future UN operations. Given that the United Nations will most likely be charged with humanitarian and OOTW type interventions in underdeveloped regions of the world, there are two policy questions for Canada. First, what type of capabilities is Canada prepared to develop which would best serve the missions that will probably fall to the United Nations? Second, what is Canada prepared to do to enhance the United Nations' ability to conduct coalition operations?

Answering the first question requires an assessment of the likely force requirements for typical UN intervention operations. From this study, Canadian policymakers could develop a force model for Canada, one that would likely include diplomatic, military, police and NGO capabilities, as the basis for a national strategy to support the United Nations. Military planners, of course, have other imperatives that drive force development, and they are spelt out in the government's *Defence 1994*. Leaning the military force model too far toward the United

Nations' needs might compromise other national defence requirements. By the same measure, leaning too far toward "battlespace" warfare and the RMA might compromise Canada's ability to aid and support the United Nations, or to join "nonconventional operations" where stealth and people are central to plans. The goals and their consequences for force development need to be reconciled within a national strategy for acting through coalitions.

Helping the United Nations is never easy. If it were easy to change patterns of international behaviour through the United Nations, then the worthy recommendations of dedicated people would have accomplished this task years ago. The United Nations is not beset by a puzzle in search of a solution. Rather, it is a political institution and functions along political lines much as its founders anticipated. However, once it has decided to intervene in some dispute or crisis, then there are ways that might enhance the effect and efficiency of that intervention. Many, if not all, of the important recommendations have been made, in some cases many times. But a crucial area frequently criticized in Canada and elsewhere concerns the command-and-control of UN forces or, rather, the weaknesses or absence of UN command-and-control capabilities. Here is an area where Canada might put forward ideas and an effort to make a difference in the interest of the both the United Nations and the Cansadians deployed under it.

Military leadership in the diplomacy of UN-mandated multinational operations is a prominent characteristic of recent coalitions. Few diplomats or soldiers were prepared for this outcome when international interventions began to multiply and developed into quasi-military campaigns unlike any operation of the Cold War-peacekeeping era. Whether in the Gulf War, or in Bosnia, Kosovo, Zaire, Rwanda, Haiti or East Timor, senior military officers and commanders have been asked or required to take decisions far outside the usual expected range of military matters.[18] Officers, including Canadians at times, are increasingly involved in political, legal and ethical questions that they must help to decide to ensure coherent coalition operations. Many officers are frustrated by the confusion that swirls out of the United Nations, NATO's political committees and national capitals. They are frustrated not only by orders and counter-orders, but also by the restrictions that follow from the notion that leaders of intervention forces ought to be "even-handed," no matter the circumstances or the activities of people in the region.

General Wesley Clark, supreme allied commander Europe (SACEUR) and overall commander of allied forces during the Kosovo war, recalls that he was asked to "use forces, not force," but to do so following "the principles of any military operation." But what was especially vexing to Clark as he dealt with conflicts in the Balkans was the idea taken from United Nations mandates that commanders would be given "unlimited obligations" to protect civilians, deliver aid, secure safe areas and so on, but "very limited authority" to accomplish any of these things.[19] The formulation is a perversion of military principles and commanders' expectations, and it contributes greatly to coalition difficulties in the field. This difficulty is exacerbated when the belligerents know that commanders of UN and other coalition forces are greatly restricted from using the military force placed under their command. Hostile political leaders, some of whom are former military officers, have little respect for, and are unlikely to be swayed by, international commanders who they know have no right to act on their own initiative. What Clark suggested should be the policy in all allied coalitions, and what he was seeking under the Dayton Accords was "to limit the obligations of the military — you can't do everything with military forces — but to give the commander unlimited authority to accomplish these limited obligations."[20]

Where Clark was frustrated in a NATO context, other officers, like General Romeo Dallaire, were ironically and tragically blocked by the same constraints under UN rulings from accomplishing the missions the United Nations sent them to do. In Dallaire's case, it led directly to the catastrophe in Rwanda. Better to send no commanders than to send officers prohibited from commanding their forces to accomplish their missions.

Sorting out the new civil-military relations between nations in coalitions and "international commanders" is an essential part of building clear methods for acting through coalitions. Canada might take the initiative in sorting out this messy relationship in international affairs by first qualifying its own policies with regard to Canadian Forces officers deployed and assigned as coalition commanders.

Coalitions of the moment

Typically today, many international interventions take place outside the Cold War, allied framework and the "Cyprus model" of peacekeeping. "Coalitions of the willing" in Somalia, the Balkans, Africa and East Timor have brought together nations and armed forces that are often strangers. One thing that is known about these new types of operations is that not much is known in advance of their assembly and deployment. There are few principles or rules or decisions in place before a crisis occurs, and Canadian officers and officials and strangers are forced to cobble together operating procedures in the midst of a crisis. Furthermore, because each operation tends to be unique in important ways, these same officials are often forced to define a Canadian position and an operational response in haste at home and overseas. Repeatedly, they and others have complained about the inadequacy of the machinery of national government and international organizations in these circumstances.

At the end of the Cold War, some believed that the new international order could be managed mainly through the offices of the United Nations and the willing cooperation of states in various types of coalitions. Certainly in Canada, the Axworthy doctrine followed these lines and encompassed a greater vision in which states, great and small, would increase the breadth and depth of interventions around the world. Canadians could not, according to Axworthy, "ignore the problems of others, even if we wanted to," and he asserted that Canadians did not want to ignore others "in an increasingly interconnected world."[21] One can argue, as Sokolsky and Jockel have, that the "human security agenda" rescued Canadian defence policy from "military irrelevance and strategic sterility."[22] Others, however, would now say that military leaders have largely gotten past Axworthy's challenges and fashioned a new (old) relevance for themselves in *Strategy 2020*, leaving foreign policy behind in the race to situate a post-Axworthy national foreign and defence strategy.

While Canadians debate the legacy of foreign ministers, the *sine qua non* of the human security agenda, interventions "in the problems of others" may be becoming increasingly problematic, not only for Canada, but for other major states as well. President Clinton's policies encouraged the American armed forces to launch in 1997-98 a mission to "shape the world through engagements." America's *National Military Strategy* defines shaping as helping to "shape the international environment primarily through [the] inherent deterrent qualities and through peacetime military engagement" of the armed forces. This effort would help "foster the institutions and international relationships that constitute a peaceful strategic environment by promoting stability; preventing and reducing conflicts and threats; and deterring aggression and coercion."[23] The strategy is controversial, not only because the goals are difficult to delimit, but also because some theatre commanders believe it cuts into resources dedicated to the armed forces' primary missions. The greatest concern, however, is the way "shaping" can turn against the "shaper" and entangle the United States in unresolvable regional and domestic conflicts. Rarely can anyone, least of all military officers and their units, help solve others' problems without taking sides or making choices that seem to favour one faction over another.[24]

The British government's defence policy directed "Britain's armed forces [to] make a major contribution to peacetime diplomacy" as part of a general policy to "be a force for good" around the world. The "key aim of these activities [known in the UK] collectively as 'Defence Diplomacy' is to teach other countries how Armed Forces should operate in support of and be accountable to, a democratically elected government. In this way [Britain] aims to reduce the potential for tension and conflict and make the world a safer place."[25] Heightened emphasis on the armed forces' traditional role of "showing the flag" is a display of British willingness to join international efforts to construct a peaceful world. Engagements in the Persian Gulf region, Kosovo, Macedonia, Sierra Leone and now especially in Afghanistan, demonstrate a British willingness to lead coalitions and to use military force to achieve these national and international ends.

Nevertheless, a certain weariness with these conflicts and a disillusionment with the outcomes of costly engagements may be signalling that the idea that "others' problems are our problems" may have run its course. The Bush administration is in the midst of a defence and foreign policy review, but no one expects President Bush to commit the United States to coalitions of the moment unless they are formed to address significant American national interests, as in the war on terrorism. The British government is widely reported as being on the verge of substantially cutting major military capabilities, and this policy will undoubtedly curtail any opportunity to "do good in the world" outside Europe. Early in/early out may not describe a change in Canadian foreign policy, but the practical effect is that Canada, too, has lowered its horizons to a very few commitments in regions tied closely to Canada's traditional national interests.

Partnership with the United States

Under bilateral and multilateral agreements, Canada has enjoyed a long and beneficial security relationship with the United States in what a colleague called "the only coalition that matters to Canada." Despite the upheavals in international relations since the end of the Cold War, few political leaders in either country expected the fundamentals of this relationship to change any time soon. Although a number of influential American leaders were questioning what they saw as Canada's failures to honour its obligations to the coalition during this period, they did not carry much weight in Washington or Ottawa. Generally, Canadian officials successfully deflected American criticisms of Canadian efforts, and most Americans felt that its armed forces provided adequate defence in North America. For the most part, the currency in the Canadian/American coalition was Canadian political support for US security interests and activities outside North America.

The recent terror attacks on the United States and the growing apprehension that they will increase in number and ferocity has changed fundamental assumptions about the Canada/US defence and security relationship. This change will be all the more dramatic if Americans believe "the longest undefended border" must be defended — according to American standards — because Canada cannot be trusted to take the necessary actions to deter and prevent terrorists from entering the United States.

Thus, Canada's most important coalition may be headed for radical transformation, from one based since about 1950 on a threat of over-the-pole air attacks and from 1989 on no threat at all, to an overwhelming, all-encompassing concern for the security of the homeland. In this circumstance, the United States will undoubtedly look to Canada to share the burden of homeland security in hitherto unimagined ways, which will impose considerable tangible and intangible costs on Canadians. Should Canada hesitate or seek to avoid these new obligations, it seems likely that the United States will blockade its northern border, undertake covert intelligence operations in Canada and act unilaterally to defend itself by deploying its armed forces in Canada whenever the president deems it necessary. Canada faces no greater foreign and defence policy challenge than finding an appropriate and credible way to reassure the United States that Canada can live up to the 1938 Roosevelt-Mackenzie King agreement under which the prime minister assured the president that no attack on the United States could come through Canadian territory.

Canada must, however, view its coalition relationships with the United States and those outside North America more broadly. It is difficult to see how NATO, the United Nations or any important coalition of the moment could succeed without American political and logistical support and, at times, its armed forces. Nor should one exaggerate so-called neo-isolationism or unilateralism in American foreign policy. Careful policy planners in the United States tend to agree that coalitions are essential — for advanced basing of military forces, for instance — if the United States is to exploit its technological superiority in any regional conflict. They agree also that the American public expects a collective response and a sharing of burdens between the United States and traditional allies. But perhaps the main reason why the United States may usually seek to act through coalitions is because coalitions "are reassuring to others [states] and may contribute more to stability than attempts by the world's only superpower to unilaterally impose deterrence [and conflict resolutions]" on the rest of the world.[26] The challenge, therefore, is not to get the United States to act within coalitions, but to shape coalitions in ways that serve each partner's interests, needs and constrains.

While there is no doubt that the United States would act to defend its interests when necessary and seems now likely to look for allies in such situations, it is not as certain that the United States will always eagerly join coalitions devised by other states for other purposes. This observation is the clear lesson of the Rwandan affair and in some diplomatic contexts. But Canada and other states whose foreign policies are closely associated with the United States and which depend, more or less, on American soft and hard assets, cannot usually wait for a happy coincidence of their goals and American interests. Therefore, it would be especially useful to find ways in which Canada could help keep the United States continuously engaged in global security issues beyond those that directly affect America's vital interests.

There are, of course, scores of initiatives aimed at bolstering America's international "engagement." However, those that attempt to embroil the United States in every regional conflict outside America's definition of its vital interests might simply defeat the general intent and particular operation. First, overdependence on the United States in coalitions can appear to Americans as though allies are "in effect 'taxing' the American public" to the detriment of the United States.[27] This perception might only fuel the rhetoric and opinions of those in the United States who believe that "entangling alliances" are essentially wrong-headed. Moreover, where these types of engagements result in American casualties, then the negative effect of coalition-building with the United States can be greatly exacerbated, as the American experiences in Somalia demonstrate.

On the other hand, involving American armed forces in UN operations and coalitions of the moment can unintentionally worsen already tense situations and lead to conflicts between coalition forces and local inhabitants. Americans, for many reasons, are international targets that some people wish to fire on simply to gain attention at home and abroad. Thus, putting Americans in situations — especially on the ground — where inhabitants might see them as opponents could turn a manageable situation into a hostile situation. Arguably, this was the case in Somalia and may now be the case in Kosovo. Allies and coalition-builders ought to carefully consider the consequences of organizing coalitions around American assets and armed forces before they devise policies that assume that engaging Americans in multilateral coalitions is universally beneficial to national and international security interests.

What, then, might Canada do in these circumstances? Ironically, the best policy might be to support those Americans who argue for the restricted engagement of the United States in coalitions formed for missions outside America's direct interests. Doing this, however, would require others, including Canada, to pick up the American burden to relieve the United States from leading and underpinning every coalition in NATO, the United Nations and elsewhere. Specifically, Canada and the other states would have to build the requisite command and control mechanisms, develop armed forces, especially army units and formations, and accept the costs these policies would entail. They would also have to lead willingly when crises arise and to sustain their efforts until some reasonable outcome is achieved. In fact, some might contend that this is the key element in Britain's security relationship with the United States now.

This logic is behind much of what is happening with a greater European defence entity, and it is expressed in the British deployment to Sierra Leone. However, keeping the United States engaged by giving it room for disengagement will fail if the political will and effective soft and hard assets are not forthcoming. What can Canada do to enhance its foreign policy through coalitions? Canada could begin the long process of building a credible Canadian capability to lead and support multilateral coalitions at levels commensurate with its traditions, wealth, international position and global responsibilities.

A new realism?

Policy planners may be facing a "new realism." In this environment, the human security agenda may be put to new tests aimed at showing relevance to specific national interests. Being my brother's keeper may no longer serve as a basis for national foreign policies no matter the value such homilies hold for individuals. According to some observers, "the pendulum is swinging the other way now, above all in Washington...NATO is under political and logistical strain. Each western-brokered accord has entailed policing, and almost every intervention has turned out to be for the long term...making for deep reluctance to back diplomacy with military muscle."[28] Is there a truer example of this realism than the grudging reluctance of states to use force to implement stability in Macedonia and the consequent limited NATO commitment to the state that many described as the fuse in the Balkan powder keg?

If it is true that "[T]hose who cause fire will be left to burn," then what future is there for Canadian foreign policy through coalitions? The focus on traditional friends and allies will not change, but less will be expected in terms of commitments to "out-of-area" operations. The United Nations will remain a stage for pleas from some and condemnation from others, but it will be less likely to react to either simply because it does not have the resources to act on its own. Coalitions of the moment will come and go, but they too will be restricted to putting out small fires in otherwise controllable neighbourhoods. Embargoes, threats of force and gunboat diplomacy in the name of any cause, no matter how noble and necessary, may be the exception rather than the rule except where states and their vital interests are directly threatened or attacked.

Perhaps the new realism is fuelled, ironically, by a reverse CNN-effect already at play. Once media broadcasts stimulated governments to act for a good cause. Now the media may be unintentionally dissuading governments from acting through their ever-present images of Hobbes's "worst of all [worlds of] continual fear, and danger of violent death; and the life of man, solitary, poor, nasty, brutish, and short" and of autocrats in thugocracies, self-centred and unwilling to be reasonable in any circumstance. Who would send their children to soldier for collective security and the United Nations on the assumption that peace can be found in such places?

These pessimistic "mean times" and "complex political emergencies" may be driving good people from the scene, but this result is not the only predictable consequence of Canada's experiences in coalitions.[29] One can flee in fear from Hobbes' world, or one can act courageously to mitigate its worse aspects where one can do so. Saving some — by reinforcing success, not failure — means making choices about where to spend scarce resources. It also means making decisions aimed at creating the conditions for success.

International policy ought to address the following needs: (1) the establishment of an international regime to guide states acting together in stability and humanitarian coalitions; (2) the creation of interoperable assets that multiply the effects of national capabilities when they are brought together; (3) the organization of permanent, international, diplomatic, military, police and NGO headquarters and staffs dedicated to planning and controlling coalition operations both with and, where necessary, without the United Nations; (4) the creation of NATO-like military cultures and procedures in other regions of the world to enable states to rapidly bring military forces into being in coalition operations outside Europe; (5) the establishment of the terms and relationships of international military leaders before crises occur; and (6) the examination of the relationship between national leaders and international diplomats and military leaders assigned to conduct coalition operations. Perhaps the most difficult duty for political leaders is to explain to domestic audiences and ruthless leaders that people will not always be left to burn, and to commit themselves and their people to finding ways to bring a degree of stability to regions where there is a fair chance of success.

The Canadian Armed Forces: Current Capabilities, Future Aspirations

What capabilities — how ready?

Measuring power and relating it to influence is a tricky business. Nevertheless, there are conspicuous relationships between a state's resources and its ability to shape international affairs in its own interests. The capabilities of a state's armed forces are an obvious element of power, but the term is not often well developed, especially in discussions in Canada about its armed forces and its foreign policy. Any discussion about armed forces and coalition operations must begin from a common understanding of what is meant by the term "capabilities" and the relationship between capabilities and their state of "operational readiness."

Too often military capabilities are described and seen simply as pieces of equipment, "platforms" used to deliver weapons to targets. In truth, however, a military capability is not simply a piece of kit or even many pieces. A capability cannot be defined simply as a ship, an aircraft or a gun. No matter the general rhetoric in defence circles about the wonders of some new piece of equipment, the finest modern ship tied to the dock for lack of a crew is no capability at all.[30]

Rather, a capability must be measured as an inseparable combination of weapons and equipment; trained personnel; adequate supporting equipment and logistics, such as ammunition; command support, that is, trained officers and commanders to bring the capability into operation; doctrine matching various capabilities; and sustained logistics and personnel reserves. If any one of these elements is missing, then the capability is degraded or nullified. Therefore, when policy-makers consider how to sustain the foreign policy tradition of acting through coalitions, and before they commit Canada to any coalition involving armed forces, they should look critically for a comprehensive assessment of the capabilities of the Canadian Forces in the near-, middle- and longer-term futures.

Timeliness and the readiness of forces are critical factors that rest at the boundary of soft and hard assets. Hard assets, no matter their sophistication, are of little value unless they can be brought to bear in a crisis quickly, efficiently and credibly. Operational readiness is usually defined as "the state of preparedness of a unit...to perform the missions for which it is organized and designed." Operational effectiveness, on the other hand, is defined as the "degree to which a unit is capable of performing its duty."[31] Policies built on coalitions, or that place the Canadian Forces into coalitions, must carefully consider beforehand both the readiness of the Canadian Forces and the degree to which committed forces will be operationally effective in the circumstances.

The issue is not merely whether the armed forces (and other assets) are ready to deploy — although that capability is significant in both soft and hard cases — but whether they are reasonably ready to be employed. In the Gulf War case, for instance, the government seemed willing enough to deploy the Canadian Forces, but overly hesitant to employ them in the theatre once they arrived there. When Canada chose to join the East Timor coalition, the government committed the Canadian Forces to an operation for which it was not ready, in a region where there was little expectation that any unit — because of lack of experience and training there — would be operationally effective. The slow and costly deployment and the considerable operational and logistical problems the forces encountered in the theatre are evidence of a commitment made without a clear appreciation of the situation.

Readiness can be expressed in degrees — high, medium or low — but the state of readiness of any asset is mostly a political responsibility and a policy choice. Each choice carries its own costs and benefits. Generally, forces at high degrees of readiness can react to orders quickly, but the costs of keeping units at this level may be correspondingly high as well. Forces at low levels of readiness are less expensive to maintain (they do not need, for example, dedicated transportation units held ready for them at all times), but they may be very slow, too slow, to react to emerging crises. The operational readiness of forces, therefore, is fundamentally a political matter, not a military matter. Governments must decide where, when and how they wish to participate in coalition operations, and these decisions will drive second-order decisions about what forces are developed and at what state of readiness they are kept. Governments, then, ought to pay the price for their decisions.

But perhaps the relationship between the state of readiness of assets and the government's flexibility in a crisis is the factor that should most concern governments. Specifically, governments could delay critical decisions on deployments and their associate political and other costs if they have forces available at high states of readiness. On the other hand, governments may be forced to commit themselves early to costly public decisions or to deploy ill-prepared forces at the last minute when they have no recourse but to units and other assets that are in low states of readiness.

The present force vs. the future force

When policy planners discuss Canada's military capabilities, there are two armed forces they should consider: the present force and the future force. The present force, obviously, includes all the people, equipment and infrastructure now within the defence establishment. The future force includes those resources that are already funded for tomorrow and those that resources planners hope to add to the inventory over time. The present force is real, while the future force is anticipated, problematic and often merely a distant hope depending on the resources dedicated to that future.

Much of defence planning is a constant battle between the present force and the future force. For example, should funds be spent on readiness for current operations — ammunition and stores stockpiled for contingency operations — or should leaders allow readiness to decrease to find funds to recapitalize the forces? Should money go to holding people in the armed forces or should people's quality of life be sacrificed for equipment, on the assumption that equipment takes a long time to come on-line and people are more readily available?

If funds are scarce and demand for equipping the armed forces is high, then planners tend to hedge toward the future force, and current capabilities and readiness decline. If current foreign policy depends on the capabilities and readiness of the Canadian Forces but defence planners are hedging toward a future force, then foreign policy might suffer. On the other hand, if in the same circumstances defence planners favour the present force — because of overcommitment, for instance — then the future force suffers. Thus, foreign policy planners may eventually find that the defence cupboard is bare, leaving Canada unable to join any coalition, no matter how important this might be to Canada's national interests. Indeed, in the worst of all worlds (for people responsible for such things, at least) Canada could face both situations at once: few resources for current operations and a steadily declining future capability.

Although there is some debate about the particular capabilities of the Canadian Forces and their state of readiness from time to time, there is no disagreement in the defence community (even within NDHQ) that the

Canadian Forces have few modern capabilities or that the armed forces are at a low state of readiness. The Chief of the Army Staff, Lieutenant-General Jeffery, described the army of 2001 as "outdated" and "fragile."[32] Moreover, the hunt for funds is causing considerable difficulty for planners, who are forced to cut into "core capabilities" or to sacrifice one set of capabilities for some other.

When asked, "Will the Canadian Forces have what they need when they need it?" Brian MacDonald answered flatly, "No." He assessed that several critical capabilities were at, or approaching, the end of their life cycle. The basis for sustained army operations was the most severely affected. Other capabilities, though due to retire in the 2005-10 time frame, had been "extended" by various patches and partial upgrades. Macdonald concluded, "As things stand now [winter 2000-01], future foreign policy will be determined by budgetary decisions forced on the Canadian Forces about what equipment will not be replaced."[33]

The auditor general of Canada in his 2000 *Report on the Department of National Defence* gave an explicit warning about the state of the Canadian Forces. He emphasized that to meet the national defence policy to maintain a "modern multi-purpose force" requires "$11 billion in capital funds over the next five years [from 1999] but [that DND] would receive only $6.5 billion, resulting in a $4.5 billion shortfall."[34] He went on to report that the defence establishment was "out of manoeuvring room." The auditor general concluded that the defence budget needed an additional $1 billion for capital funding annually, just to allow the Canadian Forces to maintain an "even smaller force while modernizing, revitalizing infrastructure and maintaining readiness."[35] In 2001, the Canadian Forces are in reality becoming smaller, letting people go and beaching older pieces of equipment in the struggle to find money for the elusive superior future force.

Recently, the House of Commons Standing Committee on National Defence and Veterans Affairs (SCONDVA) began hearings on the operational readiness of the Canadian Forces. The early reports of this government-dominated committee are pessimistic and warn that Canada finds itself "in the hapless situation of either renting [equipment] or relying on allies." More ominously, SCONDVA can find little hope for early redress to the many capabilities and readiness problems facing the Canadian Forces. For instance, the committee notes that "2000-2001 estimates indicate that the planned spending for 2002-2003 is actually lower than the budget allocated for this fiscal [2001] year." While estimates indicate higher spending for 2003-04, "this increase may not be enough given all the demands being put upon the defence budget."

SCONDVA is especially concerned about the army's diminishing capabilities. The committee reported that

> Despite the rhetoric about technology and the probability that the RMA will increase our combat effectiveness, we need to remember that peacekeeping is done by individuals on the ground. Since the onus of peacekeeping falls disproportionately on the army, it is imperative that their levels remain at the full complement.[36]

The committee was convinced "that a complement of 60,000 personnel is the *minimum* required for the Canadian Forces to remain an effective force"[37] However, the Canadian Forces today have fewer than 58,000 "effectives," and the numbers seem to be falling rapidly.[38] As a result of the continuing decline in its strength, the Canadian Forces were forced to change policies on voluntary release from the armed forces and to launch a high-profile campaign to recruit personnel. Both these measures may, in the medium term, further decrease the readiness of the armed forces, as experienced members are withdrawn from units to train the 7,000 recruits that planners hope will come to the colours annually over the next few years. Some predict that the new personnel policies will cause morale to fall further, and that the recruiting and training bills will remove more money from other vital programs, unless the government increases the defence budget to cover them.

The defence program response. Departmental and Canadian Forces' plans to overcome these and other serious weaknesses are reflected in the department's Defence Plan (DP), published yearly after the federal budget and departmental estimates are tabled.[39] The DP allocates funds to Level 1 agencies in DND and the Canadian Forces (essentially senior general officers and assistant deputy ministers at National Defence Headquarters), and sets priorities for their expenditures in relation to defence tasks and change initiatives, or makes changes to previously approved programs. The Long-Term Capital Program is updated yearly and lists (mainly) equipment acquisition requirements and planned acquisition decisions over the next 15 years.

However, there is a great deal of subjectivity embedded in the department's programs, caused mostly by the tentative nature of federal budgeting. It is difficult for planners to control decisions when many of the components of the planning process are in the hands of people outside DND and the Canadian Forces, or when decisions are subject to reviews and second-guessing in the central agencies. The Long-Term Capital Program, for instance, is necessarily spread over several years, but actual funding is rarely assured for more than one year in advance. More worrisome is the surprise factor in defence planning.

Over the past 10 years, the Canadian Forces have been deployed to places and on duties that no one would have predicted and for which no plans existed. New demands — or more correctly, deferred demands — to improve the quality of life in the armed forces, to increase pay and allowances, and to revamp the recruiting system have placed unexpected pressures on scarce dollars. The effect is to impose a "random management system" on NDHQ, characterized by incremental responses to demands and events beyond the control of any putative rational policy process or response.[40]

Neither the essential randomness of the defence-planning process nor the criticisms or warnings of outsiders is news within NDHQ. Departmental "Level 1 Business Plans" are published yearly in response to the Defence Plan, and in them senior officers and officials report on the current and foreseeable situation they face in trying to meet government policy declarations. This year, these officers and officials reportedly acknowledged the stresses on the Canadian Forces as planners and commanders try to maintain even the core capabilities that are the backbone of the multipurpose combat-capable force and future defence and foreign policies.

Each officer and official raised serious concerns in their internal reports caused mainly by the differences between their defence tasks and change initiatives and the resources they have been allocated. According to some officials, the shortfalls in the Level 1 demands equal some $1.2 billion per annum just for operations and maintenance functions. When this demand is added to the capital shortfalls identified by the auditor general, the total amounts to some $5 to $6 billion over the next 10 to 15 years.

This level of funding shortfall assumes, of course, that the armed forces can get by with 60,000 or fewer personnel and that further labour costs can be held to 2001 rates. Neither assumption is dependable. If personnel costs rise and budgets are held more or less constant, then the pressures on operations and maintenance and the capital programs will increase and one or the other will have to be curtailed. The faint hope within NDHQ is that technology can in many ways replace people.

Replacing people with technology provides little escape from this funding dilemma, and it is a questionable policy from a foreign policy perspective given that, as SCONDVA emphasized, "peacekeeping is done by individuals on the ground."[41] In other words, foreign policy based on coalitions demands both high technical capabilities *and* more people, not trade-offs between these two critical components of a modern military capability. But defence planners, apparently, cannot avoid this dilemma. Thus foreign policy may be restricted, unless governments provide the funds necessary to change declared policies into actual policies. The tradition of foreign policy by coalition may be nearing an effective end.

Evidence of this approaching foreign policy crisis was expressly provided in recent NDHQ business plans. Senior officers, individually and together, reported specific shortfalls that might severely and negatively influence Canadian foreign policy options, especially coalition-related decisions, in the next few years and thereafter. Although many specialists believe that the navy is best prepared for operations, the chief of the naval staff reportedly noted that, among other problems: (1) the navy will not be able to deliver its mandated level of maritime defence capability without additional resources; (2) the increased cost of fuel, combined with no flexibility in operating budgets, will lead to a reduction in fleet operations; (3) maintaining a balance between sustaining current capabilities at a minimum level, investing in "quality of life," generating savings for the future and implementing change remains elusive; (4) the navy faces serious personnel shortages in a number of trades and specialities; (5) force development studies for major ship and system modernization are in jeopardy due to lack of funds; and 6) aircraft fleet reductions, national procurement reductions and a shortage of personnel will have a direct impact on the navy's ability to conduct surveillance of, and control, Canadian territory. Overall, he reported, these problems have the potential to severely degrade fleet operational readiness and effectiveness.

The air force is a technical service, but it too is caught in a current operations-future development bind. The following points were made by the air force chief of staff in his business report:

- The high operational tempo, numerous change initiatives (not always well coordinated), and significant fiscal and human resource limitations contribute to an increased stress level for personnel at headquarters, wings and squadrons.
- The air force is "one deep" in many areas and has lost much of its flexibility, redundancy and ability to surge (that is, to rapidly concentrate forces for critical missions).
- The air force faces significant personnel shortfalls. For pilots, the situation is extremely serious and will likely get worse over the next three years. The result will be a loss of capability. Other air force classifications are below the "preferred manning level," also leading to loss of capability.
- One of the most difficult challenges in the period 2001-04 will be achieving resource reductions while meeting DND and government performance expectations.
- The elimination of important parts of modernization programs appears to be the only potential area of savings.

The army chief of staff, however, delivered the most pessimistic message for foreign policy planners. According to interviews, he noted, for instance, the following points:

- The army is overdrawn on its human capital account, in both the physical and psychological senses.

- The army is not sustainable under the current circumstances.
- The ability to generate only subunit-sized force packages does not meet government expectations.
- Structural changes such as army transformation will not ultimately resolve the resource dilemma.
- The level of commitment in Bosnia is not sustainable.
- Specific problems facing the army of today include personnel fatigue, stress, and regular and reserve unit strengths and leadership cadres that have fallen to critical levels. There should be no illusions as to the size and cumulative impact of rotation stresses on Canada's small army.
- Either a resource infusion will sustain existing force levels, or force levels will be reduced to match projected resource levels.[42]

The future armed forces and consequences for foreign policy. Although one can predict a crisis in foreign policy based on the assumption that the army provides the mainstay of Canada's current coalition strategies and the army is failing, senior defence planners, including the past and present chief of the defence staff (CDS), seem to disagree. They are, evidently, planning a different future for the Canadian Forces, one that will inevitably take the armed forces away from army-based operations toward a war-fighting strategy grounded in the concepts of the so-called revolution in military affairs.

"To succeed in the battlespace of the 21st century" is the central message of *Shaping the Future of Canadian Defence: A Strategy for 2020.* The "battlespace" is defined as being global in scope and ranging from the sea floor to space. The emerging cyberspace environment adds another dimension to the battlespace. Military operations will be conducted at an accelerated pace, requiring rapid coordination of political and military objectives and increasing dependence upon information. The portability, range, precision and lethality of weapons will continue to improve, while the effective lifespan of sensors and weapons systems will decrease due to the rapid pace of technological change. Many emerging threats such as cyber- and bio-terrorism will tend to be asymmetric.[43]

The "force structure" envisioned under *Strategy 2020* might shape Canada's hard assets and therefore its foreign policy choices for years to come. Specifically, senior military leaders hope to "field a viable and affordable force structure trained and equipped to generate advanced combat capabilities that target leading-edge doctrine and technologies relevant to the battlespace of the 21st century." This is a worthy aim, of course, if the intent is to be prepared to engage leading-edge opponents in the twenty-first century battlespace. But, is this the most likely conflict scenario that Canadian politicians and diplomats will face in the immediate future or even in the next 10 to 15 years?

Given the continuing tensions between federal budget allocations to defence and present operational demands and the need to revitalize the Canadian Forces, going the high-tech route will likely result in a smaller armed force equipped with a few sophisticated capabilities suited intentionally for battlespace operations. The vision set out in *Strategy 2020* is transformed into fact in *Defence Plan 2001*, the DND "business plan for fiscal year 2001/2002." The deputy minister (DM) and CDS message proclaims that:

> Optimizing force structure...[is] an ongoing and never-ending challenge for us given the changes in the international environment, technology (including the Revolution in Military Affairs), and the need to live within a finite resource base. Our efforts here will focus on, among other things, the modernization of our fleets, of Auroras and CF-18s, the acquisition of new equipment such as the replacement for the Sea King helicopters, and an ongoing review of our force structure.[44]

The resource allocations within DND and the Canadian Forces detailed in *Defence Plan 2001* reflect these priorities. Certainly Canadian defence planners need to acknowledge the changing environment of warfare brought about by technical advances. But before they buy into the RMA in any substantial way, they must also consider the consequences of such decisions on operations, and especially OOTW, that demand less technology in the sky and more skilled people on the ground.[45]

The principal documents for defence planners also describe the defence establishment's appreciation of Canada's future coalitions or, perhaps more accurately, their coalition assumptions and preferences. As they see it, Canada will hold on to and reinforce its traditional alliances and must, therefore, build armed forces with the capabilities to support these allies. A directing vision within DND and the Canadian Forces strategy and a five-year target is to "manage our interoperability relationship with the United States and other allies to permit seamless operational integration at short notice."[46] To meet this target, "defence must keep pace with new military concepts, doctrine, and technological change," all of which, by any assessment, originates in the United States.[47] Moreover, this strategic orientation is supported by a putative demand from our North American Aerospace Defence (NORAD) and NATO allies who "*want defence to be a competent partner* capable of playing a meaningful role in combined operations: therefore, our armed forces must be interoperable with our main defence partners in the United Nations, NATO and coalitions operations."[48]

Policy is not self-imposing. Mere statements from ministers, officials and senior officers are often no more than declared policy. Actual policies are made in the rough-and-tumble of bureaucracies where the opinions, assumptions, interests and preferences of those with discretionary powers and authority to decide usually carry the day. The central role of the concept of the RMA, battlespace capability planning and interoperability with the United States armed forces as enunciated in *Strategy 2020* became actual defence policy because several senior officers and some officials decided that these notions provided the best interpretation of government policy. They have tenaciously held to this strategy for Canada against all criticism.

The former CDS, General Baril, told the Calgary Chamber of Commerce in December 1999 that "since all Canadian Forces deployments will be in cooperation with the forces of other nations, *a natural objective* is to strengthen our military relations with our allies...Central to this, of course, is adopting equipment, doctrine, and communications that are compatible with our NATO allies and, in particular, with the forces of the United States."[49] The recently retired vice- chief of the defence staff, Vice-Admiral Garnett, is a firm advocate for a strong Canada/United States military coalition based on the inevitability of the RMA. In his view the "RMA has in fact arrived at an ideal time for the Canadian Forces — a time when change must be embraced if we are to sustain our combat capabilities."[50] This central planning notion requires the Canadian Forces to "implement a prudent investment strategy for technology, with the key objective being the maintenance of interoperability with our allies [and to] achieve across the whole Canadian Forces...the level of interoperability with our allies that our navy has achieved with the U.S. Carrier Battle Groups."[51]

When his views and assumptions about the priorities for force development in periods of continuing funding constraints were challenged, Garnett repeated his RMA theme. But he directly attacked his critics' "attrition mentality of the 1970s and 1980s" (presumably held by those who favour an increase in military personnel strength) while advancing his idea for "a 'smarter' model of both equipment and people."[52] He again predicted that new technologies "will be the backbone upon which we can grow new joint Canadian Forces capabilities." In this environment, as Garnett honestly admits, "our army [has] and will continue to undergo the greatest degree of change."[53] But here too, he sees salvation in a new high-tech world in which the Canadian Forces will become "light and lethal."

Both General Baril and Admiral Garnett, and their successors by some reports, acknowledge that the Canadian Forces will be called upon to undertake humanitarian and OOTW commitments, but these tasks are second-order duties that will be met, in their view, by "improved war-fighting capabilities." But again, in the absence of funds for all capabilities, which military capability choices best serve Canada's foreign policy needs and who is to make these decisions? Is it unreasonable for the minister of foreign affairs to have opinions on Canada's military force development?

Defence planning documents do refer to "peacekeeping operations." The DND and Canadian Forces' mission is "to defend Canada and Canadian values and interests while contributing to international peace and security." Traditional peacekeeping duties and so-called OOTW are interwoven within the vision of "combat-capable armed forces," but there is no specific attention given to building a force for these purposes. The worry among some experts and former senior military persons is that the concentration on building a force to fight and win the envisaged "battlespace" battle is diverting resources and people from the actual battles that are now harassing, and will most likely continue to harass, the international community. Experience and analysis envisions small-scale, drawn-out conflicts among low-skilled persons and low-tech armed forces. Experience suggests that responding to these types of conflicts requires not only RMA-type standoff forces, but also highly skilled, well-trained people on the ground who can be sustained for long periods of time. If Canada wished to join these types of operations in coalitions (depending on their frequency and intensity), it must be prepared to train, and hold ready, substantial numbers of professional soldiers.

Generally, if actual experiences and assumptions drawn from them drive defence policy, then Canada might be developing a highly trained, sustainable, rapidly deployable and larger army. If foreign policy preferences and assessments were driving defence policy planning, then a similar force structure might be on the books. However, if the vision of future conflict involves war between major powers, each with comparable high-tech capabilities, and if Canada is always a partner to decisions taken by major allies, *Strategy 2020* and its adherents offer the best choice for force development and for defence policy. Given that *Strategy 2020* is the only strategy propelling defence choices and that its conceptual assumptions are embedded in significant segments of the defence establishment, then it follows that Canada's foreign policy may be moving away from participation in OOTW and humanitarian coalitions.

Indeed, if recent appointments to the most senior decision-making posts in the Canadian Forces have any relationship to the *Strategy 2020* approach and the preferences of its adherents, then one should expect a tightening of the battlespace vision over the next four or five years.[54] By then, there may be little opportunity for foreign policy planners to reverse or modify decisions taken in NDHQ. Once the money is spent, such as on submarines and space-based defences for example, then barring a substantial increase in defence spending directed at other capabilities over a long period, foreign policy choices that call on hard assets may have been settled already.

The Waiting Agenda

Strategy has many definitions, but the most useful for policy-makers is the notion that strategy is the result of sets of decisions joining ends to means taken by people with the authority to decide and to oversee the implementation of those decisions. Although foreign policy is the province of the minister of foreign affairs, that policy is, in fact, dependent in many cases on resources that belong to others. Designing a Canadian foreign policy strategy which has at its centre the idea of achieving Canadian goals through military coalitions must necessarily involve people of authority from other departments and agencies of the Canadian government led by politicians. If ministers directed officials, military officers and other authorities to bring forward a national strategy aimed at advancing Canadian interests through coalitions, what issues would be placed on the agenda before an interdepartmental forum?

The national interest

Doubtless, the primary matter would be to discover what "national interests" could be, and should be, advanced in coalitions. Because coalition dynamics invariably require compromises, it would seem that coalition politics ought to be restricted to those things that Canada cannot achieve on its own. However, coalition politics might also be used as an avenue along which Canada could and should involve itself in the affairs of others and of the world community in general. For instance, Canada might not always have access to negotiations between the United States and Latin American states, but maintaining a connection to the United States and Latin American states through the Western Hemispheric coalition the Organization of American States does give Canadian policy-makers access to aspects of these important relationships by right of association. The basic question for the strategic forum is, however, what is the connection between Canada's strategic imperatives and its strategic choices, with each coalition, old and new?

Why, where and with whom?

Quite naturally, the second item on the agenda should bring forward a discussion of the national parameters that would guide a Canadian coalition strategy. In other words, as discussed in this paper, why would Canada decide to join or to maintain a particular coalition? Where in the world can Canada best achieve its purposes most efficiently and with whom would Canada seek and accept an alliance? This is not an easy policy framework to design and build, especially in a political culture that seems accustomed to being a partner in every UN undertaking. But if there are no boundaries to where, when and with whom Canada will act, there will be no limit to demands for resources from departments, nor any way to contain Canadian expectations in matters of international affairs.

The Canadian rules

Canada cannot join other states or take on obligations that flow from coalitions without regard for national laws, costs, domestic politics and policies, and the need to maintain public support for foreign policy. It is critically important, therefore, that policy planners and individuals who lead Canadians in coalition operations have at hand a basic national regime for coalitions — Canadian rules of the game — to guide their actions and decisions. The next question on the agenda for Canadian leaders is this: What are the explicit terms under which Canada will join and support coalitions in international affairs?

A national mechanism for managing coalition dynamics inside government

Politicians, officials, scholars and others have long criticized the federal bureaucracy and the Canadian Forces for failing to coordinate foreign and defence policies more effectively. The usual complaint is that policies are too often separately conceived and administered, and that this habit compromises more or less the national interest. But as General Baril's reports from the Zaire operation and investigations such as *The Inquiry Into The Deployment of the Canadian Forces to Somalia* make obvious, the present critical conclusion is that missions have failed or been ineffectively cobbled together and people endangered, whenever Ottawa is confused or unsure.

The present difficulties in formulating and managing a national security policy for Canada extend beyond the defence and foreign policy establishment and into the wider structure of government and its departments and agencies. Proponents for greater coordination argue that if Canada is to protect itself from, and help redress, the many security and humanitarian problems of the world, then it must offer credible hard, as well as soft, national security options to Canadians and the international community. But reaching this objective will require a concerted and unified assessment of Canada's sometimes conflicting foreign, defence and internal security policies and the "machinery of government" related to these policy areas.

There are many reasons why Canada's planning for and execution of coalition operations seem awkward, and political neglect and naivety in matters of foreign and defence policy are a central factor. But another chief reason lies in the structure of the federal system, especially with regard to international relations and coalition-building and operations. Successful internal and external operations ought to be based on a single concerted security policy built on its own foundation, not on the hope that success might simply appear from several separately conceived and administered departmental policies. Nevertheless, it is departmental politics that dominate policy-making in Ottawa.[55]

Some believe that a more efficient committee system might lead to better policy coordination. They offer, for instance, a "national security council" as a device to bring responsibility for policy and operations now resident in several departments and agencies into harmony.[56] But these types of suggestions might only complicate an already complex bureaucratic mesh. Rather than committees, federal planners need direction and standards upon which to build an operating system for a world in which states seek results through coalitions.

A single meeting or even several routine meetings of officials will never satisfy the need for Canada to continuously anticipate, plan for and manage coalition politics in Canada's interests. Perhaps the most difficult coalition, for policy-makers, is the one they must fashion in Ottawa between the myriad players who supposedly have a stake in foreign policy formulation and outcomes. However, forging some mechanism beyond ad hoc interdepartmental committees to control coalition policies is a decisive matter. The new mechanism ought to take a Canadian perspective and thus should not be composed of "representatives" sent by departments to ensure that the home team is protected from centralist decisions. Rather, the mechanism should be especially designed to build coherence between intentions and outcomes. In this regard, the mechanism might best be situated under a minister who has responsibility for the resources that change coalition intentions into fact, as long as the minister's decisions are carefully guided by a strong national strategy.

This argument is not made in favour of some abstract rationality — for instance that a putative perfect system would produce perfect decisions and outcomes. Rather, the recommendation to review, at least, the way Ottawa decides is based on practical concerns about effective governance of vital national security issues raised by the incoherence in operations noted in this paper. Moreover, the present security decision-making process, though briefly overturned by then Foreign Minister John Manley's Cabinet committee on national security (now defunct), is premised on managing episodes rather than continuing operations which now define the security environment Canada faces. If strategy follows structure, then the extant structure is unlikely to accommodate or formulate an appropriate strategy for acting through coalitions in the twenty-first century.

Joining ends to means

Michael Ignatieff, commenting on the need to use force to defend human rights, concluded, "If we will the ends, we had better will the right means. For the means we select may betray our ends."[57] The warning is germane to Canada's situation and the growing disparity between what Canadians wish to do in international affairs, what they think Canada can do, and what capabilities are really available now and may be available in the future to do anything meaningful. The arguments between military experts may not be comprehensible to everyone, but even an informed casual observer would understand that in the long term — say to 2020 — if budgets remain constant relative to today, the Canadian Forces will have fewer resources and fewer people to deal with a world that is most likely to be more, not less, turbulent. Notwithstanding that some capabilities will certainly be greatly enhanced and "more lethal," it is not certain that they will be especially suited to the usual pattern of international security affairs; that is, to situations short of conventional war.

A national foreign and defence strategy for 2020 must join ends to means and allocate resources appropriately between strategic imperatives and strategic choices. Care must be taken to avoid the allure of "double-hatting" assets (assigning multiple duties to the same resources) to cover gaps in capabilities because it leads to the assumption, which will invariably be proved false in a crisis, that all contingencies are covered and that one person or one unit can be everywhere and do everything all the time. Matching Canadians' will to national means would be a critical item on any agenda to craft a coherent coalition strategy for Canada.

Canada, helpful fixer or helpful follower?

In 1963, Robert Sutherland — a respected senior defence analyst writing in a paper on national strategy that he had prepared for then defence minister Paul Hellyer — declared:

> There is...a distinct limit to how far one can define a Canadian position in advance of discussions with our allies. In the course of such discussions it must be anticipated that Canada's position would be necessarily subject to reconsideration and redefinition...the most that is possible at the present time is to define an "initial" Canadian position, accepting the fact that this position might require substantial revision in the course of discussions.
>
> From the point of view of the Department of National Defence, it would be highly advantageous to discover a strategic rationale which would impart to Canada's defence programs a wholly Canadian character. Unfortunately, such a rationale does not exist and one cannot be invented.[58]

Hellyer rejected this "little Canada" concept and tried, but failed, to win support in Cabinet for a fuller, more mature strategy for Canada in the 1960s. Though many things have changed since that report was written, some might say that not much has changed at all in the way some Canadians think about Canada's place in international affairs and multilateral coalitions.

Certainly, Canada cannot expect to lead the major powers. But is it true that Canada can only act on its own behalf in coalitions led by others? This is an assumption that needs to be tested. In some respects, tagging along in multinational coalitions diminishes Canada's place in the world, betrays foreign policy traditions, lowers other states' esteem for Canada and exposes Canada to cynical criticisms as a country that wants to play in the big leagues but dodges the attendant responsibilities. Moreover, public support for national goals may be harmed because many Canadians resist this assumption, especially when they believe that it involves an unthinking and uncontrollable surrender of national sovereignty to more or less unilateral acts by the United States. Neither Canada's interests nor those of the international community can be well served if the major powers, especially the United States, must lead every international coalition or if Canadians think of themselves as merely helpful followers.

Other wealthy states like Canada are well situated to take the initiative in some multinational coalitions, as Australia did in East Timor, and to use their less imposing powers to bring nations together in common causes. Canada did just this in 1956 and failed in the attempt in 1996. Unfortunately, the failure, the result of a betrayal of ends and means, seems to have reinforced the assumption that Canada cannot lead the international community and did not promote a determination to build the means necessary to retrieve Canada's lost legacy. A forum on a new national strategy must decide if Canada is to again become a leader or to just remain a cheerleader in international affairs.

New players, new methods

Coalitions today, as usual, are created around states and their diplomats, armed forces and other agencies. But multinational coalitions now also include various mixes of nontraditional allies and entirely new allies from national and international nongovernmental organizations (NGOs) and from international organizations. Diplomatic and military leadership may come from states or from international organizations, principally the United Nations and NATO. Arrangements, therefore, within coalitions are seldom sure at the outset and often ambiguous in the field, especially where NGOs are important actors. Nevertheless, these arrangements can have a significant impact on Canada's interests, domestic and foreign policies, and on the lives of members of the Canadian Forces. But arrangements are complicated in some new coalitions because they are predicated not only on sovereign states and their rights and laws, but also on the assumed rights of nonstate actors and the international standing of various multijurisdictional entities. Officials and officers preparing a national strategy for Canada must consider in their deliberations the terms and conditions that will underpin future Canadian commitments to multinational/multijurisdictional coalitions and the rules governing Canadians assigned to such coalitions.

Trusting Canadians

Public support for foreign policy is of paramount importance to the successful implementation of such policies over the longer term. This fact is especially pertinent whenever Canada acts through coalitions in an environment where every step in the field may be recorded and broadcast immediately by the media. An agenda for a forum on a national coalition strategy for Canada must include some consideration of how Canadians will be informed of the choices Canada faces in international relations. This may be a daunting assignment in a crisis, if the public and the commentators are ignorant of Canada's real capabilities and the circumstances in which Canadian diplomacy is played out.

Too often, Canadians seem to have higher expectations of foreign policy than the circumstances suggest. For instance, many Canadians, including most members of Parliament, believe that Canada is an important participant and a leader in international peacekeeping missions worldwide. They appear convinced that Canada has "influence" in NATO and the United Nations because of the commitments made there; but the reality is different.[59] Consequently, the public may be disillusioned when they discover a more sober truth, as many did when they found Canada outside the Contact Group directing NATO operations in the former Yugoslavia.

A forum on a national strategy to guide coalition policy ought to address three items concerning public support. First, to redress any public misunderstanding of Canada's capabilities to act through coalitions, politicians should forthrightly explain the state of Canadian diplomatic and military assets and the situation of prominent Canadian-based NGOs. Second, leaders should organize a public campaign to describe to citizens the complexities of the "new world disorder" and the consequences it brings to Canadian foreign policy. Third, politicians should describe the opportunities available for Canadians to take the lead in some types of multinational coalitions and the costs such efforts might entail. The public might then appreciate that while Canada could build coalitions of the willing around soft assets where risks are low — as in specific arms control areas and international judicial matters — they might also lower their expectations of its ability to act in coalitions where hard assets are needed and high risks are anticipated. Alternatively, Canadians might decide to assemble the means needed to match the vision they have of Canada in the world.

Although attempts in the late 1980s and early 1990s to construct policy from public debates, consultations and so on do not inspire much confidence in the method and cause deep frustrations in the bureaucracy, arguably the fault is in the application of the concept, not in the idea itself. Too much detail was introduced in the early open forums and participants expected too much influence. A new round based, perhaps, on meetings assembled to present citizens with technical information and reasoned options and their consequences, as officials and politicians see them, might help Canadians express their concerns and preferences. At the same time, meetings with limited aims might overcome some of the weaknesses of earlier attempts to develop foreign and defence policy through public debates. Certainly, the two 1992-93 open, wide-ranging, special joint committees of the Senate and the House of Commons appointed to inform citizens and to solicit their opinions on these matters might be a useful model to consider in the future.

A framework document for acting through coalitions

An officials' forum on a national strategy for acting through coalitions ought to produce for political leaders a framework document to govern "coalitions and Canadian foreign policy." This document (perhaps even a Cabinet White Paper, given the continuing emergency) ought to provide a comprehensive, coherent and authorized statement of intent and an indication of the resources needed to achieve it. It should be written to inform the public, to guide and control the policy discretion of officials and Canadian Forces officers and to bring order to the ends and means of foreign and defence policies.

Throughout its history, Canada has looked to "the mother country" — first Britain, then the Second World War allied leaders, NATO and the United Nations — to provide strategic direction to Canada's foreign and defence policy. But now Canada is "home alone" and must decide these things for itself, simply because candidate mothers are too busy with their own chores and interests to lead Canada any longer. Canada must resist the habit of merely lending troops to others, leaving them unattended to serve some communal interest while assuming it is a common interest.

Although Canada could sit still, leaving international responsibilities to others, it would then risk sliding out of sight in international affairs. Canadians would then have to accept that other states, willing to take the risks and pay the price, would set the agenda and receive any resulting benefits. Canadians would also have to set aside a legacy of sacrifice and compassion and a willingness to champion values that have defined Canada at home and abroad. On the other hand, an ambitious document might introduce Canadians to a road toward a new horizon and to a national policy that would place Canada in the vanguard of the gathering movement toward international peace and security through multinational coalitions.

But a crusade fuelled only on rhetoric will go nowhere, even if led by a new generation of political leaders. To regain the prominence it once held in the international community, Canada ought to heed the words and courage of the man who did so much to create it long ago. Lester Pearson believed "that the maintenance of an overwhelming superiority of force on the side of peace is the best guarantee today of the maintenance of peace," and Canadians were willing then to back his words with their own efforts. Few could credibly argue that Canadians today are less willing to back sound policies aimed at bringing greater freedom and security to the international community, but they need trusted national leaders to inspire them to make the sacrifices this end demands. This much is clear in the aftermath of the September 2001 attack on the United States and Western liberal democracies.

Canada has a respected international tradition to uphold. But Canadians, in their own interest, also have a responsibility to allies and the global community to help organize international affairs on a foundation built on peaceful change and security for all while defending liberal democracy at home and promoting it reasonably abroad. Diplomatic and security coalitions will continue to be a central instrument — an organizing principle — through which Canada achieves these related aims. Coalitions, fortunately, can also be the most productive means for explaining, guarding and realizing Canada's national interests. The challenge, therefore, is to collect our national thoughts; construct a national consensus on Canada's place in the world to guide politicians, military officers and officials; and then build the machinery of government and the soft and hard assets that will turn policy visions into policy outcomes.

Notes

1. Thomas (2000, p. 79).

2. Trudeau (1969).

3. As quoted in Gibler and Vasquez (1998, p. 787).

4. Ward (1982, p. 74).

5. As quoted in Hennessy (2001, p. 20). For another perspective, see Johnson (2001, p. A14).

6. Hennessy (2001).

7. See, for example, Kasurak (2001, pp. 8-12).

8. For a detailed review and assessment of the "policy process" leading to the commitment in Zaire, see Hays (1999).

9. Commission of Inquiry into the Deployment of the Canadian Forces to Somalia (1997).

10. See, for example, Department of National Defence (1999*a*).

11. For a recent Canadian example, see Brigadier General Robin Gagnon (2001, pp. 19-24).

12. Delvoie (2000, pp. 13-24).

13. Department of National Defence (1999*b*).

14. Bland (1995, pp. 214-24).

15. Delvoie (2000, p. 13).

16. Cellucci (2001).

17. The conflict in Colombia is not simply an internal security affair, as it has already spread more or less to neighbouring states. Latin American attempts to address the crisis with the assistance of the United States under various policies, including Colombia's "Plan Columbia," will probably touch Canada in the next few years. Who will lead this venture is an open question, especially if the United States Congress remains suspicious of the United Nations. However, the OAS is a possible candidate, and it will look to Canada for support and perhaps a diplomatic or armed commitment. What rules to govern such a coalition must be in place before Canada would act?

18. See Gagnon (2001).

19. General Wesley Clark (2001, p. 59).

20. Clark (2001).

21. Department of Foreign Affairs and International Trade (1999).

22. Jockel and Sokolsky (2000/2001, p. 1).

23. Jordan et al. (2000, p. 4).

24. For an historical perspective see Maloney (2001, p. A18).

25. Ministry of Defence (1999*a*, p.2). See also Ministry of Defence (1999*b*, pp. 187-202) and Tim Edmunds et al. (2001).

26. Manwaring (2001, pp. 60-71).

27. Haglund (2000, p. 92).

28. *The Times* (London) (2001).

29. The terms come from Stein et al. (1999).

30. See for example House of Commons (2001*b*, pp. 12-5).

31. Commission of Inquiry into the Deployment of the Canadian Forces to Somalia (1997, pp. 671-73).

32. Lieutenant-General Mike Jeffery as quoted in Mooney (2001, p. 10).

33. MacDonald (2001*a*, p. 32). See also MacDonald (2001*b*, pp. 14-7).

34. Auditor General of Canada (2001, ch. 16).

35. Auditor General of Canada (2001, para. 125).

36. House of Commons (2001*a*, p. 7).

37. House of Commons (2001*a*). Emphasis in original.

38. The term "effectives" means the personnel strength *less* the chronically sick, those excused from duty, and personnel in basic training and on retirement leave. Generally, the strength of the Canadian Forces usually reported includes some 10 percent who are "noneffectives."

39. Department of National Defence (2001).

40. Bland (1987, pp. 175-82).

41. House of Commons (2001*a*, p. 7).

42. These descriptions of the Department of National Defence 2000, Level 1, Business Reports were developed from confidential interviews in Ottawa and other sources.

43. Department of National Defence (1999*b*, p. 4).

44. Department of National Defence (2001, p. iii). Emphasis added.

45. See for example, Bland (2001*b*, pp. 19-35).

46. Department of National Defence (1999*b*, p. 10).

47. Department of National Defence (1999*b*, p. 3).

48. Department of National Defence (1999*b*, p. 3). Emphasis added.

49. General Maurice Baril (2000). Emphasis added.

50. Vice Admiral Gary Garnett (2001*b*, p. 5).

51. Garnett (2001*b*, pp. 8-9).

52. Garnett (2001*a*, p. 6).

53. Garnett (2001*a*, p. 7).

54. The positions of CDS, vice chief of the defence staff (VCDS), and deputy chief of the defence staff (DCDS) have been filled by air force and naval officers who, by all public reports, are firmly wedded to the *Strategy 2020* and the underlying revolution in military affairs approach to force development and resource allocations. The most senior army officer in NDHQ is in the second rung and well removed from decisions concerning the overall distribution of present or future resources for the Canadian Forces.

55. Bland (2001*a*, pp. 40-7).

56. Boulden (2000).

57. Ignatieff (2000).

58. As quoted in Bland (1995, p. 226).

59. Bland (1999, pp. 34-5).

References

Auditor General of Canada. 2001. *Report on the Department of National Defence*. Ottawa: Public Works and Government Services.

Baril, Maurice. 2000. "Canadian Defence Force." Delivered in Calgary, November 23, 1999. In Department of National Defence. *Vital Speeches*. Ottawa: Public Works and Government Services.

Bland, Douglas L. 2001*a*. "Defence And Security: The Next Generation." *Policy Options*, 22, no. 2 (March): 40-7.

——. 2001*b*. "The RMA: Managing An Idea." In *Managing The Revolution In Military Affairs*, edited by Ron Matthews and John Treddenick. Basingstoke, UK: Palgrave.

——. 1999. "Parliament, Defence Policy, and The Canadian Armed Forces." *The Claxton Papers* no. 1. Kingston: Queen's University, School of Policy Studies.

——. 1995. *Chiefs of Defence: Government and the Unified Command of the Canadian Armed Forces*. Toronto: Canadian Institute for Strategic Studies.

——. 1987. *The Administration of Defence Policy in Canada: 1947 to 1985*. Kingston: R.P. Frye.

Boulden, Jane. 2000. "A National Security Council for Canada?" *The Claxton Papers* no. 2. Kingston: Queen's University, School of Policy Studies.

Cellulicci, A. Paul. 2001. "Canada/U.S. 2001." Delivered to the Pacific Northwest Economic Region and the Council of State Governments of the West, Whistler, British Columbia, July 26.

Clark, Wesley. 2001. *Waging Modern War*. New York: Public Affairs Publishing.

Commission of Inquiry into the Deployment of the Canadian Forces to Somalia. 1997. *Dishonoured Legacy: The Lessons of the Somalia Affair*, vol. 4. Ottawa: Minister of Public Works and Government Services Canada.

Delvoie, Louis. 2000. "Canada and International Security Operations: The Search For Policy Rationales." *Canadian Military Journal* 1, no. 2 (summer): 13-24.

Department of Foreign Affairs and International Trade. 1999. "Notes for an address by the Hon. Lloyd Axworthy, 'Kosovo and The Human Security Agenda,'" delivered in Princeton, New Jersey, April.

Department of National Defence. 2001. *Defence Plan 2001: The Department of National Defence and Canadian Forces Internal Business Plan for Fiscal Year 2001/2002*. Ottawa: Ministry of Public Works and Government Services Canada.

——. 1999*a*. *The Minister's Monitoring Committee on Change in the Department of National Defence and the Canadian Forces*. Ottawa: Ministry of Public Works and Government Services Canada.

——. 1999*b*. *Shaping the Future of Canadian Defence: A Strategy For 2020*. Ottawa: Ministry of Public Works and Government Services Canada.

Dorman, Andrew. 2001. "Reconciling Britain To Europe in the Next Millennium: The Evolution of British Defence Policy in the Post-Cold War Era." *Defence Analysis* 17, no. 2 (August): 187-202.

Edmunds, Tim, Andrew Cottey and Anthony Forster. 2001. "Defence Diplomacy In Central and Eastern Europe: Challenges to Comparative Public Policy." Swindon, UK: Economic and Social Research Council Research Programme on Future Governance.

Garnett, Gary. 2001*a*. "The Canadian Forces, More Capable in an Unpredictable World." *On Track* 6, no. 2 (June): 4-7.

——. 2001*b*. "The Canadian Forces and the Revolution in Military Affairs: A Time for Change." *Canadian Military Journal* 2, no.1 (Spring): 5-10.

Gagnon, Robin. 2001. "Multilateral Intervention Forces." *Policy Options* 22, no. 2 (March): 19-24.

Gibler, Douglas, and John Vasquez. 1998. "Uncovering The Dangerous Alliances, 1495-1980." *International Studies Quarterly* 42, no. 4 (December): 785-807.

Haglund, David. 2000. "Allied Force or Allied Forces: The Allies' Perspective." In *Alliance Politics, Kosovo, and NATO's War: Allied Force or Forced Allies*, edited by Pierre Martin and Mark Brawley. New York: Palgrave.

Hays, John. 1999. "Conditions of Influence: A Canadian Case Study in the Diplomacy of Intervention." Occasional paper no. 19. Ottawa: Norman Paterson School of International Relations.

Hennessy, Michael. 2001. "Operation 'Assurance': Planning a Multi-National Force for Rwanda/Zaire." *Canadian Military Journal* 2, no. 1 (spring): 11-20.

House of Commons, Standing Committee On National Defence and Veterans Affairs. 2001*a*. "Report On Plans And Priorities." Ottawa: House of Commons.

——. 2001*b*. "Chairman's Remarks To SCONDVA: Operational Readiness of the Canadian Forces." *On Track* 6, no. 2 (June): 12-5.

Ignatieff, Michael. 2000. *Virtual War: Kosovo and Beyond* New York: Henry Holt.

Jockel, Joe and Joel Sokolsky. 2000/2001. "Lloyd Axworthy's Legacy." *International Journal* 56, no. 2 (winter): 1-18.

Jordan, Thomas, Douglas Lovelace, and Thomas-Durell Young. 2000. "'Shaping' the World Through 'Engagement': Assessing The Department of Defence's Theatre Engagement Planning Process." Carlisle, PA: US Army War College, Strategic Studies Institute.

Johnson, R.W. 2001. "Rwanda Redux: Don't Blame the US." *National Post*, September 10, p. A14.

Kasurak, Peter. 2001. "Sorting Out the Essentials of Readiness: A Look at the Canadian Forces." *On Track* 6, no. 2 (June): 8-12.

MacDonald, Brian. 2001*a*. "Will the Canadian Forces Have What They Need When They Need It?" *Policy Options* 22, no. 2 (March): 32-8.

———. 2001*b*. "The Capital/Capabilities Gap: The Final Rustout Decade of the Canadian Forces." *On Track* 6, no. 2 (June): 14-7.

Maloney, Sean. 2001. "Mideast 'Observers' Would Be Scapegoats." *National Post*, August 16, p. A18.

Manwaring, Max, ed. 2001. *Deterrence in the 21st Century*. Portland, OR: Frank Cass.

Mooney, Paul. 2001. "Army More Capable but Fragile: Commander." *The Maple Leaf* 4, no. 20 (May): 10.

Ministry of Defence. 1999*a*. *Defence Diplomacy*. London: Ministry of Defence.

———. 1999*b*. "Policy Paper - Paper No.1, Defence Diplomacy." London: Ministry of Defence.

Stein, Janice Gross, Michael Bryans, and Bruce Jones. 1999. "Mean Times: Humanitarian Actions in Complex Political Emergencies — Stark Choices, Cruel Dilemmas." *Report of the NGOs in Complex Emergencies Project*, vol. 1, no. 3. Toronto: University of Toronto, Centre for International Studies.

Thomas, James P. 2000. "The Military Challenges of Transatlantic Coalitions." *Adelphi Paper* no. 333, International Institute For Strategic Studies.

The Times (London). 2001. "The New Realism: Warnings for Arafat, Albanians, and More," *The Times* (London), August 14, 1.

Trudeau, Pierre. 1969. "Address to the Alberta Liberal Association." In *Speeches and Statements*, no. 68/9.

Ward, Michael. 1982. *Research Gaps in Alliance Dynamics*. Denver: University of Denver Press.

6

The Canadian Forces and the Doctrine of Interoperability: The Issues

Danford W. Middlemiss and Denis Stairs

Introduction

In June 1999, there issued from the Department of National Defence (DND), over the signatures of Deputy Minister Jim Judd and Chief of the Defence Staff General J.M.G. Baril, an unusually ambitious disquisition: *Shaping the Future of Canadian Defence: A Strategy for 2020.*[1] Not a White Paper, officially it had no formal government approval, certainly no Cabinet imprimatur. It was intended instead as a *dirigiste* inspiration from "the senior leadership team" for the department's working professionals. More precisely, it was described as "a strategic framework for defence planning and decision-making to help guide the institution well into the next century." It would be "updated periodically and used to guide [the DND's] planning, force structure and procurement decisions, as well as our investments in personnel, education and training."[2] Or so its readers were told. And so it was subsequently to be.

For a document with such impressive aspirations, its contents were surprisingly unremarkable. Indeed, they were typical of bureaucratic think pieces written under the dark shadow (we can only assume) of the Treasury Board. Its language was safely anchored in the leaden vocabulary and conceptual abstractions common to the faddish constructs that pass in our time for "theories" of organization management. In faithful obedience to what is now everywhere *de rigueur* in Canadian government, there was a discussion of the "defence mission" and of the "Canadian values" that it was intended to embody and preserve.

The values were identified as "democracy and the rule of law; individual rights and freedoms as articulated in the Charter; peace, order and good government as defined in the Constitution; and sustainable economic well-being."[3] There was also a "stakeholder analysis," and an assessment of the "emerging strategic environment" (as conceived in geopolitical, military, socio-economic and organizational terms). "Strategic imperatives" were identified, imperatives requiring that the strategy itself be coherent, that it foster pride in the defence establishment among those who labour within it, that it make possible effective collaboration with other departments of government as well as with like-minded nations abroad, that it maintain a relevant force structure, and that it ensure a balanced use of available resources.

After these came a list of the attributes — 11 of them — the strategy was intended to exhibit, along with a statement of the vision by which it was to be guided.[4] The pursuit of the vision required in turn the strengthening of "distinctive competencies in five principal domains" (command and leadership; multiskilled people; doctrine, technology and training; modern management practices; and special relationships with principal allies). These domains were said to build in turn upon the eight long-term strategic objectives of existing defence policy and the various five-year targets associated with each of them, respectively. In relentlessly earnest spirit, these were listed, too. And at the end, there was the assertion that success would be achieved only if the effort were sustained by unity, continuity, resolve and partnership.

The exposition was only 12 pages long, but it is easy to imagine the accomplished Canadian scholar-diplomat John W. Holmes complaining, ever so gently, of its lamentable display of "hardening of the categories" — a disease, however, that Holmes himself associated less with policy-makers than with academics whose judgment he thought had been corroded by scholasticism.

Nonetheless, there could be discerned in the plethora of largely empty abstractions at least one concept that had potentially significant practical implications, one quietly revealing display, as it were, of bureaucratic code. Its designation was "interoperability"; and it cropped up in several places in the document, often accompanied by reference to the need to take advantage of new military technologies, and to do so in the company of "our main defence partners in the UN, NATO and coalition operations."[5] This clearly meant, above all others, the United States of America.

Taken at face value, the interoperability concept might at first be regarded as no more remarkable than the document as a whole. In practice, Canada has always operated militarily overseas in coalition with others. This has been true in the conduct of war. It has been equally true in the making, and sometimes the enforcing, of peace — that is, in peacekeeping of both the first- and second-generation kinds. Indeed, ever since 1949,[6] and arguably since 1940,[7] most of Canada's standing (or contingency) defence arrangements and commitments have been institutionalized in lockstep with the Americans in particular. This has been effected prominently through North Atlantic Organization (NATO), but also by way of a host of highly integrated mechanisms for the direct defence of the North American continent. Obviously, not all of these arrangements have entailed technical interoperability for the Canadian forces (CF) at the practical (that is, operational, or combat) level, but they have certainly encouraged it (see below). That being so, the declared dedication to interoperability in *Strategy for 2020* could easily be regarded as no more than a restatement of traditional practice and continuing necessity. Such restatement might be regarded as particularly timely, given recent operational experiences in the Persian Gulf and elsewhere, as detailed below.

Having said that, however, it is also clear that the new language reflected an intensifying Canadian defence problem, a problem arising in considerable measure from the contraction of the defence budget, and hence of the country's store of real military assets, particularly when placed in the context of Canada's proximity to the world's only transcendent superpower. Some might not regard this as a "problem" at all, but rather as the happy consequence of the end of the Cold War. Peace was at hand, and its dividend could therefore be enjoyed. But for practising military professionals, the difficulty was how to sustain a significant capacity for contributing to the resolution of "pointy-end" security challenges, at whatever level, at a time when their assets were so badly rundown.

Part of the answer might lie with an advancing military technology — more bang, more accurately delivered, for fewer bucks. But the rest of it could well require operating not just in tandem with others, but as a wholly integrated (or at least integration-capable) component of the defence establishments of coalition partners. In practice, once again, this might really mean the defence establishment of the United States. As the leadership document itself innocently observes, "At its core, the strategy is to position the force structure of the CF to provide Canada with modern, task-tailored, and globally deployable combat-capable forces that can respond to crises at home and abroad, in joint or combined operations. The force structure must be viable, achievable and affordable." Hence, one of the strategy's "critical attributes" is elaborated to mean a strengthening of "our military relationship with the US military to ensure Canadian and US forces are interoperable and capable of combined operations in key selected areas."[8]

Such a stratagem might make perfect sense for a professional military community searching desperately for ways in which to preserve its operational salience and effectiveness. But in principle it could have larger consequences, too, among them consequences for the degree of latitude actually available to Ottawa in the deployment of the CF. Clearly, this would depend on what "interoperability" actually meant in practical terms. That is the central question this paper is intended to explore. The need to consider it, as we have argued, has been brought home further by the circumstances — not to say the confusions — surrounding CF deployments in the wake of the September 11 terrorist assaults on New York and Washington. Those circumstances provide a clear and present demonstration of realities that might otherwise have been consigned by policy planners and independent observers alike to that vaguely nagging (but not practically salient) category of conceivable, yet improbable, possibilities.

Our discussion begins with a brief exploration of interoperability as a concept. We then attempt to place it, briefly and broadly, in its longer-term historical context, so as to lend perspective to judgment. Our observations on earlier incarnations of the interoperability phenomenon as a feature of Canadian policy are followed by a consideration of the various additional catalysts in support of it that have emerged from more recent operational experiences in the field, as well as from other developments. We conclude with an assessment of the initiatives that are now underway and an attempt to identify some of their potential implications, both good and bad, for Canadian foreign and security policies in the broader sense.

Interoperability: The Concept and the Rationale

Like many of the common concepts of military discourse, the term "interoperability" seems relatively straightforward in principle, but its practical implications, if not carefully delineated, can be disarmingly — and confusingly — ambiguous. In the real world, moreover, they can be immensely difficult to put into practical effect.

The standard definition — the one used by the military establishments of the NATO allies, including Canada and the United States — holds that interoperability is the "ability of systems, units or forces to provide services to and accept services from other systems, units or forces and to use the services so exchanged to enable them to operate effectively together."[9] This points to the fact that the concept has both technical and operational dimensions, both of which must be in place if interoperability among coalition militaries is to be fully achieved.[10]

Students of defence literature can easily identify a bewildering array of other military terms that can be used synonymously (and often confusingly) with the interoperability concept — among them "compatibility," "interchangeability," "commonality" and "standardization." In the simplest terms, interoperability is best conceived as lying near the middle of a continuum between basic compatibility at the low end (where systems and forces can operate, as it were, side-by-side without interfering with one another's functioning) to complete integration at the high end (where there is an ineluctable element of functional interdependence between systems and forces acting together).[11] Precisely where any particular relationship of military interoperability actually lies along this continuum, of course, can become a sensitive and sometimes intensely debated political issue, as well as a source of vexation for the military themselves — a point to which we will return below.

It follows from the foregoing that, in the most general terms, the ultimate goal of interoperability is not to ensure that all the contributors to a given coalition will deploy identical types of military systems and units, but simply to achieve a more practicable level of cooperation among them. As the DND's current strategic capability planning document puts it, "The capability to work seamlessly with our most important allies in an operational setting ensures that we can participate effectively in those crises most likely to affect our vital interests."[12] From the military point of view, the overarching objective is thus to make a militarily relevant and effective contribution to multinational security efforts at the maximum possible level of efficiency. Normally, of course, such a contribution will impart (or at least will be intended to impart) political benefits, too: among them, an entitlement to participate in decisions that affect the purposes, the scope and sometimes even the practical conduct of the hostilities; improved access to the intelligence resources of coalition partners (the result of establishing, through direct military involvement, a need to know); and a discernible, if intangible, enhancement of diplomatic influence (a potential by-product of bringing cosmetically useful, as well as practically significant, military assets to the table).

More concretely, interoperability seeks to overcome a number of obstacles to the more effective functioning of multinational forces that have been established on the assumption that their various national components will act in concert. These obstacles are commonly grounded in such factors as disagreements or misunderstandings over mission goals, priorities and rules of engagement (ROEs); the reliance of different coalition contributors on different types of equipment, or on similar equipment with different specifications; the commitment of the various national forces involved to incompatible tactical, organizational, leadership or other professional doctrines; the involvement in coalition campaigns of units that have been exposed to unrealistic and/or insufficient training and preparatory exercises; and a variety of other "soft factors," ranging from different organizational "cultures" to outright policy disagreements at the highest levels of national decision-making. Left unattended, such sources of behavioural divergence can create havoc in the field, particularly when many of the national contingents involved are not large enough to be logistically, and in other respects, self-sufficient — precisely the circumstance in which Canadian units have found themselves in every operation since the Second World War, with the possible exception of ground force units in certain phases of the Korean War.[13]

It is instructive to note that the pursuit of interoperability usually emanates in the first instance from military establishments that are attempting to deal with problems that from the purely professional standpoint are no more than technical-cum-operational. That being so, the issue often seems not to be very central to the concerns of politicians who are engaged in planning the broader direction of their national policy. In practice, however, interoperability can seem to be very much a two-edged sword when considered from the political point of view. By enabling a country with only modest military capabilities (like Canada) to contribute in a meaningful way to multinational operations, it can lend a certain visibility to the willingness to make a substantive commitment to the resolution of the conflict at hand, a visibility that can earn political dividends from audiences at home as well as from coalition partners abroad.

This is one of the reasons, rarely advertised in public, for Canada's preference — demonstrated in a variety of multinational operations from the Persian Gulf War to the present — for mounting contributions of more or less self-contained "task groups," rather than individual troops, ships or aircraft.[14] Furthermore, for countries (again like Canada) that proclaim their eagerness to promote an institutionalized and rule-governed international environment, gaining a solid reputation for possessing a modern, fully capable and interoperable military establishment can generate additional opportunities for constructive participation in order-enhancing multinational coalitions. Indeed, as recent events from the Gulf War onward have demonstrated, the number of states that can operate effectively in this way with the "big battalions" is actually very small.

Having said all that, the very notion of working with larger and more powerful allies in military coalitions, as opposed to operating unilaterally, raises the prospect of some — or even a complete — loss to the smaller powers of autonomy in decision-making, a circumstance that carries with it the risk of generating perceptions both at home and abroad of a decline in national prestige, and a reduced capacity for acting independently in the national interest. A preoccupation with the danger that interoperability will go hand-in-hand with perceptions of political dependency has thus been a prominent theme in the calculations of successive Canadian governments from the earliest days of Confederation to the present.

Another concern, and a recurrent one for Canada in recent years, is that being numbered among the few countries that can actually operate "seamlessly" with the United States may make it more difficult for Ottawa to refuse requests to contribute to American-led operations, or alternatively to participate in operations in which the Americans are not involved. The inability of the government to negotiate satisfactory terms for Canada's participation in the United Kingdom-led International Security Assistance Force in Afghanistan, and its apparent pursuit of a "second-best" arrangement in conjunction with US combat forces in Kandahar, provides a recent case in point. That being so, the political dimensions of military interoperability with the United States may now warrant more attention than they have tended in the past to receive.

But before considering these and other implications of current and recent developments, it may be helpful to place the issue as a whole somewhat more fully, although still very briefly and selectively, in its longer-term historical perspective.

Interoperability: The Historical Context

It should be clear that, in any military alliance, interoperability is primarily an issue for the lesser powers. This is because it is the lesser powers that must deal with the military equivalent of "keeping up with the Joneses." Nowhere has this been more starkly revealed than in NATO, where all the members, save in some degree the United Kingdom and France, have found it a perennially daunting challenge to maintain military forces that can operate effectively with the vastly superior military establishment of the United States. Such was the case even in the earliest days of the alliance. The difficulty has come not so much from the superiority of American forces in quantitative terms as from the lead they hold in qualitative terms, which in recent years has intensified in tandem with accelerating (and increasingly expensive) advances in modern military technology.

The underlying roots of the problem, however, precede the technology gap. This is partly because the militaries of the various NATO members, including Canada, have evolved in very different geopolitical, historical, cultural and economic circumstances. Hence, they frequently exhibit significantly different characteristics in relation to size, composition, equipment, training, roles and competencies. As a result, no single military establishment is an exact clone of any other. Indeed, in the early days of NATO's doctrine of "balanced collective forces," Canada, for one, looked upon the hodgepodge of diversities among the various member forces with approval, since it seemed to hold out the possibility of "an international division of labor in the area of collective defence" to which each individual partner could contribute what it was best able to provide.[15] But in practice, this idea of niche roles and forces for alliance members was never widely embraced, even by Ottawa. The result was that the most broadly prevalent pattern took the form of a seemingly endless struggle on the part of the smaller contributors to obtain at least some measure of standardization and interoperability with the forces of their dominant military partner — the United States.

In the case of interoperability, therefore, as in the case of most other military concepts, it can be argued that there is little that is genuinely new under the martial sun. Indeed, Canada's experience historically in this field has been characterized by repeated attempts — variously motivated and certainly with varying degrees of success — to accommodate its military equipment, doctrine, training, and sometimes even its objectives and roles, to those of a patron power. The process began first in relation to the British. Later, starting roughly with the onset of the Second World War, the focus shifted to the United States, where it has remained ever since.

From Confederation to the end of the First World War, Canada's military preparedness — to the extent that it amounted to preparedness at all — reflected Ottawa's primary reliance on Great Britain, as the parent imperial state, for Canadian defence. With the impressive expansion in the power of the United States, however, and given London's primary preoccupation with potential challenges in other theatres, the British in this period were becoming increasingly leery of any prospect of significant military entanglements in North America on the ground.[16] Hence they came to rely more and more on the new dominion to provide for its own defences by raising a local militia. Alas, from the military point of view the results were less than impressive. When circumstances posed the prospect of military operations in South Africa in the late 1890s, the British War Office initially made no attempt to acquire military contingents from the colonies. It was simply assumed that it would take too long to train them to British standards.[17] Similarly, while the British navy might be counted on, *in extremis*, to extend its protective umbrella to Canada, it became increasingly apparent in the years leading up to Sir Wilfrid Laurier's Naval Service Bill of 1910 and his subsequent defeat in the "reciprocity" election of 1911 that London ultimately expected Canada to ante up ships and sailors, not for local defence, but for the far more compelling cause of defending the Empire itself.[18] In practice, then, the presumed purpose of the Canadian military was to supply troops to British-commanded units, and to operate in accordance with British doctrine, procedures and training standards — all in the service of the greater glory of the British Empire.

At the outbreak of the First World War, Canadian military units were woefully unprepared for the conflict. Ottawa, moreover, lacked any independent vision that could be regarded as pertinent to the conduct of the hostilities. It was simply taken for granted that Canadians would fight alongside the British, under British leadership, and that they would follow British tactics and procedures in battling their common foe.[19] Here was interoperability with a vengeance! As the war progressed, of course, and as the size and stature of the Canadian contribution grew in parallel with disillusionment over the quality of British generalship, there was evidence of a "Canadianization" of the Dominion's involvement.[20] But the basic premise of the operation was nonetheless starkly clear from the very beginning.

The interwar period was characterized essentially by the demobilization and retrenchment of Canada's armed forces. With the failure of the League of Nations, however, and with storm clouds gathering in the Far East as well as in Western Europe, Ottawa began in the late 1930s to show signs of moving toward a closer defence relationship with its continental neighbour. Indeed, in many respects, the reciprocal pledges of mutual defence support that were issued by President Franklin Delano Roosevelt and Prime Minister Mackenzie King in 1938 constituted the basic foundation for Canada-United States defence co-operation, a foundation that endures to the present day.[21]

These lofty rhetorical reassurances from the summit later found concrete expression in the Ogdensburg Declaration of August 1940, through which the president and the prime minister together authorized the creation of a Permanent Joint Board on Defence (PJBD) in recognition of "the beginning of a new era in Canadian-American relations."[22] As one analyst later noted, this declaration "provided the fundamental condition for military co-operation: a full and systematic exchange of military information upon which joint plans, operations and logistics could be based."[23] North American defence preparedness, in all its forms, had now become what it would henceforth remain: a matter of mutual concern.

The same spirit of cooperation was soon extended to the related area of defence production. In the Hyde Park Declaration of April 1941, it was agreed "as a general principle that in mobilizing the resources of this continent each country should provide the other with the defence articles which it is best able to produce...and that production programs should be co-ordinated to this end."[24]

Here was niche interoperability writ large, and from these two surprisingly informal declarations of intent flowed a host of more specific arrangements, both during and immediately following the Second World War. Taken together, they had the cumulative effect of tying both the military establishments and the defence industries of the two states into a seemingly irrevocable web of closely coordinated interconnectedness. As time went on, interoperability became an increasingly prominent feature, and consequence, of this intricate construct.[25]

It was the 35th recommendation of the PJBD, adopted by both countries in 1947, that most clearly pointed the way to greater interoperability in the post-war period. It committed the two governments to the "[a]doption, as far as practicable, of common designs and standards in arms, equipment, organization, methods of training and new developments to be encouraged, due recognition being given by each country to the special circumstances prevailing therein."[26]

This was followed almost immediately by a public "Joint Statement" on February 12, 1947, that was intended in part to quash rumours to the effect that the United States was insisting on large-scale defence projects, including military bases, in the Canadian north, and more generally to respond to ever-present Canadian sensitivities over the erosion of sovereignty. "[I]n the interest of efficiency and economy," the statement asserted, the militaries of the two nations should "continue to collaborate for peacetime joint security purposes" on the basis of five principles:

- Interchange of select individuals to increase the familiarity of each country's defence establishment with that of the other country
- General cooperation and exchange of observers in connection with exercises and with the development and tests of material of common interest.
- Encouragement of common designs and standards in arms, equipment, organization, methods of training and new developments, etc.
- Mutual and reciprocal availability of military, naval and air facilities in each country, applied as may be agreed in specific instances. Reciprocally each country will continue to provide with a minimum of formality for the transit through its territory and its territorial waters of military aircraft and public vessels of the other country
- As an underlying principle all cooperative arrangements will be without impairment of the control of either county over all activities in its territory[27]

These principles soon found concrete expression in the announcement of a nine-station weather network to be built across the Arctic Circle; the interchange of officers at several Canadian, American and British training schools and colleges; initiatives aimed at standardizing weapons and equipment; the regular exchange of observers and technical information; and, by early 1949, joint naval exercises and full-scale combined army and navy amphibious assault manoeuvres.[28]

While some of these initiatives may seem prosaic and routine to the casual observer today, for military planners of the immediate post-war period they were the bedrock of interoperability and effective operations, and they represented significant progress. Indeed, one such measure, the tripartite "Screw Thread Convention" of 1948, has been heralded as no less than the "germinal act of engineering statesmanship" that eventually led to a wide array of additional, and equally vital, standardization agreements "in non-material areas such as doctrine, organization, operational procedures and communications, in logistic matters such as technical procedures for supply and maintenance, and in researches [sic] and investigations."[29]

From the perspective of command and control, the North American Air [later Aerospace] Defence (NORAD) Command formed in 1958 represented the zenith of Canadian-American military interoperability. Almost from its inception, however, NORAD also raised concerns among some Canadians over loss of autonomy in decision-making.

At its inception, NORAD was but the latest in a series of bilateral air defence arrangements that were designed to provide a more rapid and effective joint response to the commonly perceived threat to North America from Soviet long-range bombers carrying nuclear weapons.[30] This response initially included plans for destroying incoming bombers, but as the bomber threat receded in importance following the advent of intercontinental ballistic missiles, NORAD's primary function became that of monitoring North American airspace so as to provide adequate warning time to flush aloft the retaliatory forces of the US Strategic Air Command, and thereby reinforce the deterrent premise of Mutual Assured Destruction (MAD).

The formal agreement by which NORAD was established observed that "the air defence of Canada and the United States must be considered as a single problem" and declared that this problem "could best be met by delegating to an integrated headquarters the task of exercising operational control over combat units of the national forces made available for the air defence of both countries."[31] It further stated that "the agreed integration is intended to assist the two governments to develop and maintain their individual and collective capacity to resist air attack on their territories in North America in mutual self-defence."

In practice, the command and control arrangements that NORAD embodied meant that a US military officer would have "operational control" over Canadian interceptor squadrons in both peace and in war, although care was taken to ensure that adequate prior "consultation" would be maintained and that each government would have the final say over its own allocation of personnel and units to the NORAD structure.

Given this degree of integration, there were, and indeed there still remain, points of friction for Canada, as the smaller partner in an organization so unequally composed. One of these became very evident during the Cuban Missile Crisis in October 1962, when Canadian components of NORAD were automatically placed on the same alert status as their American counterparts, even though the Diefenbaker government initially regarded such action as both unnecessary and potentially provocative.[32] A similar incident occurred during the Middle East crisis of 1973, when US forces were placed on higher alert, in an action that again affected the Canadian contingent at NORAD headquarters. Since then, procedures have been developed to allow US personnel to replace their Canadian colleagues should the two governments disagree on the need for a particular alert.[33]

Another continuing issue has been posed by the prospect that Canada's participation in NORAD could embroil it more generally in security policies with which it disagrees, such as the abrogation of the Anti-Ballistic Missile (ABM) Treaty of 1972; in potentially destabilizing military projects, like the Strategic Defense Initiative (SDI a.k.a. "Star Wars") of 1983; or, most recently, in the National Missile Defense (NMD) program that the Bush administration is now quietly pressing Canada to endorse. The point worth noting here is not that Ottawa has no choice on these matters. Clearly it has at least some room for manoeuvre, and in the past it has made use of it both to underscore its commitment to the ABM Treaty and to decline to participate directly in SDI. The point, instead, is that continued participation in the kind of interoperability arrangement that NORAD represents entails certain costs, some of which are political and others financial.

Even among the staunchest of Canadian nationalists, moreover, there is sometimes recognition that being a full partner in a binational organization like NORAD, inherently unequal though the partnership may be, can afford Canada a uniquely "inside" view of the intentions of the larger partner. That being so, it may even make possible from time to time the exercise of a moderating influence. For example, in accordance with Canada's wishes, a joint recommendation of the PJBD led to a realignment of NORAD's regional boundaries to coincide with national borders in 1983.[34] How real and how common the use is of this kind of influence, and whether it is worth the erosion of decision-making autonomy that goes with the opportunity to try it on, is a matter of political judgement. But it can be argued that the alternative — opting out of the partnership altogether — is almost a sure-fire guarantee that Canada's interests will be benignly neglected or even deliberately ignored by the United States. The costs, moreover, of "going it alone" would then almost certainly rise to prohibitive levels. Supporters of interoperable arrangements are thus easily drawn to their rhetorical question: What, in such circumstances, would be the true "value added" for Canadian sovereignty?[35]

After the Cold War had ended, of course, the first real test of Canada-US interoperability came with the war in the Persian Gulf in 1990-91. In that example, far from undermining Canada's autonomy, the CF's closely knit "connectivity" with American forces enabled Ottawa to parlay a relatively minor deployment into a major foreign policy advantage, one that was consistent with the nation's traditional multilateralist preferences. It therefore deserves more detailed attention as we come now to consider some of the experiences of the past 10 years, experiences that have accelerated further the drive to enhance

Canada's ability to participate effectively in joint military operations, most notably with the United States.

Interoperability: Recent Catalysts

Over the decade or so since the collapse of the Soviet Union and the demise of the Cold War, domestic political demands for peace dividends in the non-US NATO partners have led the governments involved to degrade their respective military establishments to such a significant degree that the United States now spends more on its armed forces in absolute terms each year than all of the other allies combined. This growing disparity has developed in spite of the fact that there have been similar pressures for reductions in defence spending within the United States itself. One consequence has been that the US has been able to devote a much larger proportion of its overall military expenditures than its allies have to equipment modernization and training improvement.[36] This, in turn, has made it increasingly difficult for the other partners to keep pace with the qualitative improvements in American capabilities. The holy grail of interoperability within NATO is thus becoming notoriously hard to achieve. This is in spite of the fact that even the most autonomy-conscious of the allies are now forced to concede, in principle at least, that reaching it ought to be among their highest priorities.

While the argument for interoperability may seem particularly strong in the case of the smaller powers, it should be noted that the Americans themselves have come to share the same aspiration. This is partly for reasons that are peculiarly their own. Among them is their desire to encourage a capacity for effective co-operation in the field among their own individual services, a capacity they call "jointness." Almost all major military establishments have to cope with inter-service rivalry, but in the United States there has been a particularly long and vigorous tradition of warring among the various branches of the armed forces over their respective budgets, and hence more generally over doctrine, procurement and turf.[37] In addition, however, the Defense Department in Washington now recognizes the importance, even for the United States, of developing better mechanisms for working with the armed forces establishments of coalition powers. This is because the US has increasingly, if somewhat reluctantly, been compelled for diplomatic as well as practical reasons to ally with other players in conducting multinational military operations. This circumstance has applied not only to operations established under the aegis of the United Nations and/or NATO, but also in the context of ad hoc coalitions like the current campaign in Afghanistan against the forces of terrorism and their patrons.[38]

For Canada, field experiences in the Gulf War in 1990-91, in several NORAD and NATO training exercises, and in the Kosovo air campaign of 1999, have all reinforced the lesson that the effectiveness of multinational military operations now more than ever depends on greater attention being given to the interoperability imperative. From the professional military point of view, this is particularly urgent in relation to the larger allies, of which the United States is by far the most significant.

In the case of the war against Iraq, which involved operations in and around the Persian Gulf (Operation FRICTION), Canadian naval and air components both had considerable success in establishing effective interoperability "up" to the forces of the United States and "down" to those of other coalition members. Moreover, astute Canadian commanders in situ were able to parlay the benefits of a reputation for reliable performance, naval vessels that were highly interoperable (especially in communications), and the intangible attribute of "understanding the way they do business" into a highly visible, combined logistics command subcommand of the coalition's naval operations — a distinction unique to Canada during the Gulf conflict.[39]

Nonetheless, there were some serious problems and weaknesses. For example, Canada's A/B model of the CF-18 fighter aircraft, less modern than the C/D and E/F models used by the Americans, was found deficient in its tactical air communications equipment. Specifically, the Canadian CF-18s lacked the Link-4 ship-to-aircraft computer communications system, which, for American units, made it possible to sustain secure transmissions with the AEGIS anti-aircraft cruisers that were guarding the coalition surface fleet. Although the problem was ultimately rectified with the help of upgrade arrangements hastily negotiated with American military authorities, the deficiency raised, at the beginning of the operation, the unsettling prospect of "blue-on-blue" (friendly fire) incidents in what had quickly become a congested theatre, to say nothing of the possibility that engagements might be launched in error against innocent civilian aircraft. So serious was the problem as a source of hesitation at the operational level that, at times, it even delayed the process of intervening against the hostile military aircraft of the Iraqis.[40]

Similarly, following the Mulroney government's subsequent authorization of "sweep and escort" missions for the CF-18s, the lack of secure voice radio communications, such as "Havequick," forced a change in coalition bombing missions from lower to higher altitudes. The decision was required in the first place to afford Canada's fighters greater protection against enemy missiles and anti-aircraft artillery, and in the second place to reduce the need for violent evasive manoeuvres — which, in the absence of continuous voice-radio contact, would have been potentially dangerous to other aircraft.[41]

Other limiting problems for Canada's CF-18s included: inadequate strategic refuelling capability; the lack of a precision-guided munitions (PGM) capability;

and doctrinal differences with coalition partners that necessitated a "crash" retraining regimen to ensure that everyone involved was adequately prepared for high-altitude bombing procedures.[42]

Many of the shortcomings of Canada's aging CF-18s resurfaced during Operation ECHO, Canada's contribution to the coalition air campaign against Serbian forces in Kosovo, as well as in Serbia itself in 1999.[43] For example, Canada's pilots lacked both night-vision goggles and helmet-mounted bombsight "cueing" systems. While these deficiencies were also evident in other coalition aircraft, including even some that were deployed by the United States, there is no doubt their lack restricted the safe manoeuvring and bombing effectiveness of the Canadian aircraft. The complete absence, moreover, of any organic Canadian strategic air-to-air refuelling capability, a deficiency first encountered during the Gulf conflict almost a decade earlier, forced a near-total dependence on scarce US Air Force air tankers, and considerably lengthened the island-hopping transit times for CF-18s flying in and out of the Canadian base in Aviano, Italy. In addition, only 12 of the 18 Canadian CF-18s were fitted with "Nite Hawk" Forward Looking Infrared (FLIR) pods, a circumstance that to some degree limited the overall bombing capacity of the Canadian contingent.

During the course of the campaign, the fighters were also confronted with shortages in the supply of the more accurate, laser-guided bombs. An American Air Force officer, Lieutenant Colonel Samuel J. Walker, has also reported that the Canadian CF-18s did not have the stores-management systems or weapons interfaces required by current generation air-to-air weapons and many PGMs. Finally, perhaps the most widely cited shortcoming of the Canadian CF-18s operating in the Kosovo air campaign was their lack of secure, interoperable voice communications, data link interfaces, and "friend-or-foe" identification equipment that were up to the American "Havequick" and Link-16 standards. To some extent these latter deficiencies, which were not confined solely to Canadian aircraft, forced a "dumbing-down" of the way in which operations in the air were conducted, so as to accommodate the lowest common capability denominator. Doing so significantly increased the combat risks involved.[44]

Interoperability: The Way Ahead for Canadian-American Practice

As we indicated at the outset, the June 1999 DND document, *Shaping the Future of Canadian Defence: A Strategy for 2020*, represents the department's detailed "guidance" for strategic decision-making in the immediate future, given in particular the lessons learned from recent operations as well as from assessments of newly emerging security challenges.[45] To some degree, of course, the latter constitute a moving target, as the events of September 11 have served to demonstrate in such tragically dramatic form. Indeed, it can be argued that one of the immediate impacts in the Canadian context has been not only to intensify the drive to interoperability as a matter of principle, but to broaden its implications to include a potential integration of command and control arrangements across the board and at the highest level. The 9/11 assault may also have had an important political consequence — perhaps positive from the professional military point of view — in creating a climate of receptivity within Canada for the further enhancement of interoperability arrangements. This may be true even if such arrangements come to imply an additional loss of autonomy and freedom of manoeuvre in relation to potential Canadian force deployments (see discussion below).

In any case, it is now worth emphasizing again that one of the main pillars of *Strategy for 2020* is the emphasis it gives to reinforcing Canada's special relationship with its principal allies, most notably the United States. As the document puts the argument in its most explicit form:

> A vital distinctive competency is the ability to work together with our allies. Our most important ally now and for the future is the United States where our strong relationship has long benefited both countries. We must plan to nurture this relationship by strengthening our interoperability with the US Armed Forces, training together, sharing the burden for global sensing and telecommunications and pursuing collaborative ways to respond to emerging asymmetric threats to continental security.[46]

"Objective 5: Interoperable" is listed as one of the defence establishment's eight long-term strategic objectives. The goal is to "Strengthen our military to military relationships with our principal allies ensuring interoperable forces, doctrine and C4I (command, control, communications, computers, and intelligence)." Three "five-year targets" are identified as prerequisites for achieving this end, and they require the department and the CF to:

- Manage our interoperability relationship with the US and other allies to permit seamless operational integration on short notice
- Develop a comprehensive program to adopt new doctrine and equipment compatible with our principal allies
- Expand the joint and combined exercise program to include all environments and exchanges with the US[47]

Skeptical old hands might be inclined at first to suspect that there is more rhetoric than substance in this elaborate exposition. Statements of lofty intent, common though they may be to the hierarchical bureaucracies of government, often remain empty and unfulfilled unless they are buttressed by explicit, authoritative directives and are closely monitored to enforce compliance with their aims. But a closer examination would lead even the most hardened observer to conclude that the authors of the document had both said what they meant and meant what they said. This is because *Strategy for 2020* was soon followed by other documentation that was clearly designed to put flesh on the bones, to give the high-blown phrases practical definition.

More precisely, specific policy-planning guidance for greater interoperability can be found in two major in-house treatises. The first is "Defence Plan 2001" (a transitional document) which replaces the former annual "Defence Planning Guidance for the CF" as a whole, along with the Level 1 Business Plan documents for the individual environmental service commands. The second is the "Report on Plans and Priorities 2001-2002."[48] Taken together, these two publications provide ample evidence that the department has fully embraced interoperability as a critical objective for the immediate future.

What is particularly notable about all of the new policy-planning frameworks, in fact, is the degree to which they demonstrate that the concept of interoperability is now seamlessly, consistently and centrally woven into both the thinking and the fabric of the defence establishment. This is evident at the highest level of strategic direction for change, as manifested in *Strategy for 2020* itself (e.g., "Objective 5: Interoperable").[49] It is also reflected in the five so-called core priority areas of focus for major initiatives, as outlined in the *Report on Plans and Priorities 2001-2002* (e.g., "Optimizing Canada's Force Structure" and "Fostering Canada's Defence Relationships").[50] And it crops up prominently once again in the specific, short-term priorities that are identified for DND in the current fiscal year (e.g., "2.2 Develop DND/CF's C2ISR capability to ensure we can operate effectively in the information age with our allies").[51]

Absent a dramatic change in the receptivity of the external environment (that is, the United States), or an unexpected contrary intervention on the part of the political leadership, this tight interconnection of defence planning documents should help to ensure that the various capability enhancements identified for the CF will have the full weight of the department behind them. Moreover, a Joint Capability Requirements Board (JCRB) has recently been created with a mandate to "review proposals, challenge the issues and provide direction for the development of multi-purpose CF capabilities including the long term capital plan."[52]

Given that the Canadian forces lack the capacity to achieve their mission objectives by themselves when deployed abroad, yet another key DND document, "Strategic Capability Planning for the Canadian Forces" (SCP), also assumes that Canada will have to act with its major allies. Hence, it notes that

> [T]he fundamental asset that the CF requires for international operations (also a key contributor to domestic responsibilities) is what may be termed a tactically self-sufficient unit (TSSU). It follows that TSSU's must be capable of integrating into a Combined Force package as a "task-tailored" component. The consequences of [this] requirement...is that TSSU's must be modular and adaptable, capable of integrating with other international and national forces that are likely to be involved in a joint and combined operation. The most likely coalition leader for CF TSSU's is the US, which leads to the emphasis placed on interoperability with US forces by CF leaders. A corollary of the coalition nature of Canadian military operations is that decisions regarding commitments of Canadian TSSU's are vitally important, highlighting the importance of military strategic level command capability in the CF.[53]

In addition, however, and with commendable candour for an in-house planning publication, the SCP also offers a warning about the possibly adverse political consequences of the growing emphasis on greater Canada-US military interoperability:

> This trend raises concern over the degree to which CF units and Canadian political authorities can retain the ability to make autonomous decisions in future fast-paced combat situations. The trend toward integrated operations and interoperability may create an unintended interdependency if CF units become too enmeshed in Alliance controlled network systems that require automatic linkages of sensor and weapon systems for effective tactical operation.[54]

One way in which this could occur would be through the Canadian Navy's ability to "plug in" to a USN "net-centric" battle group via the evolving US concept of "cooperative engagement capability." Employing dedicated data links, ships and aircraft can share secure data gathered from a variety of different sensors to gain a single, composite battlespace management picture. This can then be used by a central command authority to command, fire and control the weapons of a subordinate, third-party unit to attack a target never seen by the firing entity. But as one Canadian officer has warned, there may be a high political price to be paid for such a far-reaching interoperability arrangement:

> Integration into a carrier battle group operating like this implies a level of commitment to force goals and therefore to the political objectives that underlie them. There will be no time to check national rules of engagement and 'opt

> out' on a case-by-case basis or refer difficult decisions to national headquarters. Buy-in on operational decisions will be absolute and the consequences shared accordingly. Today, the USS Vincennes must take the full blame for misidentifying and engaging an Iranian 747 aircraft — tomorrow she could do it again, but with a missile fired from a Canadian ship.[55]

There is one final, but centrally significant, indicator of the commitment of both the DND and the uniformed establishment to the implementation of the interoperability doctrine. Interoperability is partly about the sharing of operational practices. It is also about hardware. The equipment procurement programs currently planned for the CF thus help to tell the tale. And from the vantage point of even the most well informed analyst, the list of new additions and improvements that have interoperability as their primary justification is bound to look impressive. The package as a whole is an unambiguous demonstration that Canadian defence planners have been moving toward greater interoperability with the United States in a manner that is far more comprehensive and rapidly paced than most Canadians realize.

The best single account of these accumulating procurement projects, at least in the public domain, has been compiled by Sharon Hobson of *Jane's International Defense Review*. In a recent article, Hobson quotes Brigadier-General Marc Dumais, chief of staff for the deputy chief of defence staff, on the challenge that now confronts Canadian defence planners in deciding where to invest the DND's limited capital funds. Dumais observes that the United States is "at the leading edge of many developments, and from a budgetary perspective, it's evident that we can't keep up in all areas, so we have to be very careful as to which areas we ensure that we keep up in, and get the most bang for our buck in terms of focusing future interoperability requirements."[56]

To a large extent, this is a reflection of the "catch-up" problem facing Canada as it strives to strengthen its defence relationship with the United States at a time when the much-discussed "revolution in military affairs" (RMA) is threatening to leave behind all but the biggest of the defence spenders in the Western alliance. As the DND's *Report on Plans and Priorities, 2001-2002* notes, "the US is leading the RMA, and thus, by remaining interoperable with US forces, Canada will remain interoperable with any major, future coalition force led by the US."[57] Ultimately, according to General Dumais again, "it comes down to resources, and leveraging our partnerships and getting the most out of collective defence. Because clearly we wouldn't be able to afford providing our own defence for our sovereignty, for example, or something of that nature."[58]

If "going it alone" is not, then, a viable option for the CF, where exactly is the DND placing its interoperability bets? As noted earlier, "Defence Plan 2001" stresses the need for the CF to give priority to their command, control, and communications as well as their intelligence, surveillance and reconnaissance capabilities "to ensure that we can operate effectively in the information age with our allies...[and] to enhance Canada/US interoperability."[59] This will require, among other things, the completion of the Omnibus Canadian Military Satellite Communications (CANMILSATCOM) Project already underway, at a budgeted cost of $480 million. Under this project, the DND will participate in the US Advanced EHF satellite system, which is scheduled to be operational in 2006. At present, the CF have no dedicated long-range communications system of their own and must rely, instead, on commercial satellite services. They can only operate with American forces, however, if they have a capability in, and guaranteed access to, the type of system upon which the Americans themselves rely. The CF will also continue to cooperate with the United States through the Canada-US Space Co-operation Working Group on initiatives like the Joint Space Project.[60]

On land, *Report on Plans and Priorities 2001-2002* notes that the Canadian Army will "work to develop and integrate interoperability benchmarks into the design and implementation of the 'Army of Tomorrow.'"[61] The army is currently restructuring itself to become a lighter and, therefore, more mobile force by acquiring 651 General Motors LAV III armoured personnel carriers. But army planners hasten to add that becoming a "medium-weight force does not equate to a niche force." The army's planning concept is to "structure expeditionary forces to be of maximum strategic value to a joint force and potential coalition partners, particularly the US."[62] To this end, a top priority project for the army is the Land Force Omnibus ISTAR project. This will provide a capability for integrating and distributing information from a variety of army land- and air-based surveillance and target acquisition systems.[63]

Similarly, the Canadian Air Force is planning a comprehensive Omnibus CF-18 Incremental Modernization Project (IMP) to upgrade its reduced fleet of CF-18 fighter aircraft to US model C/D configuration standards. This project, if funded to the full estimated cost of $1.226 billion, will encompass 10 individual projects designed to overcome certain of the key interoperability problems experienced during the Gulf and Kosovo air campaigns, and will extend the operational life of the CF-18s to about the year 2020, thereby ensuring that they can play potentially significant roles in any future US-led coalition operations.[64]

With respect to the CF's strategic airlift capabilities, it is interesting to note that the DND has evidently decided against adopting a fully integrated solution. At one point, the air force had considered buying as many as six Boeing C-17 aircraft for this purpose and then "loaning" them to the United States when they were not required by the CF. In return, the US would share the operating costs and would ensure that six of their own C-17s would always be available for

Canadian military missions. According to Pat Dowsett, project manager for the Future Strategic Airlift Project, this "swap" arrangement has fallen afoul of a variety of "complicating factors" — factors having to do in part with deciding which country's colours would adorn the aircraft and who would crew them. There were "political sensitivities involved in that you might see American airplanes with Canadian crews doing Canadian jobs and the Americans might not be too happy with that," especially if Canada were using the C-17s to transport aid to Cuba, for example.[65]

As a stop-gap alternative, Ottawa is now planning to convert two of its CC-150 Airbuses into strategic refuellers, an initiative that has doubtless gained political support in the wake of the embarrassing delays the CF have faced in getting 750 Canadian troops and their 12 Coyote reconnaissance vehicles to Afghanistan. While the refuellers will extend the range of Canada's existing Hercules tactical transport fleet, the Americans reportedly are still keen on Canada acquiring the C-17s. Dowsett explains that "should we acquire C-17s and should we have an excess capacity during peacetime, the Americans are very interested in...tapping into that excess capacity." In that model, Canada would not actually lend the aircraft to US forces, but rather "on a case by case basis, we would satisfy an airlift task." In return, "there would be an offset, and some advantage to us in some other areas."[66]

The navy is examining a slightly different approach in attempting to enhance the CF's currently limited strategic lift capability. Specifically, it is planning to procure modern replacements for the two elderly fleet-replenishment ships that it now has in operation, ships absolutely essential to the support of its two naval task groups. It is estimated that the Afloat Logistics and Sealift Capability (ALSC) Project will cost between $1.446 and $2.3 billion (depending on the number of ships involved), and it has been fast-tracked to generate an initial operating capability in 2005. In addition to supporting the task groups with fuel, food, spare parts and so on, the ALSC is also expected to provide some sealift capability for troops and heavy equipment, headquarters command and control facilities, and support capacity for joint forces ashore. It is also slated to operate independently of a jetty, a requirement that has brought interoperability considerations into play. Commander Dave Harper, ALSC project director, confirms that the ship chosen will have to be compatible with the US Navy, and reports that "we're wrestling with what does integration [sic] mean and we believe that if we were involved in an allied effort, it may be wise to be able to accept a US Navy hovercraft. We would simply design the well deck in such a way that we would have the capability of accepting one."[67]

These examples — and there are many others — represent genuinely innovative approaches to achieving the interoperability objective, and they underscore the extent to which the military is factoring it into its long-term re-equipment plans in order to help defray some of the costs of maintaining an ability to project power abroad. To a very large degree, and in the absence of a sudden and uncharacteristic impulse on the part of Ottawa to increase Canadian defence spending dramatically, interoperability with the Americans is the only game in town.

Interoperability: Some Observations on the Wider Implications

From the purely military point of view, the rationale for pursuing the goal of interoperability seems unassailable. It maximizes military returns and minimizes combat risk. It holds out the prospect of operating with cutting-edge military technologies and in tandem with the most sophisticated military machine ever constructed in the history of humankind. It seems to be consistent, moreover, with the pattern of Canada's past experiences with combat operations abroad, operations that have always been conducted in the context of coalitions with allies. That being so, it also offers what Canadian diplomacy wants most from the military game, which is the enhancement of its political credit where it really counts: in the capitals of the foreign powers that matter most to Canada. And at the most mundane level, it makes the best of a bad budget.

But there are other potential implications that may also warrant the close consideration of Canadians as they reflect on a process which, as we have tried to demonstrate, is now very far advanced. As recent reports of continental defence integration across the board and at the command level suggest, it is also a process that may shortly be taken to new heights. Interoperability in the operational sense is not the same as the integration of structures for command and control, but the two are closely related and, clearly, mutually reinforcing. That being so, we may be dealing here with a heavy train rolling downgrade without brakes. We are led, therefore, to offer the following as issues that may warrant public attention and debate.

First, is it necessarily the case that advancing the objective of interoperability, as its defenders have sometimes argued, will enhance Canada's role as a potential facilitator or enabler for other states wishing to "plug into" American-led operations, and to reap such rewards of multilateral influence as this may bring?[68] It is certainly true that full interoperability with the Americans will also carry with it a greater degree of de facto interoperability with other potential allies. It is also true that Canada had some success in precisely this way in the context of naval operations during the Gulf War. Despite the rumblings of certain European allies in support of an ESDI (European Security and Defence Identity), moreover, there are indications that even the Europeans recognize the need to remain reasonably interoperable with American forces.[69] (On the other hand, to the extent that they do it themselves, they will have less need of Canadian bridges to US contingents.)

But having said all that, the reality is that there is now increasing anecdotal and other evidence that in recent years the Europeans have come to assume that Canada has so fully integrated with the United States — economically and diplomatically as well as militarily — that it can no longer be regarded as a useful interlocutor, much less as an independent player. The recent European reaction to the Canadian desire to participate in the peacekeeping operation in Kabul is a telling demonstration of the point. To the extent that this is true, the Canadian position may be rendered more, rather than less, difficult as the interoperability process continues to advance.

Parenthetically, it should be noted here that, from the US perspective, interoperability with potential coalition allies is more a matter of choice than of necessity. If the allies further advance their interoperable capabilities, their doing so may on occasion be helpful. But from the American point of view, such co-operation is not essential, and if the problems involved in promoting it become more trouble than the results are worth, US forces will not allow such complications to retard their own military exploitation of cutting-edge technologies.

Second, will greater Canada-US interoperability have the effect in practice of narrowing Ottawa's options with respect to participation in future operations? Is it possible that Canada will be left with little choice but to go when — and only when — the Americans go, or not go at all? And is it conceivable that in some cases the second of these two options will be ruled out as well, given the kinds of pressures that may emanate from Washington as a result of the expectations of automatic support that will be created there by Canada's eager pursuit of further integration with the American structure?

These questions presuppose among other things that Ottawa remains unwilling to foot the cost of gaining a truly independent strategic lift capability. Whether or not this assumption proves to be well founded will depend in part, of course, on Canada's near-term decisions in connection with the acquisition of dedicated strategic lift capabilities in both the air and sea environments. But that, in any case, may be only one dimension of the problem, and it points to a much larger issue having to do with the scale of Canadian defence expenditures more generally. In concrete terms, one of the purposes (although certainly not the only purpose) of interoperability is to compensate for the unwillingness of the political leadership, and perhaps ultimately of the electorate, to allocate a larger portion of public funds to the armed forces. It is quite likely, however, that there is a direct functional connection between interoperability "on the cheap" and loss of autonomy. To put the point another way, a fully funded interoperability arrangement might still leave Canadian decision-makers with at least some military options of their own because it would not deprive them of the capacity to operate independently of their larger partner. By contrast, a lightly funded arrangement could force them to act in lockstep with American initiatives whether they like them or not, and it might even prevent them from responding to requirements and opportunities (generated, for example, through the United Nations) that are regarded as important in Ottawa, but as trivial, distracting or politically inconvenient in Washington.

This is a variation, of course, on a very old theme, a theme reflected many years ago in Nils Ørvik's well-known aphorism that one of the primary purposes of the Canadian armed forces in the North American theatre was to provide a "defence against help." Their function, in other words, was to ensure that the Americans would not feel the need, in their own security interest, to defend Canada whether Canadians liked it or not, and to intrude on Canada's sovereignty in the process. You get what you pay for. You may also get what you don't pay for, but not in a form you like.[70]

Third, will greater interoperability ultimately have the effect of weakening the operational autonomy of Canadian theatre commanders? This question bears not so much on general policy issues of the "to go or not to go" variety, but more immediately on the position in which Canadian commanders may find themselves once their forces have actually been committed. Traditionally, Canadian in-theatre commanders, and through them the authorities in Ottawa, have always retained the final say — i.e., command — over Canadian contingents operating under the overall control of foreign commanders. This was certainly the case in relation to the Gulf War, Somalia, the Aviano and Adriatic deployments, the individual Canadian warships operating with both NATO's Standing Naval Force Atlantic (SNFL) and the West Coast US Carrier Battle Group, and the Kosovo air campaign.[71] Currently, it applies also to the case of Operation APOLLO in the Arabian Sea and Afghanistan. Decades ago, it was the practice as well in the context of NORAD, where in both 1962 and 1973 the Canadian government opted not to put Canadian Forces on higher states of alert that were commensurate with the decisions taken by their US counterparts.[72]

Arguably, there has been no loss of effective Canadian autonomy in any of these instances or in others like them. But it is now much less clear that this is still the case, not least because of the accelerating integration of Canadian and American forces on the one hand, and changes in the character of the threat to the security of both North America and targets elsewhere on the other. Jeffrey Simpson, for example, has recently reported that during the NORAD alerts that were stimulated by the 9-11 attacks, the Canadian government was not consulted before the decision was made, nor even notified of it afterwards.[73] If his report is accurate, it may be writing on the wall.

Fourth, is it likely, as the integrative process proceeds, that there will be an increasingly unavoidable linkage between technical and operational interoperability with the Americans on the one hand, and broader security policy matters on the other? And if so, will the trend towards greater interoperability lead to a much more intensive focus on bilateralism in Canadian foreign policy at the expense of the more traditional emphasis on multilateralism as the preferred diplomatic strategy? As an example of a broader security issue that might be affected in this way, the National Missile Defense initiative comes immediately to mind. If acceding to the American position on NMD were a serious problem for the political leadership in Ottawa, either because it was internally divided on the issue or because it feared an adverse public reaction that it was unwilling to confront, Canada would probably still have the freedom to refrain from active participation, particularly since the Americans have not been asking for a significant financial contribution. But given a greatly elaborated system of interoperability at multiple levels, it would be much more difficult for the government to express open opposition to the American preference. Such opposition, in any case, would be inconsistent on its face with the underlying premises of the interoperability doctrine, and would enormously complicate the defence community's attempt to maintain credibility with its American counterpart upon which the full implementation of the doctrine ultimately depends.

With regard to the potential implications of interoperability for the Canadian preference for multilateralism, John Manley has argued, in effect, that this is not a problem,[74] but there may be more wishful thinking and cosmetic politics than serious judgment in the assertion. Among other things, it ignores the relatively recent experience of Manley's immediate predecessor as minister of foreign affairs, Lloyd Axworthy, who clearly had to hold his nose while his government agreed to commit Canadian Forces to the American-led coalitions in both Bosnia and Kosovo. These were, of course, NATO-sanctioned missions, and even the current deployment in Afghanistan has at least the partial endorsement of both NATO and the UN. But the ultimate policy "driver" in all these cases has been the United States, and this reality may repeat itself, while becoming increasingly transparent, as time goes by. At the time of writing, for example, Prime Minister Chrétien has been implying that Canada would not be likely to join with the United States were it to open up a new round of hostilities against Iraq. But with Canadian Forces operating closely with American units in the theatre, and given the general context of what the Americans have been pleased to call their "war against terrorism," he might well find in the practical event that his capacities for abstention are more limited than he currently appears to assume.

Finally, does Canada really have a choice? Is there, in fact, an alternative to the interoperability option, which is already far advanced? In principle, there obviously is. However, the difficulty — one which the DND has clearly concluded is insurmountable — is that any serious alternative would demand a far greater expenditure of public funds than the workings of Canadian politics are likely to make possible. Certainly, this is true of any alternative of a kind that would maintain a significant remnant of Canada's position as a useful contributor to security affairs. This difficulty might be overcome by a powerful and determined display of firm commitment from the political leadership, but such a display is improbable given that it would entail opportunity costs in the public policy sector (e.g., in health care, education, social welfare, or whatever) that could be politically damaging to whomever happens to hold public office.

It might be argued that alternatives of a somewhat different sort are available. One of them would be a constabulary-only force that would be dedicated primarily to the protection of Canada itself. But this would entail a price of its own — notably the near-total evaporation of any Canadian capacity for exercising even marginal influence over the course of international security affairs. Another option would be to concentrate on the formation of a niche specialization that would lead to Canadian units acting as single-purpose components deployed as complements to the multipurpose forces of the United States (and perhaps occasionally those of other allied powers, too). But as the DND's senior policy planner has recently pointed out, his department's ability to predict the particular niche that would best fit future circumstances, given that these are unknown, "is very bad."[75] Any option of this sort, therefore, risks dooming Canada to a place of irrelevance on the international stage.

In summary, as their views are manifested in their politics, Canadians appear to want their country to be an active and contributive player in a wide variety of contexts around the world. On the other hand, they also appear reluctant to allocate the resources that are appropriate to their aspiration. The consequence, as interpreted thus far by a policy community that is attempting to respond to both of their preferences, is uncomfortably clear: in order to play a significant role on the world stage, Canada has to get into bed with the United States. For some, this may be a sobering thought. Others may not mind it at all. But whatever the reaction, surely no one would think that the process itself should proceed by stealth, or even by osmosis. It warrants a closer public look.

Notes

1. DND (1999).

2. DND (1999, foreword).

3. DND (1999, p. 2).

4. The so-called "critical attributes" included modernization, deployability, interoperability, force structure, domestic capability, jointness, capital program, command and control, engage Canadians [sic], human resources and pro-activity. DND (1999, p. 6).

5. DND (1999, p. 3).

6. The North Atlantic Treaty Organization was officially established in April of that year at a signing ceremony in Washington. Lester Pearson was in attendance. He later reported that the military band of the US Marines, apparently innocent of irony, offered a background musical repertoire that included "two selections from Porgy and Bess: 'I Got Plenty of Nothing' and 'It Ain't Necessarily So.'" Munro and Inglis (1973, p. 37).

7. The year in which President Roosevelt and Mackenzie King agreed to establish the Canada-United States Permanent Joint Board on Defence, an institution that persists to this day.

8. DND (1999, p. 6). Emphasis added.

9. Vice-Chief of the Defence Staff (2000, p. 28). See also, North Atlantic Treaty Organization (2002, p. 221), Department of Defense (2001) and the US Joint Chiefs of Staff (1997, p. 352).

10. National Research Council (1999, p. 1, ch. 2).

11. Faughn (2001, p. 5).

12. Vice-Chief of the Defence Staff (2000, p. 8).

13. It needs to be recognized, of course, that self-sufficiency is itself a matter of degree, and the issue is therefore given explicit attention in the DND's current internal discussions of capability planning. We will refer to these discussions in more detail below. For present purposes, however, the central point is clear enough: the less self-sufficiency there is, the greater will be the need for smoothly working interoperability arrangements and routines, and the more necessary it will be to have them in place at multiple levels of operational activity. The obvious corollary comes also to mind: the capacity to act autonomously in the field, and by extension in the offices of higher-level decision-making as well, is inversely related to the degree to which an armed forces establishment is capable of deploying and operating its military contingents on a stand-alone basis.

14. For more on the evolution and practice of the Canadian Naval Task Group concept see, Haydon (2000) and Morse (2000). Also, Thomas (2001).

15. Lester Pearson, quoted in McLin (1967, p. 18).

16. Morton (1992, p. 111).

17. Morton (1992, p. 113).

18. A cause often shared by Canadian political leaders, as evidenced by Laurier's unhesitating support for British-led conflicts when he stated in 1910: "When Britain is at war, Canada is at war. There is no distinction." Morton (1992, p. 130).

19. Morton (1992, pp. 145-51). Morton also notes that men of the Canadian Expeditionary Force (CEF) enlisted as "imperials" under the British Army Act and that officers carried temporary British commissions and came under the authority of the British War Office as part of the Imperial army. Morton (1992, p. 145) concludes, "From strategy to finance, Britain's authority was supreme and unquestioned."

20. Morton (1992, pp. 147-51).

21. During an August 1938 visit to Canada, President Roosevelt pledged "that the people of the United States would not stand idly by if domination of Canadian soil is threatened by any other empire." Two days later, Prime Minister King responded by assuring the United States that "we, too, have our obligations as a good friendly neighbour, and one of these is that, at our own instance, our country is made as immune from attack or possible invasion as we can reasonably be expected to make it, and that should the occasion ever arise, enemy forces should not be able to pursue their way either by land, sea or air to the United States across Canadian territory." Quoted in Eayrs (1965, p. 183).

22. Government of Canada (1940).

23. Smith (1963, p. 303).

24. Government of Canada (1941).

25. For a brief overview of these cooperative developments, see Middlemiss (1982, pp. 86-114).

26. Government of the United States (1946, pp. 65-7).

27. Government of Canada (1947).

28. Willoughby (1951, pp. 682-3).

29. Smith (1963, pp. 307, 318). On the Screw Thread Convention itself, see Government of Canada (1948).

30. For a comprehensive account of these arrangements leading up to the creation of NORAD, see, Jockel (1987).

31. Government of Canada (1958).

32. It should be noted that, while Canada's air force was not formally placed on the higher alert level for some 42 hours, Canadian naval vessels were "discreetly" dispatched into the NATO Atlantic region to "back-stop" US naval forces that were being withdrawn from their regular NATO patrol areas to enforce the quarantine around Cuba. For more on this little appreciated example of interoperability, see Haydon (1993).

33. House of Commons (1986, p. 25).

34. Once Canada's two Regional Operations Control Centres (co-located at North Bay, Ontario) became fully functional on August 15, 1984, this reconfiguration of air defence boundaries meant that Canadians now exercised full operational control of all NORAD surveillance, identification and interceptor control activities in Canadian airspace. House of Commons (1986, p. 22).

35. These and other pros and cons of embracing the "continentalism" implied by interoperability with the United States are dutifully trotted out and publicly rehearsed each time the NORAD agreement comes up for renewal. But there is nothing really new or unique about Canada's continuing angst over sharing its own continent with so overpowering a neighbour. The same sorts of concerns about American domination arise for the European powers, as well as for Canada, in NATO's European theatre. They even emerge in the various agencies of the United Nations. In these latter fora, as in the case of NORAD, the revealed preference of successive Canadian governments has been to choose to work for reform from inside, not outside, the institution in question.

36. In 2000, the United States spent 70 percent of the total NATO expenditures on equipment for that year. Stockholm International Peace Research Institute (2000).

37. See for example, "Joint Doctrine Story." available online at URL: http://www.dtic.mil/doctrine/joint_doctrine_story.htm, See Department of Defence (2000).

38. See especially Faughn (2001, pp. 7-17).

39. This important story is detailed by Lieutenant-Commander Gimblett (1996, pp. 231-43).

40. Lieutenant Colonel Walker (2001, pp. 259-61). See also, Major Morin and Lieutenant Commander Gimblett (1997, pp. 131, 158-9).

41. Walker (2001, pp. 259-61).

42. Walker (2001, pp. 262-3); Morin and Gimblett (1997, pp. 159-75).

43. These examples are drawn primarily from the analysis in Walker (1997, pp. 269-71).

44. This was confirmed by General Raymond Henault, CDS, in his testimony to the House of Commons (2001a). See also the testimony of the senior US and NATO leadership to this effect following the Kosovo air campaign, cited in Faughn (2001, pp. 12-3), plus the remarks of Canada's own top airman, Lieutenant General Kinsman (1999, pp. 7-10).

45. See the testimony of Dr. Kenneth J. Calder, Assistant Deputy Minister, Policy, DND. Calder (2001).

46. DND (1999, p. 8).

47. DND (1999, p. 10).

48. In order to reduce workload and the number of planning documents produced yearly, and to improve transparency through a single authoritative document that is valid within the DND as well as for external audiences, including parliament's "Defence Plan 2001," will soon be replaced by a single planning document for the DND, *Report on Plans and Priorities*. See DND (2000a, i).

49. DND (1999, p. 10).

50. DND (2001b, pp. 24-5, 28-30).

51. DND (2001a, pp. 5-4 to 5-5). Note, C2ISR refers to command, control, intelligence, surveillance and reconnaissance.

52. DND (2001a, Annex A to ch. 7).

53. Vice-Chief of the Defence Staff (2000, p. 18). Emphasis added.

54. Vice-Chief of the Defence Staff (2000, p. 8). Emphasis added.

55. Tunnicliffe (2000, p. 5).

56. Hobson (2001).

57. DND (2001b, p. 28).

58. Quoted in Hobson (2001). Emphasis added.

59. DND (2001a, p. 5-2).

60. These examples are drawn from Hobson (2001); and DND (2001b, p. 29).

61. DND (2001b, p. 29).

62. Hobson (2001).

63. Hobson (2001).

64. Walker (2001, pp. 272-4).

65. Quoted in Hobson (2001).

66. Quoted in Hobson (2001).

67. Quoted in Hobson (2001). Note that the CF have no plans to acquire hovercraft or any other amphibious capability of their own.

68. For example, Hobson (2001) quotes Commodore Dan McNeil, director, Force Planning and Program Co-ordination, as saying that "we also see ourselves as a facilitator for smaller nations as well...other nations can plug into us, and not have to worry about plugging into the Americans. We can be enablers and facilitators."

69. Testimony of Colonel (Ret'd) MacDonald to the House of Commons, Standing Committee on National Defence and Veterans Affairs. House of Commons (2001b).

70. Ørvik (1983, pp. 3-7).

71. See, for example, the recollections of General (Ret'd) MacKenzie (2001, p. A17).

72. On the post-9-11 NORAD rules of engagement for shooting down "hostage" civilian airliners, see the testimony of General Henault to the House of Commons, Standing Committee on National Defence and Veterans Affairs. House of Commons (2001a).

73. See Simpson (2002, p. A18).

74. McCarthy (2002, p. A4).

75. See the testimony of Dr. Calder (2001).

References

Calder, Kenneth J. 2001. *Evidence* 7 (November).

Department of Defense. 2001. *Department of Defense Dictionary of Military and Associated Terms* (as amended through October 15, 2001), Joint Publication 1-02. Washington, DC: US Government Printing Office.

Department of National Defence. "Joint Doctrine Story." http://www.dtic.mil/doctrine/ joint_doctrine_story.htm (Accessed 2 January 2002).

——. 2001a. "Defence Plan 2001." Ottawa.

——. 2001b. *Report on Plans and Priorities 2001-2002*. Ottawa.

——. 1999. *Shaping the Future of Canadian Defence: A Strategy for 2020*. Ottawa.

Eayrs, James. 1965. In *Defence of Canada: Vol. II, Appeasement and Rearmament*. Toronto: University of Toronto Press.

Faughn, Anthony W. 2001. *Interoperability: Is It Achievable*? Boston: Harvard University.

Gimblett, Richard H. 1996. "Canadian Coordination of the Persian Gulf Combat Logistics Force." In *Multinational Naval Forces*, edited by Peter T. Haydon and Ann L. Griffiths. Halifax: Centre for Foreign Policy Studies, Dalhousie University.

Government of Canada. 1958. Canada Treaty Series, no. 9.

——. 1948. Canada Treaty Series, no. 21.

——. 1947. Canada Treaty Series, no. 43.

——. 1941. Canada Treaty Series, no. 14.

——. 1941. Canada Treaty Series, no. 14.

Government of the United States. 1946. *Foreign Relations of the United States*, vol. 5.

Haydon, Peter T. 2000. "The Evolution of the Canadian Naval Task Group. In *Canadian Gunboat Diplomacy: The Canadian Navy and Foreign Policy*, edited by Ann L. Griffiths, Peter T. Haydon, and Richard H. Gimblett. Halifax, NS: Centre For Foreign Policy Studies, Dalhousie University.

——. 1993. *The 1962 Cuban Missile Crisis: Canadian Involvement Reconsidered*. Toronto: Canadian Institute of Strategic Studies.

Hobson, Sharon. 2001. "Canada Aims for Defense Interoperability with U.S." *International Defense Review*, online ed. (January). http://idr.janes.com.

House of Commons. 2001a. Standing Committee on National Defence and Veterans Affairs. *Evidence*, meeting no. 23 (October 16).

——. 2001b. Testimony of Brian S. MacDonald to the House of Commons Standing Committee on National Defence and Veterans Affairs, *Evidence*, meeting no. 37, 27 (November).

——. 1986. *Canada-U.S. Defence Co-operation and the 1986 Renewal of the NORAD Agreement*. Report, 1st Session, 33rd Parliament, February 14.

Jockel, Joseph T. 1987. *No Boundaries Upstairs: Canada, the United States and the Origins of North American Air Defence, 1945-1958*. Vancouver: University of British Columbia Press.

Kinsman, David. 1999. "Canada's Air Operations in the Balkans." *National Network News* 6, no. 2 (fall/winter): 7-10.

MacKenzie, Lewis. 2001. "We're in This Fight Together." *The Globe and Mail*, January 31, A17.

McCarthy, Shawn. 2002. "Manley Defends Deployment." *The Globe and Mail*, January 9, A4.

Middlemiss, Dan. 1982. "Economic Defence Co-operation with the United States, 1940-63." In *An Acceptance of Paradox: Essays on Canadian Diplomacy in Honour of John W. Holmes*, edited by Kim Richard Nossal. Toronto: Canadian Institute of International Affairs.

Morin, Jean, and Richard H. Gimblett. 1997. *Operation Friction, 1990-1991: The Canadian Forces in the Persian Gulf*. Toronto: Dundurn Press.

Morse, David. 2000. "The Canadian Naval Task Group." In *Canadian Gunboat Diplomacy: The Canadian Navy and Foreign Policy*, edited by Ann L. Griffiths, Peter T. Haydon, and Richard H. Gimblett. Halifax: Centre For Foreign Policy Studies, Dalhousie University, .

Morton, Desmond. 1992. *A Military History of Canada: From Champlain to the Gulf War*, 3rd ed. Toronto: McClelland and Stewart.

Munro, John A., and Alex I. Inglis, eds. 1973. *Mike: The Memoirs of the Rt. Hon. Lester B. Pearson — Volume Two: 1948-1957*. Toronto: University of Toronto Press.

National Research Council, Committee to Review DOD Plans and Programs. 1999. *Realizing the Potential of C4I: Fundamental Challenges*. Washington, DC: National Academy Press, .

North Atlantic Treaty Organization. 2002. *NATO Glossary of Terms and Definitions*, AAP-6.

Øivik, Nils. 1983. "Canadian Security and 'Defence Against Help.'" *International Perspectives* 12, no. 3 (May/June): 3-7.

Pearson, Lester. 1967. *Canada's Changing Defense Policy, 1957-1963: The Problems of a Middle Power in Alliance*. Baltimore: The Johns Hopkins Press.

Simpson, Jeffrey. 2002. "What Matters is Defence Integration, Not the Eggleton Sideshow." *The Globe and Mail*, February 2, A18.

Smith, Foster Lee. 1963. "Canadian-United States Scientific Collaboration for Defense." *Public Policy* 12.

Stockholm International Peace Research Institute. "Tables of NATO Military Expenditure: NATO Distribution of Military Expenditure by Category, 1991-2000." http://projects.sipri.se/mitex/mex_date_index.html (accessed February 2, 2002).

Thomas, Doug. 2001. "The Canadian Naval Task Group: What Is It, What Can It Do?" *Maritime Affairs* 1, no. 2 (summer/fall).

Tunnicliffe, Mark. 2000. "The Revolution in Military Affairs and the Canadian Navy in the 21st Century." *Maritime Affairs* 2, no. 2 (spring/summer).

US Joint Chiefs of Staff. 1997. *The Joint Doctrine Encyclopedia*. July 16.

Vice-Chief of the Defence Staff. 2000. "Glossary for Strategic Capability Planning for the CF." *Strategic Capability Planning for the Canadian Forces*. Ottawa: VCDS, Department of National Defence.

Walker, Samuel J. 2001. "Interoperability at the Speed of Sound: Modernizing the CF-18 Hornet." In *Over Here and Over There: Canada-US Defence Co-operation in an Era of Interoperability*, edited by David G. Haglund. Kingston: Centre for International Relations.

Willoughby, William R. 1951. "Canadian-American Defense Co-operation." *Journal of Politics* 13, no. 4 (November): 682-3.

7

Sailing in Concert: The Politics and Strategy of Canada-US Naval Interoperability

Joel J. Sokolsky

At the dawn of the 21st century, the US Navy is seeking to define its missions across the AORs and how it will carry out those missions...US policy statements make it clear that whenever possible, US forces will seek to respond to requirements for military force in concert with other countries. These responses may take the form of ad-hoc coalitions or bilateral actions with other countries, and may or may not have mandates or consent from the United Nations, NATO, or other international bodies. This being the case, a particularly important question in defining the US Navy's role is how will it operate with its key allies? The fact that navies will be less engaged in classical blue water operations and instead will be more involved in littoral operations enhances the possibility of improvised multinational naval cooperation. Under these circumstances, achieving interoperability becomes much more complicated, for it is no longer simply a question of standardization and compatibility. Interoperability also involves issues of political will.

Kenneth Gause, Center for Naval Analysis[1]

Accepting...that Canada is more likely to conduct overseas military operations in concert with other like-minded nations...it is also advantageous to both the US and its allies that the US not act unilaterally as the "world's policeman." If recent experience is an indication, there will continue to arise any number of situations in which naval forces of medium powers such as Canada can make a difference by working in combination with the USN...Opponents of closer allied cooperation will argue that integration into US and NATO naval formations undermines Canadian sovereignty, but that is not necessarily so. Rather, because each mission is a function of choice, it tends to strengthen Canadian sovereignty. This issue, of the potential effect of closer allied cooperation upon independence, is a matter of national strategy. For the navy, it must be noted and considered while implementing the requirement identified in Strategy 2020 for enhanced interoperability.

Leadmark: The Navy's Strategy for 2020[2]

Introduction

Dispatching the fleet again: a "traditional" response to a nontraditional threat

While some pundits in Canada derided the commitment as insufficient and inappropriate, it was not surprising that Ottawa's initial response to the American call for allied assistance in the war on terrorism was to dispatch several of the Canadian Navy's most advanced surface ships. Practically and politically, ships were the easiest units to send right away. To this extent, the deployment was fully in line with traditional Canadian responses dating back to the Korean War, through the Gulf War and during the heightened "new" peacekeeping of the 1990s. It could be said that the dawn of the twenty-first century finds the Canadian Navy where it has been since its founding at the beginning of the twentieth: sailing far from home in concert with its principal allies as part of multilateral maritime coalitions.

But it is also the case that relative to other branches of the Canadian Forces (CF), the navy has gone furthest in seeking interoperability with the armed forces of the United States. And this is why the fleet sailed so quickly, once again, in concert with the United States Navy. Indeed, today, more than a decade into the post-Cold War era and despite the disappearance of the Soviet threat, Canadian defence policy is more closely linked to that of the United States than at any other time in our history. Ottawa seems to be "glued" to its seat at the allied table and as a consequence, its defence posture, especially at sea, seems even more fused with that of the United States.

This paper examines the strategy and politics of Canada-US naval interoperability. It begins with a discussion about sea power in the post-Cold War era. Since 1990, the political and strategic nature of multilateral sea power has changed, shifting from a focus on securing the seas to that of the projection of power ashore. This is followed by an examination of interoperability in the context of overall American national security policy in a "unipolar world." The changing nature of sea power combined with the implications of unipolarism has meant that the United States Navy (USN) has placed greater emphasis upon interoperability with a host of traditional and nontraditional allied navies. The paper then turns to consider what this has meant for the Canadian Navy given the role of Canada's maritime forces in the country's overall foreign and defence policy and national interests. The paper argues that interoperability is an inescapable and highly beneficial fact of life for the Canadian Navy if Ottawa wishes to maintain a "global" defence policy. Moreover, despite its size relative to the USN, the Canadian Navy can make an effective contribution to Western maritime multilateralism. To a certain extent, the USN's interest in interoperability with the Canadian Navy is not fundamentally different than its fostering of similar ties with other navies. At the same time, it is argued here that from the Canadian standpoint (and perhaps from the American after September 11, 2001) there are unique considerations. Above all, there are intangible factors of sovereignty and autonomy. If the immediate naval response to the war on terrorism was traditional, then so too are the questions that can be raised about the strategy and politics of interoperability in the post-Cold War era. For however tactically and operationally sensible, it raises old questions about Canada's ability to conduct an independent national security policy or its ability to determine, on its own, the extent and character of its contributions to American-led multinational collective defence and security operations. Another consideration is that until now, bilateral interoperability has been focused on the application of naval power overseas, to secure regions and allies in Europe, the Middle East and Asia. Canada's "region," North America, was not a priority for the USN. What will "homeland" defence mean for the strictly bilateral dimension of Canada-US naval cooperation?

Sea Power in the Post-Cold War Era: The Influence of Sir Julian on History

If, for the USN, the passing of the Cold War has brought about what is already a decade's worth of intellectual effort to redefine the role of sea power, for the former Soviet, now Russian, navy, the past 10 years have been a continual and not very successful effort simply to stay afloat.

The end of the Cold War has occasioned a similar challenge to the meaning and role of sea power as that which took place at the end of the Second World War. At that time, the advent of atomic weapons appeared to make navies obsolete and superfluous in any future war against the Soviet Union. "How could enough time be allowed for sea power to take its effect, where war was characterized by strategic bombing with nuclear weapons?" asked leading naval theorist Bernard Brodie. Nations, their land and air forces, as well as their economies would "disappear in the first blows of the nuclear war."[3] Not only did atomic weapons appear to undermine the need for sea power, but American sea power seemed so absolute that Admiral Chester Nimitz worried that it would be taken for granted.[4]

The Cold War atomic era did not see the eclipse of sea power, but quite the opposite. By the early 1950s, the USN had developed a carrier-based nuclear strike capability. The last years of that decade saw the advent of the nuclear-powered ballistic missile submarine (SSBN) which, with its submarine-launched ballistic missile (SLBM), became the capital ship of the new age. It was upon this third and secure leg of the nuclear triad that the credibility of deterrence rested. The development first of nuclear-powered attack submarines (SSNs) and later of sea-launched cruise missiles (SLCMs) made it evident that nuclear propulsion and especially the deployment of nuclear weapons at sea had endowed navies, particularly the USN, with a power and strategic significance unmatched even when Britannia ruled the waves.

Had strategic nuclear deterrence been the sole, or even dominant, role of sea power in the Cold War, then the USN and other Western navies would never have attained the size and sophistication that they did. But navies continued to be concerned with traditional roles, protection of the sea-lanes of communication (SLOCs), the projection of force ashore, gunboat diplomacy and naval presence. Even in the absence of a comparable rival such as the Soviet high seas fleet, sea power maintained relevance in the global balance of power. Indeed, Samuel P. Huntington argued in 1954 that the USN's monopoly of the seas and Soviet land power in Eurasia had resulted in a new kind of navy — a "transoceanic" one. The USN's role was not to prepare for a Mahanian fleet-on-fleet struggle for the high seas but to apply power on the "narrow lands and the narrow seas which lie between" the "great oceans on the one hand and the equally immense spaces of the Eurasian heartland on the other."[5]

This was especially the case for the NATO alliance. From its earliest days, the alliance focused on securing the seas immediately adjacent to Europe. Moreover, while it was the case as Huntington argued that the USN and its allies dominated the high seas, in the "narrow seas" around Western Europe the Soviet Union could, even in these early years, deploy sea denial forces (principally submarines) that would have made the immediate projection of force ashore difficult. In later years when, due to the emergence of a more powerful and high-seas-capable Soviet fleet along with a considerable land-based naval aviation capability, NATO grew increasingly apprehensive about its ability to protect the transatlantic SLOCs upon which the strategy of flexible response rested. Sea power had been an essential component of collective defence.[6]

In what turned out to be the last years of the Cold War, NATO took specific and deliberate steps to address what was viewed as a growing maritime threat.

In 1981, the Defence Planning Committee (DPC) adopted a "Concept of Maritime Operations" (CONMAROPS) which stressed the importance of containing Warsaw Pact forces through forward operations, of defence in depth, and of gaining and maintaining the initiative at sea. Although differing in some respects from CONMAROPS and the cause of considerable controversy, the USN's much heralded, and much maligned, Maritime Strategy of the 1980s also drew attention to the need to provide a more effective counter to growing Soviet naval capability.

But it was never the various formulations of maritime strategies that defined the role and significance of sea power in NATO during the Cold War. Nor was it, fundamentally, only the naval balance of power which determined the need for maritime forces. Allied naval plans and the forces acquired to implement them were simply reflections of the overall goal of NATO, which was to provide for collective defence and deterrence ashore in Western Europe. The Cold War had wrought many changes in international strategic relations, but it did not change the true essence of sea power, which remained the ability to secure, deny to the enemy, and utilize the oceans for the projection and sustainment of military power ashore in peace and war. The ultimate objectives of naval forces, even in the nuclear age, have been ashore, because it is there that organized political communities exist. As Colin Gray has observed, "The sea, like the air and like space, has strategic meaning only in relation to where the human race lives, the land."[7] Accordingly, the measure of the effectiveness and significance of naval forces rests in their ability to influence the situation ashore. Whether it was the USN's SSBNs and carriers reinforcing extended nuclear deterrence or the combined NATO fleet's support for conventional deterrence through the maintenance of a flexible response capability, the ultimate objectives of allied sea power were ashore. It was this reality which, despite the consternation over the future of sea power at the dawn of the Cold War, made the NATO navies major contributors to the final victory in that "long twilight struggle."

To this extent, it is evident that while in the Cold War era, and especially in the 1980s, the USN said it was relying upon the theories of Mahan, it was in fact more closely following the writings of Sir Julian Corbett. The global dispersal of the USN seemed to suggest that commanding the sea meant commanding the land and that the great challenge to the USN was a rival fleet able to challenge it for control of the seas. But until well into the 1970s, and even then, no country had a fleet capable of matching the American Navy. Yet the USN was essential to the American and allied strategic posture.

All this was described by Corbett, who argued that "the ultimate purpose of naval forces working in the service of the government's wider policy objectives was to pressure the enemy in various ways, assist the army, the diplomats, the country's allies." The USN's emphasis upon joint operations "also reflects an acknowledgement of his statement that the idea in tactics remains the same in strategy because 'The Army' was the primary objective around which all dispositions turned."[8]

By the early 1970s, it appeared that a rival fleet was emerging to challenge the USN and its maritime allies. For most of the early Cold War period, the Soviet navy had also been following the principles of Corbett, insofar as its prime mission and resulting capabilities were directed toward the defence of the Soviet Union and its position on the ground in Eastern Europe. With its heavy emphasis on attack submarines and land-based naval aviation, this was a sea-denial force meant to thwart the power projection capabilities of the American "transoceanic" navy. Here the Soviet Union had the advantage. For in the waters off the European littoral, despite its overall quantitative and qualitative inferiority, it still presented a formidable challenge to any power projection efforts by the USN and its allies. Limited as its range might have been, the Soviet navy could control waters vital to any allied war plans that envisioned a protracted conventional conflict along the central front or the northern or southern flanks. To be sure, the USSR acquired a sea-based atomic delivery capability, and thus the ultimate power-projection weapon. But here too, defence and sea denial were important, and it devoted significant assets to the protection of its SSBN fleets in their northern bastions.

As the Soviet navy grew, it also took on aspects of a true blue-water fleet, sailing more widely and seeking out port and base facilities far from the homeland. As with the USN, the Soviet navy was to be an instrument of global presence, protecting distant allies. In the Mediterranean, the Indian Ocean and, with the establishment of a base in Vietnam, in Southeast Asia, the Soviet navy challenged America not so much for control of the seas, but rather for global political influence. As reflected in the USN's vaunted "Maritime Strategy" of the mid-1980s, the United States now believed it finally faced a Mahanian challenge worthy of its steel. But with the abrupt ending of the Cold War, this (potential) challenge vanished.

For the USN and its allies, however, because their sea power had always fundamentally been about power projection, the new era did not mean the disappearance of the need for maritime forces. And here too, it has been the Corbett, as opposed to Mahanian, approach that has made allied sea power relevant to the post-Cold War era. The USN articulated its post-Cold War strategy in *From the Sea* in 1992. It is an aptly named document for it constituted a shift in focus from the sea of the 1980s maritime strategy to the land, where the real objectives of sea power have always been. Command of the sea is meaningless unless it can allow for the projection of force from the sea to the land. "Derived from" the

National Security Strategy of George H. W. Bush's administration, which emphasizes peacetime presence and engagement, promotion of stability, thwarting of aggression mobility and flexibility in meeting regional, rather than global threats to American interests, the USN's strategic direction was described as

> a fundamental shift away from open-ocean warfighting on the sea towards joint operations conducted from the sea. The Navy and Marine Corps will now respond to crises and can provide the initial "enabling" capability for joint operations in conflict — as well as continued participation in any sustained effort. We will be part of a "sea-air-land" team trained to respond immediately to the Unified Commanders as they execute national policy.[9]

With the coming into office of the Clinton administration, new guidance was provided for the role of military forces, one which reflected the new roles and missions, especially in the areas of peacekeeping that the US military had undertaken.[10] Given the shift in emphasis toward the "new dangers" posed by "aggression by regional powers," it was necessary to again review naval strategy. In November 1994, the USN published *Forward...from the Sea*. The document notes that while naval forces "are designed to fight and win wars, our most recent experiences...underscore the premise that the most important role of the naval forces in situations short of war is to be engaged in forward areas, with the objectives of preventing conflicts and controlling crises."[11] As the chief of naval operations put it, the "cornerstones" of American sea power will be "forward presence, power projection, strategic deterrence, sea control and maritime supremacy, and strategic lift. Naval forces are going to come from the sea. They are going to work near land and over land."[12]

For Jan Breemer, the USN's approach to the role of sea power in the post-Cold War era marks the "end of naval strategy." In making this argument, he draws a distinction between naval strategy, which is concerned with securing command of the sea, and "maritime strategy," of which naval strategy is a subset and which is concerned with the relationship between navies and armies. Naval strategy is at an end to the extent that the USN's focus is no longer on planning for war at sea but rather on support of joint operations on land. Gone is the "Mahanian vision of naval power as the struggle for command of the sea by battlefleet," a vision which was integral to the maritime strategy of the 1980s. The then rising Soviet fleet is gone. Because the USN need no longer "look over its shoulder for the next blue water challenge," it can concentrate on "operations other than war at sea." Indeed, it can concentrate on operations other than war, on littoral operations to contain crises.[13]

Yet, as argued above and as Breemer acknowledges, in 1954 Huntington had already pointed out the importance of the Eurasian littoral as the true objective of American and NATO naval power. In the post-Cold War era, the orientation of sea power away from the sea to the land had become even more pronounced. To be sure, allied naval forces continue to prepare for article 5 operations in defence of NATO territory. To this extent, as the statement on British maritime doctrine makes clear, the principles of NATO maritime operations remain consistent with CONMAROPS; that is, "seizing the initiative, containment, defence in depth and presence."[14] The alliance cannot fully discount the possibility that it may in the future face a challenge for command of the sea. However, consistent with trends in American maritime strategy, the thrust of the allied military posture has been to support the overall objective of enhancing stability and expanding cooperation in and to Eastern Europe in addition to peace support operations out of area. It is not so much that allied naval forces need to project power ashore as to project political and military stability. And to the extent that this projection has been mainly undertaken under the leadership of and in concert with the USN, interoperability has become an important element in contemporary maritime strategies and postures, including for Canada.

Interoperability: Maritime Concerts in a Unipolar World

The interest of the United States Navy in interoperability with the Canadian Navy must be seen in the context of broad trends in American national security policy. As Michael Mastunduno has argued, since the end of the Cold War, "US officials have in fact followed a consistent strategy in pursuit of a clear objective — the preservation of the United States' preeminent global position."[15] In the pursuit of this objective, the United States does seek out others who are willing to act with it, but as a means to preserve its unipolar status. The events of September 11 have only served to make America more determined to maintain this position, one that is now clearly linked to the physical security and safety of the American people.

Reflecting on the events of the 1990s, Coral Bell noted the importance of a "pretense of concert" in American national security policy in the post-Cold War era. In the current international environment, America needs only to conform to the "pretense of concert." The Clinton administration adopted the view that "the unipolar world should be run as if it were a concert of powers."

> In a sense, the post-World War II "institutionalization" of diplomacy — through the UN, NATO, the G-7, the WTO, the World Bank, the IMF, the OSCE and so on — has more or less imposed that strategy on policy-

> makers. Resolutions must get through the Security Council and consensus must be sought in other organizations to "legitimate" the policies that are deemed to be in the US national interest. Of course, the policies could be followed without seeking their legitimation by the "international community," but the advantage of securing it are worth the diplomatic labor it takes. A resolution or consensus eases consciences both in America and abroad, and helps protect US allies from their respective critics at home (though not in Washington, of course).[16]

This is not to argue that coalitions, and the interoperability that may be necessary for them to act together, are not important for the United States. They are. It is only to emphasize that American interest is based upon perceived national interests. That being said, the USN does have an interest in fostering interoperability. It allows it to "operate with foreign navies during a crisis or conflict." During the Persian Gulf War and the Kosovo campaign, the USN found that military and political effectiveness was enhanced by the ability of certain "high-end allies" to contribute to operations. Interoperability promotes a "harmonization" of foreign naval programs with those of the USN, providing "the political benefit of shaping foreign navies." This extends the influence of the United States over military developments in old and newer allies and, in the event of joint action, may afford even further scope for effective contributions from allies. Above all, the USN supports interoperability because it is "required by US policy."[17]

As part of its fundamental efforts to preserve America's dominant position, the Clinton administration placed particular emphasis upon the continued engagement of allies and the enlargement of the number of democratic regimes with market economies. This encompassed sustaining and adapting security relationships that would be of value to the United States in the event of future crises and coalition efforts, and spreading the burden of "responsibility for international peace and stability across nations." In particular, the regional commanders-in-chief were tasked with shaping the security environment in their regions, in part by developing close relationships and working with foreign militaries. The goal was not only to help these nations improve their own security capabilities but to enhance the ability of these countries to participate "in international coalitions of varying degrees of complexity."[18]

While the George W. Bush administration came into office sharply critical of Clinton's emphasis upon multilateralism, it too sees the cultivation of security ties as important. This is particularly so with the Department of Defense's (DoD) "new planning construct." As explained, in the September 30, 2001 *Quadrennial Defense Review Report*, "the United States will maintain *regionally tailored forces* forward stationed in Europe, Northeast Asia, the East Asian littoral, and the Middle East/Southwest Asia to assure allies and friends, counter coercion, and deter aggression against the United States its forces and its allies." The objective is to allow the United States to maintain regional balances "in concert with US allies and friends with the aim of swiftly deterring attacks with only modest reinforcements and, where necessary, assuring access for follow-on forces."

> Security cooperation will serve as an important means for linking DoD's strategic direction with those of its allies and friends. DoD will focus its peacetime overseas activities on securing cooperation to help create favorable balances of military power in critical areas of the world and to deter aggression and coercion. A particular aim of DoD's security cooperation efforts will be to ensure access, interoperability, and intelligence co-operation, while expanding the range of pre-conflict options available to counter coercive threat, deter aggression, or favorably prosecute war on US terms.[19]

Drawing upon the NATO definition, the US Department of Defense defines interoperability as, "the conditions achieved among communications-electronics systems or items of equipment when information or services can be exchanged directly and satisfactorily between them and/or their users." Gause broadens this definition to include three types:

- *Technical interoperability*, like the NATO and DoD definitions, concentrates on an exchange of services
- *Operational interoperability* considers whether units from different countries operating together can complete a mission
- *Political/cultural interoperability* examines why and how each country conducts military operations the way it does

Gaps in one type of interoperability may originate in more than one type. He notes in particular, that "interoperability gaps that at first may appear to be technical in nature, upon close examination turn to have their origins in political/cultural factors."[20]

At the heart of technological interoperability is the ability of weapons platforms, ships, planes and submarines to exchange information. This allows interoperability in command, control communications, computers and intelligence (C4I) in a manner that would allow a multinational force to carry out naval missions as if it were a single national force. This would include the ability for digital exchange and satellite connectivity. Ideally, this would afford the individual contributing national units to share a "common operational picture." So linked together, the combined forces would have the capacity for surface warfare, air and anti-air warfare, undersea warfare including antisubmarine warfare and, potentially, theatre ballistic missile defence.

The technological gaps between the USN and allied navies are the result of a number of factors. The USN, as with the entire American armed forces, has put a higher priority on C4I than platforms. In the development of its C4I, the USN does not take into consideration allied use of these systems. Allies complain that they cannot keep up with American development in these systems. Adjustments can be made, but altering them too much in order to accommodate allies may undermine their effectiveness. There is the concern that attempts to integrate technologically inferior allied forces into USN units would expose the Americans to greater risks "and thus unacceptable political costs in case of casualties because weapons and systems were not used to their full capacity."

The technological gaps can, in part, be accommodated in multilateral maritime operations if allied units are assigned less demanding roles. Another alternative is to accept the idea that there are levels of interoperability. The highest is "a seamless fusion of military forces," placing great demands "across all dimensions of interoperability: technical, operational and political/cultural." Such a level would be most important in high-intensity combat operations, especially where the issue at stake is one considered vital by the United States. Therefore only navies with the requisite capabilities and whose governments are in full agreement with the objectives and the methods would be allowed to participate. In operations other than war (OOTW) such as various types of peacekeeping, peace enforcement, noncombatant evacuations and humanitarian/disaster relief operations, a lower level of interoperability would be necessary. This could involve close coordination with a combined task force with basic communication among contributing forces.[21]

Interoperability in OOTW is more feasible, not simply because of the less demanding technological requirements, but because in the final analysis, interoperability is a manifestation of political consensus. And it is easier to obtain a consensus in OOTW than at the higher end of the conflict spectrum. Thus more allies would be willing to participate in combined maritime operations, and the United States would be more willing to accommodate the technological gaps. The political dimension is crucial because it influences the technological and doctrinal approach which the USN takes to interoperability, which in turn determines not only how, but with whom and when the USN will operate.

It was the political imperatives of alliance cohesion which fostered and allowed the elements of interoperability and standardization within NATO and the development of a broad set of publications on procedures and tactics. There has evolved a generally accepted NATO maritime doctrine providing guidance on how navies "intend to operate in a collaborative environment." The USN was a major factor in the evolution of allied interoperability. However, for the USN, being able to operate in coalition with NATO and other friendly navies was in addition to its primary task of being able to operate unilaterally. In the present circumstances, the USN would like to see allies move toward its standards. "This would allow the US Navy to develop a seamless global fleet rather than an Atlantic fleet more connected to NATO allies and a Pacific fleet more unilateral or connected to Asian allies...United States is creating a multifaceted doctrine to deal with the myriad threats that a global superpower might need to confront, while its allies are creating doctrines to deal with specific practical cases in circumscribed areas."

> NATO maritime doctrine is concerned with three dimensions: (1) small-scale local conflicts (Kosovo et al.), (2) rogue states with some missile and WMD (Weapons of mass destruction) capabilities on the periphery of Europe, and (3) major regional powers with power projection capacity. Only the United States has developed doctrine to deal with all three of these threats…and in preparing for the higher end of the spectrum, the US has opened a doctrinal gap between itself and the allies. No allied navy has followed this path for reasons of domestic politics, history, perceptions of threat, national interests, and budgetary constraints.
>
> The doctrine gap has led to technology gaps. To face the spectrum of threats, the global navy of the United States has developed specific technologies for force protection and power projection intended to minimize casualties and maximize punishment. In doing this, it has adopted deterrence and denial strategies that can be used on a global scale: from a small-scale contingency in Kosovo to a major regional conflict in Northeast Asia. To remain interoperable with the US on all levels, allied navies would require a large increase in their budgets. More importantly, they would need to rethink naval doctrine. The very roles and missions of the navy (which drive the force structure and systems that are purchased and maintained) would have to be changed. This would mean changes in the domestic politics and national interest calculations that end up in the elaboration of doctrine.
>
> Technology developments in the United States have been sparked by what is "possible" for a global navy with a large budget for research, development, and acquisition. This has been nurtured in an environment of competitive internal institutions seeking to exert influence on the future direction of the navy. This is in marked contrast to allied navies. To the

extent that they desire to cooperate with the US Navy, their doctrine will be hostage to the direction of US Navy doctrine. They can attempt to follow and play niche roles within larger allied concerns by specializing in a particular type of naval warfare, or they can create an extremely small number of ships capable of interoperability with US Navy battle groups. Either way, such national choices limit the roles and missions of their navies and force them to make important decisions about doctrine.[22]

For the USN then, as for the rest of the American military, interoperability is a means to an end, the enhancement of American national security through the continued global dominance of the US military. This not a secret agenda, it is made quite clear in US statements and in the relationship of the USN with foreign navies. The USN is prepared to work with other navies and encourages them to upgrade their capabilities in order to enhance the ability to collaborate. It is recognized that the American technological lead has and will create gaps, especially at the higher end of the conflict spectrum. The American position seems to be that it will be up to allies to bridge these gaps, and if they cannot (which will be the situation in many cases), this will restrict their ability to operate with the USN. Fundamentally, the USN is postured to be able to support American policy unilaterally. The record of the post-Cold War era suggests that many nations, including Canada, are anxious to achieve some level, however modest, of interoperability with the USN and make whatever adjustments are necessary to achieve this goal.

Between Strategy and Sovereignty: The Interoperability Dilemma

It is doubtful that the Canadian Navy would describe its planning and posture as being "hostage to the direction of US Navy doctrine." Yet, there is little doubt that the navy, as with the rest of the Canadian military, has been looking into the strategic future through the prism of American unipolarism. Given this, it would be easy for critics to charge that once again the Canadian military is placing allied ties — in the navy case, the old American-organized international "fraternity of the blue uniform" — ahead of the national interest. But in this, the navy (and the rest of the military) has not parted company with the country's political leaders. The foundation of the Canadian Navy's interest in interoperability with the USN is the continuation of a broad consensus within Canada, especially among the foreign policy elite, that the country must continue to remain globally engaged. To the extent that it has involved the use of the navy as a policy instrument, Canadian overseas engagement has meant adjusting to trends in Western sea power, namely the shift in focus from "blue" to "brown" littoral waters. In its operations in those waters, the USN has been open to and has sought out ways in which its forces can operate with those allies located in the littoral states and traditional Western allies, like Canada, who can bring their forces to bear in those waters. But this, in turn, has fostered some unease about the consequences for Canadian sovereignty and autonomy of these maritime trends on the part of political leaders, whose very policies encourage the strategy of interoperability.

For students of Canadian history, the current concerns will have a familiar ring. After all, the Canadian Navy was born in 1910 amid a great national debate on how best to assist the Royal Navy (RN). There were some in Canada who feared that a national navy, closely tied to the RN, would involve the country in imperial entanglements. Whatever the original debates, the history of the Canadian Navy in the twentieth century was one of close and continual cooperation with the dominant naval powers of the era, the RN and the USN. This was so in two world wars and during the Cold War when the CN was deployed primarily in the Atlantic and postured to support the American-dominated NATO maritime forces. Historically, Ottawa has wanted a distinct national navy, but part of that naval tradition has been that the Canadian Navy has always sailed and fought alongside its powerful allies even while political leaders sometimes fretted about autonomy and sovereignty. For example, in the late 1960s, Ottawa agreed to supply ships to NATO's Standing Naval Force Atlantic (STANAFORLANT). Yet at the time it also sought assurances that it would be able to recall the ships during a time of tension if the government had misgivings about a particular operation or the ships were needed closer to home.[23]

This approach continued in the post-Cold War era. The 1994 White Paper on defence declared that the CF must be prepared to "fight with the best against the best." After half a decade of intensive operations in Eastern Europe, it may not be clear who the opposing best is, but it is clear with whom the CF wishes to fight. In *Shaping the Future of Canadian Defence: A Strategy for 2020*, this is made explicit. The CF must strengthen its "military to military relationships with our principal allies ensuring interoperable forces, doctrine and C4I." In particular, it calls for expansion of "the joint and combined exercise program to include all environments and exchanges with US."[24] Given the history of the post-Cold War decade which saw the CF deploy abroad along with the US and its principal allies, in a host of UN and especially NATO operations, this approach is the only one that makes sense for the CF.

Interoperability is the direct military consequence of accepting unipolarity, or at least American dominance. Interoperability with the US is the logical defence posture for a Canadian national security policy based upon continued global engagement. For Canada, the present international security environment, especially at sea, is favourable. To be sure, the navy has a role in protecting our natural resources at sea and backing up the civil authority against those who undertake criminal acts and environmental damage. As noted below, the navy may well have additional responsibilities in terms of the "homeland" defence of the United States and Canada. But no hostile maritime power or group of powers today seriously and immediately threaten the seas upon which Canada's security and prosperity depend. We have a stake in the safety of trade routes at sea, but over 80 percent of our trade is with the United States and the routes that link us to Japan and Europe are in no danger of being cut by a rival naval power. What this means is that a strong case cannot be made for the maintenance of a strong navy, especially a blue-water one, on the grounds that major and direct maritime threats exist.

While not escaping the budget cuts of the recent years, the Canadian Navy has emerged as the most unscathed of the three services and is relatively well-prepared to fulfill national and international roles. It has 16 major surface warships, including 12 state-of-the-art *Halifax*-class frigates. It also has 12 Maritime Coastal Defence Vessels (MCDV). One of the three support ships, which had been scheduled to be paid off in 1996, will be retained. Also retained will be the existing force of 18 maritime long-range patrol aircraft. Shortly after its election in 1993, the Liberal government fulfilled a campaign promise to cancel the program to acquire 50 new helicopters to replace the existing fleet, including those carried on the destroyers and frigates. The government is still looking for new helicopters. The navy is also going to take delivery of the *Victoria*-class submarines from the Royal Navy. It has been looking at acquiring a multirole support vessel, one capable of transporting troops and equipment, making Canada less reliant on allied sealift in peace support operations.

While the navy faces major challenges even in maintaining its present capabilities, let alone enhancing them, much has been done to achieve interoperability with the USN. This is because the navy's modernization has produced ships which, because they have independent capabilities, are of use to American forces. Indeed, among America's naval allies, the Canadian Navy is considered "high-end." With the emphasis upon command, control, communications, computers, intelligence, surveillance and reconnaissance (C4ISR), the Canadian Navy has achieved a level of sophistication that allows it to operate with the USN. As *Leadmark* notes, during the Gulf War the Canadian Task Group commander was the only non-American warfare commander because of the compatibility and interoperability of Canadian ships with those of the US and other allies. The upgrading of the command-and-control and area defence standards of the four *Iroquois* class destroyers (making them guided missile destroyers [DDGs]) allowed them to play a major role in NATO's Balkan operations and "in support of US Drug Enforcement Agency counternarcotics operations in the Gulf of Mexico." The integrated combat system of the *Halifax*-class patrol frigates has allowed these ships to integrate "seamlessly into USN carrier battle groups deployed to the Persian (Arabian) Gulf in continued enforcement of United Nations resolutions against Iraq."[25]

At its most exacting, multilateral interoperability should allow forces from different countries to engage in integrated high-intensity combat against like forces on, over and under the sea and to project power ashore against an equally sophisticated adversary. In reality, there are almost no maritime or other forces not allied with the United States which can present a peer challenge. Most of the operations in which the Canadian Forces have been involved alongside the USN have been OOTW, with the opposing naval forces either nonexistent or very limited and unsophisticated. As noted above, interoperability is easier to achieve at this level of conflict. While the Canadian Navy has achieved a significant level of interoperability with the USN and some of its assets are leading-edge, it is unclear how it would perform at the extreme level of maritime combat. At the same time, the likelihood of a major war at sea (or elsewhere) between the US and equally sophisticated forces is remote. Jeffrey Record has argued that mastery of the revolution in military affairs (RMA) "is mastery of a war that will likely never be fought." He adds, though, that "to the extent that it facilitates situational awareness it will be a plus at any level of combat."[26] Thus the sophisticated C4ISR capabilities which the Canadian Navy can bring to OOTWs and limited combat are important. This was evident in the Kosovo campaign.

The interoperability achieved by the Canadian Navy in the 1990s is consistent with the long traditions of that service. The navy has long had a truly global deployment capability. It is a navy which traditionally has been geared toward NATO maritime multilateralism, both as a national policy and in terms of anticipated operations. In the post-Cold War era, it has continued to sail on wider seas because it remained the foreign policy of the country to contribute to international stability. The Canadian Navy, which played only a minor role in the era of classical peacekeeping, has over the first decade of the post-Cold War era been called upon to support the country's contributions to peace enforcement in all its varied dimensions.

What that has meant, though, is that in the last 10 years, the focus of Canadian Navy has, in a sense, moved even further from the national waters than was the case during the Cold War. Then, in concert with the USN, the navy was vitally concerned with the maritime defence of North America, but it was also postured to secure the transatlantic bridge against a Soviet navy dedicated to cutting the SLOCs. Today, the navy deploys to the coastal waters of lands throughout the world where regional or ethnic conflict is seen as requiring combined Western intervention. For example, Canadian naval forces have been more active in the Mediterranean on behalf of NATO in the post-Cold War era than before. They deployed to Southeast Asia to support operations in East Timor. They participated in the embargo against Haiti. And now a substantial force has been sent to the Indian Ocean as part of the campaign against Afghanistan.

To this extent, the objective contained in *Leadmark: The Navy's Strategy for 2020*, to become a "medium global force projection navy that will serve Canada as a multipurpose, interoperable force capable of joint and combined operations worldwide," is consistent with international security trends and overall Canadian policy in the post-Cold War era. To achieve this,

> the Canadian Navy will continue its development as a highly adaptable and flexible force, ready to provide the government with a wide range of relevant policy options across a continuum of domestic and international contingencies. The Navy will generate combat capable forces that are responsive, rapidly deployable, sustainable, versatile, lethal and survivable. Canada's naval forces, from individual units to complete Task Groups, will be tactically self-sufficient and be able to join or integrate into a joint US or multinational force, anywhere in the world.[27]

Whether the navy will be able to achieve all its goals given the continued pressure on the defence budget remains to be seen. More will need to be done, especially in the area of C4ISR, in order to allow the Canadian Navy to expand the scope of interoperability, allowing Canadian Forces to contribute to and benefit from a common operational picture generated by advanced sensors and communications. In addition, if Canada wants its forces to be interoperable, then they must be able to sustain and supply themselves for extended operations. But there is little doubt that is the "leadmark" toward which naval planning is heading. As the document notes

> Accepting...that Canada is more likely to conduct overseas military operations in concert with other like-minded nations...it is also advantageous to both the US and its allies that the US not act unilaterally as the "world's policeman." If recent experience is an indication, there will continue to arise any number of situations in which naval forces of medium powers such as Canada can make a difference by working in combination with the USN...Opponents of closer allied cooperation will argue that integration into US and NATO naval formations undermines Canadian sovereignty, but that is not necessarily so. Rather, because each mission is a function

of choice, it tends to strengthen Canadian sovereignty. This issue, of the potential effect of closer allied cooperation upon independence, is a matter of national strategy. For the navy, it must be noted and considered while implementing the requirement, identified in Strategy 2020, for enhanced interoperability.[28]

As the statement notes, interoperability raises sovereignty concerns. It is correct in noting that participation in US-led maritime operations and the level of commitment remain discretionary. The fact that the Canadian Navy plans and postures itself to be able to achieve interoperability with the USN does not necessarily bind the government to dispatch forces when Washington decides to deploy its fleet. Even when it does decide to send forces, Ottawa can still specify what activities it will or will not undertake as part of a combined fleet. To this extent, an interoperable naval capability can serve as a useful tool. It allows an independent policy decision to join with USN forces where the government believes Canada ought to be present and to select the appropriate niche role. Moreover, the access to the common operational picture which can be afforded Canada through its interoperable C4ISR contributions can also afford Canadian policy decision-makers greater knowledge of the activities of an American-led multinational naval operation.

However, the fact that Ottawa retains a measure of discretion does not override the fundamental consideration that convincing allies to dispatch ships to join the USN, and encouraging them to be as interoperable as their budgets allow, is yet another manifestation of the American pursuit of global dominance. For the United States, unilateralism and multilateralism are simply two means to this end. It is Washington's choice as to which one will best serve overall American interests in particular circumstances. In part, this decision will be based upon the demands of a specific operation and its level of importance. In instances where vital US interests are considered to be at stake, the United States will be less inclined to undertake the political and military adjustments that coalitions demand. For example, if Canada or other countries place too many restrictions regarding roles and rule of engagement, the USN may simply not accept a contribution. Most importantly, even when allies contribute forces and there is agreement on specific tasks, there is no concomitant expectation that smaller contributors, such as Canada, will therefore share in the higher strategic and political decisions associated with the operation. Interoperability may well allow the Canadian Navy to make a useful contribution at sea, but it is not likely to permit Ottawa a greater voice or leverage in Washington.

This, and not fears about foreclosing an independent Canadian decision to participate, is the true sovereignty concern about interoperability. The choice to sail or not to sail in concert with the USN will remain a Canadian one, as will any choices regarding particular roles and rules of engagement consistent with national policy. But once those decisions have been made and Canadian naval forces are at sea, Ottawa does yield a measure of independence and its policy options are narrowed. This is as it must be, otherwise interoperability with the USN would not be to America's advantage. From the American perspective, interoperability is attractive because it allows the USN to collaborate with allies toward common objectives, not because it secures strategic and political influence on the part of allies over American decisions and direction as to how those objectives are to be achieved.

Here is the political dilemma posed by interoperability to Canada's political leaders. The country still wishes to remain globally engaged, and the navy is an extremely useful tool in support of this wish. It is also recognized that a "modest" global navy does not operate alone, and that in the unipolar post-Cold War world of the twenty-first century this means being able to operate with the USN. Having the ability to choose to operate with the American navy is, therefore, an important part of Canadian security policy and military strategy. But in both policy and practice, interoperability with the American navy, from Washington's perspective, affords Ottawa no special status or influence.

As troubling as this dilemma may be, it can be said that for Canada this is by no means a new situation. Much of Canadian foreign and defence policy since the founding of the country has involved navigating between the commitments and constraints that go along with being a global actor but not a global power. Thus it will be important for political leaders to acknowledge the American perspective on interoperability and what it means before making the decision to commit elements of the Canadian Navy to a coalition operation overseas. In theory, Ottawa can always call the ships home if it does not agree with how the operation is being conducted; such is the flexibility traditionally attributed to sea power in contrast to land forces. In reality, once Canadian ships join up with the USN, the political cost of withdrawal or of placing very restrictive conditions on their use is no different than for other elements of the CF.

But the discretion which Ottawa retains, despite adopting a naval strategy in which interoperability figures so highly, should not be discounted. This is particularly so since overseas coalition operations with the USN are themselves likely to be discretionary. As important as these operations may be in the context of regional stability, and as much as Ottawa will maintain an internationalist perspective on its security, the record of the post-Cold War era suggests that Canadian vital interests will not likely be at stake on the high seas or in the "brown" littoral waters across the oceans.

The same, though, may no longer be the case for the littoral waters of North America. The USN will remain a force whose primary focus will be the projection of force overseas. But long considered by the USN as strategic backwaters, the ocean approaches to the continent are likely to become important in the global calculus of American national security policy. For Canada, this could mean that interoperability, heretofore mainly applied to overseas operations, could take on new meaning in the American response to September 11, 2001.

Interoperability Begins at Home: The Return of Continental Defence

In one very important respect, Canada is unlike other allies and friends, new and old, who are enlisting under the war on terrorism. It is directly involved in the defence of the American homeland. As former US Under Secretary of Defense John Hamre recently told a Canadian audience, "There is no longer a way to secure the United States without securing the United States and Canada simultaneously." This is nothing new; strategic defence has long been the essence of the bilateral Canada-US defence relationship.

In 1938 in Kingston, Ontario, President Franklin Delano Roosevelt declared, "The Dominion of Canada is a part of the sisterhood of the British Empire: I give you an assurance that the people of the United States will not stand idly by if domination of Canadian soil is threatened by any other Empire." A few days later in response, Prime Minister William Lyon Mackenzie King stated that Canada too had its obligations as a "good and friendly neighbour, and one of these is to make sure that our country is made as immune from attack or possible invasion as we can reasonably be expected to make it, and that, should the occasion ever arrive, enemy forces should not be able to pursue their way either by land, sea or air to the United States across Canada."[29] The two declarations reflected the growing apprehension of Ottawa and Washington about the deteriorating international situation and the potential threat to both countries. They also bespoke the friendly feeling between the two countries and indeed between the two leaders. But they also reflected differing strategic perspectives based upon complementary, but not identical, national interests. For the United States, the problem was that Canadian weakness might endanger American security. Canada could not become a strategic liability in the defence of the US homeland. For Canada, the problem was satisfying this legitimate concern without compromising, however benignly, its own national sovereignty. This could be accomplished by taking measures on its own, to secure its territory, airspace and maritime approaches.

Although the focus was overseas, North America itself had lost the protection afforded by its ocean boundaries with the advent of nuclear weapons and long-range bombers. America could not hope to credibly extend its deterrent if the US itself were vulnerable at home. In essence, the Cold War brought about a situation wherein the North American continent was to be regarded as a "strategic unity" for purposes of defence. This made Canada, as John Foster Dulles once put it, "a very important piece of real estate."[30] Thus strategic air defence, between the Royal Canadian Air Force (RCAF) and the United States Air Force (USAF), came to be the most important dimension of the bilateral defence relationship. By the early 1960s, the focus shifted to the threat posed by land-based Intercontinental Ballistic Missiles (ICBMs) and submarine-launched ballistic missiles (SLBMs) fired from nuclear-powered ballistic missile submarines (SSBNs). The United States decided not to pursue ballistic missile defence (BMD) in the late 1960s, and this was reinforced by the 1972 US-Soviet Anti-Ballistic Missile Defense (ABM) Treaty. Since there was no protection against missiles once fired, NORAD's prime mission became surveillance, warning and attack assessment in order that the American deterrent would not be caught off-guard. As the missile threat grew, and with no BMD systems, Canadian airspace and territory became less important, as there were no missile-warning facilities on its soil.

In part because no formal overall bilateral combined command was established for North American defence, and because the US military command structure also lacked a single continental command, there were only informal linkages between the aerospace and maritime defence structures of the continent. Responsibility for the defence of the American West Coast fell to the USN-dominated Commander-in-Chief, United States Pacific Command (CINCPAC), headquartered in Hawaii. An additional factor was that for the USN, the focus was always overseas power projection. To be sure, the United States was concerned with monitoring Soviet submarines, both nuclear-powered attack submarines (SSNs) and SSBMs. Thus at sea, especially in the Atlantic, the Royal Canadian Navy and the United States Navy maintained and developed close collaboration in monitoring the maritime approaches to the continent. Much of this activity took place in the context of NATO's Allied Command Atlantic (ACLANT) which included a western (North American) region and, within it, the Canadian Atlantic Sub-Area (CANLANT). During the Cold War, the bulk of the Canadian Navy was dedicated to NATO antisubmarine warfare (ASW) roles primarily directed against SSNs. Thus continental defence missions meshed well with the NATO focus. To this extent, bilateral maritime collaboration, in contrast to aerospace cooperation under NORAD, was politically invisible. It was also politically comfortable because it was mostly part of the multilateral NATO arrangements with which Ottawa was more at ease. This collaboration, but for the level of integration, was as extensive in aerospace matters. Even in this case, however, as the Soviets developed longer-range SLBMs with the capability to hit North America from waters closer to Eurasia, the relative importance of continental maritime defence progressively diminished.

There was concern in the mid-1980s about new threats from Soviet sea-launched cruise missiles fired from SSNs, including possible use of the Arctic waters as a transit route. This potential threat prompted increased American interest in maritime security and the Arctic, which in turned raised fears about Canadian sovereignty. The proposal in the 1987 White Paper for a fleet of Canadian SSNs was meant to deal with both the security and sovereignty threats. This was a classic response to the so-called defence against help dilemma. If Canada could not secure its waters, the USN would, and it would not necessarily have to tell Ottawa how it was doing it.

The end of the Cold War brought about a decline in the importance of traditional continental defence and helped sink the Canadian SSN program. The early 1990s saw a marked scaling back of NORAD activities. To be sure, neither Washington nor Ottawa was prepared to dismantle the radar lines and disband the interceptor squadrons. Moreover, to the extent that NORAD's prime missions had become warning and assessment of missile attack and space surveillance, there was a continued role for the combined command. Thus the agreement was renewed in both 1991 and 1996. But the strategic value to the US of Canadian airspace, which had steadily declined throughout the Cold War in any case, was greatly diminished. Indeed, by the early 1990s, despite the NORAD renewals, there were suggestions in the United States that NORAD be dismantled, its missile warning and attack assessment missions handed over to United States Space Command or United States Strategic Command, and its residual air defence role given to the USAF Air Combat Command (ACC), a component command of the newly created United States Atlantic Command (USACOM), which has responsibility for defending most of the continental United States.

As to the maritime defence of North America, while the submarine threat had diminished, Russian *Akula*-class submarines were still detected patrolling in North American waters. In May and June of 1995, they conducted exercises near Kings Bay, Georgia, the east coast SSBN base for the USN.[31] According to the USN Office of Naval Intelligence, this "was the first deployment by a Russian submarine near the US East Coast since 1987." Later in 1995, an *Akula* was operating near the Bangor, Washington, SSBN base where "Russian SSNs have not been seen in recent years."[32] For over 40 years, Canada has been a participant with the United States in the Integrated Undersea Surveillance System (IUSS), which had several facilities, including the USN base in Argentia, Newfoundland. When the US closed the Argentia base because of budget cuts, Canada decided it would build a new facility to process and display information from the IUSS. In May 1995, Trinity, the Canadian Forces Integrated Undersea Surveillance System Centre, was opened in Halifax. Commanded by a Canadian officer with a staff of about 140, 30 of whom are members of the USN, Trinity is "a unit of

Canadian Maritime Forces, Atlantic, but falls under the control of the US Navy's commander, undersea surveillance, at Norfolk."[33] Nevertheless, the continuing decline of the Russian navy, whose motto would now be "backward from the sea," appeared to further reduce the strategic importance of the maritime approaches to the continent.

Although the defence of North America was on the decline in the early 1990s, by the middle of the decade American overseas operations contributed to a revival of interest in the defence of the American homeland. This was an interest that had not been seen since the early days of the Cold War, with the exception of the Reagan administration's short-lived Strategic Defense Initiative (SDI) with its proposal for an elaborate BMD system. The renewed interest in BMD was initially the result of the Gulf War, when attention was focused on Theatre Missile Defence (TMD), and the need to provide protection for deployed forces and regional allies. In its 1994 White Paper on defence, the Canadian government seemed to alter its position to a possible BMD role for NORAD:

> The Government will examine closely those areas which may require updating in accordance with evolving challenges to continental security. Canada will work towards an agreement that furthers our national interests and meets our defence needs, now and into the 21st century.
>
> Canada's potential role in ballistic defence will not be determined in isolation, but in conjunction with the evolution of North American and possibly NATO-wide aerospace defence arrangements.[34]

As noted, the impetus for the new concern with BMD came from continued American activities abroad. Richard Betts observed that US policies abroad may actually increase the danger to the American homeland: "Today, as the only nation acting to police areas outside its own region, the United States makes itself a target for states and groups whose aspirations are frustrated by US power." It is "US military and cultural hegemony — the basic threats to radicals seeking to challenge the status quo — that are directly linked to the imputation of American responsibility for maintaining world order. Playing Globocop feeds the urge to strike back."[35] This especially includes ballistic missile threats from so-called rogue states. These states were developing, or already had, chemical, biological, radiological, nuclear and enhanced high-explosive (CBRNE) weapons.

A new concern with "homeland" defence was emerging in the United States, one not seen since the 1950s. Paradoxically, these fears arose at a time when America's relative military power had never been greater. While fears about homeland defence were being pushed most forcefully by Republicans in the Congress, polls in the late 1990s conducted by the Chicago Council of Foreign Relations found that while "Americans feel secure, prosperous and confident," with "fear of armed threats from a rival superpower diminished, they are, nevertheless...alarmed by violence at home and abroad" and "support measures to thwart terrorists, prevent the spread of weapons of mass destruction, and keep defence strong." Moreover, while the vast majority of Americans did not see vital threats to US interests abroad, 84 percent regard "international terrorism" as the number one "critical threat" to American interests.[36] Here at least, the supposedly "uninformed" and "disinterested" American public "know-nothings" knew something.

In what is now an eerily prophetic comment, the United States Commission on National Security in the 21st Century, the Hart-Rudman Commission, in a report subtitled *New World Coming*, predicted in 1999 that

> America is becoming increasingly vulnerable to hostile attack on our homeland — and our military superiority will not protect us...In fact there is a school of thought that American military superiority on the conventional battlefield will push our adversaries towards unconventional alternatives. This school further postulates we are entering a period of "catastrophic terrorism" with terrorists gaining access to weapons of mass destruction including nuclear devices, germ dispensers, poison gas and computer viruses. States, terrorists, and other disaffected groups will acquire weapons of mass destruction, and some will use them. Americans will likely die on American soil, possibly in large numbers.[37]

Yet another important indication of growing American concern with homeland defence was the renaming in October 1999 of USACOM to United States Joint Forces Command (USJFCOM). In addition to its responsibility to prepare US forces for overseas deployment, USJFCOM is responsible for "homeland defence," including "providing military assistance to civil authorities for consequence management of weapons of mass destruction (WMD) incidents within the continental United States, its territories and possessions."[38] It will also "support the WMD consequence management efforts of the other combatant commands" throughout the world. In setting up USJFCOM, then Secretary of Defense William Cohen appointed an army national guard brigadier general as the first commanding general of Joint Task Force-Civil Support (JTF-CS). "The JTF-CS will ensure Department of Defense assets are prepared to respond to requests for support from a lead Federal Agency such as the Federal Emergency Management Agency (FEMA)."[39]

In the wake of the events of September 11, 2001, the US military has made homeland security and the defence of the United States "the highest priority." As the *Quadrennial Defense Review Report* notes

> The United States will maintain sufficient military forces to protect the US domestic population, its territory, and its critical defense-related infrastructure against attack emanating from outside US borders, as appropriate under US law. US forces will provide strategic deterrence and air and missile defense and uphold US commitments under NORAD. In addition, DoD components have the responsibility, as specified in US law, to support US civil authorities as directed in managing the consequences of natural and man-made disasters and CBRNE-related events on US territory. Finally, the US military will be prepared to respond in a decisive manner to acts of international terrorism committed on US territory or the territory of an ally.

As part of this new emphasis, the US intends to again review the organization of its forces within the continental United States. The *QDR* called for a continued examination of the "roles and responsibilities" of the active and reserve forces "to ensure they are properly organized, trained and equipped, and postured." It is clear, the report went on, "that US forces, including the United States Coast Guard, require more effective means, methods, and organizations to perform these missions. As part of this examination, DoD was to review the establishment of a new unified combatant commander to help address complex inter-agency issues and provide a single military commander to focus military support."[40] The United States Marines expanded the scope of their special units to deal with attacks that might take place in the United States. Senior administration officials were suggesting revision to the US *posse comitatus*, laws that restrict the use of the regular armed forces in civilian law enforcement.

This rising concern with homeland defence was already impacting upon the bilateral defence relationship and, indeed, the character of overall relations between the two countries before September 11. As noted above, the American strategic interest in Canada is that it not be a strategic liability for the United States. In the Cold War, with both countries accepting shared threat, the strategic unity of the continent ensured that this would not be the case. But the real defence of the continent lay in the deterrent capabilities of the US, not in joint measures for direct defence.

In the wake of the attack on the United States, Canada has again become a very important piece of real estate for Americans. This started minutes after the first attacks of September 11 as Canadian personnel at NORAD joined in the effort to prevent further attacks and secure North American airspace. The air defence role, which had been in decline since the late 1950s, has taken on a new focus in order to prevent a repeat of those attacks. The "undefended" border now needs to be secure if trade is to continue to flow in a timely manner. Canada has been told to reform its immigration policies and augment its internal counter-intelligence efforts so as not to make itself a security liability.

In the war on terrorism, as in the Cold War, forward-deployed offensive forces will accompany the new emphasis on homeland defence. Indeed, despite statements regarding the priority now to be attached to homeland defence, the bulk of the mighty American military posture will still be focused on the projection of force overseas. However, the relative importance of homeland security will increase. This is already evident with regard to BMD, where the events of September 11 gave President Bush all the justification he needed to finally do what his administration had promised to do: withdraw from the ABM Treaty and go forth with a BMD system. Along with the new emphasis upon air defence, this will again bring Canada's role in NORAD to the top of Ottawa's bilateral security agenda.

In the fall of 2001, USJFCOM established a new Homeland Security Directorate which was tasked with developing an organization, mission, roles and doctrine for its new mission. The focus of the new directorate is given as land and maritime defence, while for aerospace defence, it is "partnered" with NORAD and US Space Command.[41] This, though, appears to be only a temporary measure.

In January 2002, it was announced that by October 1 the US would stand up a new unified command, Northern Command, whose commander-in-chief (CINC) will "have responsibility for homeland security for the United States." The specific tasks of the command and its relationship to other branches of the US government still need to be worked out. According to General Pace, vice-chairman of the Joint Chiefs of Staff,

> From our standpoint, this individual is going to have to take NORAD and ensure that the very, very long-standing close relationship with Canada is maintained and nurtured and taken properly into the future and to figure out, is there a way, then, to add to the air defence, the land and sea defences. He's going to have to figure out, for starters, where is the best place to be? We want him near the Capital region, probably in the Capital region — maybe not, if homeland security is something that we want to be concerned about with some kind of an attack in and around Washington — and then building the staff and what types of functions: do we want this CINC to be able to perform posse comitatus; how much do we want our military to actually do or not do inside the United States?
>
> Right now we have folks who are going to be detailed to the borders of the United States in support of other government agencies. And how do we work all that? So I can't give you a precise answer yet, because we are

just in the beginning of understanding the types of capabilities that we need this country to have. And then who best should perform those functions and provide those capabilities? Should it be the states? Should it be the federal government? And if it's the federal government, should it be FEMA? FBI? The military? We need to make all those determinations.

So whoever this new CINC is going to be, come 1 October, he is going to be very busy just figuring out what questions to answer and then determining how to go about answering them.[42]

As noted above, in the US responsibility for the maritime defence of the homeland is presently divided between USJFCOM in the Atlantic and PACOM in the Pacific. Traditionally, the USN has jealously guarded its dominance in the Pacific. Whatever the specific new arrangements, since terrorists are capable of making use of the seas to bring WMDs to America, it is likely that Northern Command (or whatever it will finally be called) will take renewed steps to secure its ocean approaches. This could include, as proposed, making greater use of the US Coast Guard, perhaps bringing it under the new command. For Canada, this will mean that its traditional sovereignty protection roles, ones that will be directed against non military threats, will now take on added meaning in terms of overall North American security. The navy will still be tasked with protecting Canadian economic resources and enforcing Canadian law. But the relative importance of its other long-held tradition, collaboration with the USN in the maritime defence of the continent, is likely to become more important than it has been since the end of the Cold War.

Yet if the necessity for interoperability in overseas littoral waters raises concerns about Canadian sovereignty and independence, how much sharper are these concerns when it comes to home waters? To some extent, the fact that the ocean approaches to North America tended to be strategic backwaters for the United States and, at least in the Atlantic, handled within a NATO framework, eased Ottawa's apprehensions and the "defence against help" dilemma. Now, worries about how much bilateral maritime cooperation will be required to assure Washington may yet emerge again, even more so given the scope and intensity of American concerns. And, here, there is little discretion available. Ottawa can choose not to deploy overseas with the USN; it cannot choose to ignore American efforts to secure the maritime approaches to the continent. In addition, overall sovereignty concerns will be exacerbated if NORAD is to be subsumed within the new Northern Command, thus depriving it of its distinctive bilateral character, so important to Canadian considerations of autonomy. Indeed, the prospect of a new overall American command, whether solely US or some continental or hemispheric "Americas" command, has already generated controversy in Canada. John Manley raised concerns while Foreign Minister and former foreign affairs minister Lloyd Axworthy and former defence minister Paul Hellyer recently sounded the alarm, claiming that Canada is in danger of surrendering sovereignty at home.[43]

Once again, Canada is in familiar waters, not just because they are its own. But with regard to North American maritime efforts, sovereignty concerns can be addressed without sacrificing autonomous interests. To be sure, Ottawa will have to take care to ensure that Canadian interests and laws are respected in any new North American command structures or arrangements that are put in place. This could be a difficult task if for no other reason than the sorting out of responsibilities within the US may be complex and confusing. (Indeed, it may be easier for Washington to reach an agreement with Canada than among the American services and other federal agencies.) While any new arrangements will tie Canada closer to the US in terms of continental and, especially, maritime security, it needs to be remembered that interoperability in North America will allow the Navy to make a contribution in the face of a direct threat to Canada. After all, in these waters, as opposed to foreign seas, Canada has a vital national interest. Increased collaboration with the USN and the US Coast Guard in the defence of North America should be a priority for the Canadian Navy, consistent with the other steps Ottawa is now taking. In other words, interoperability should begin at home if the navy is to fulfill its primary role, the defence of Canada's maritime interests.

Conclusion: Sailing in Concert Abroad and at Home

It has been said repeatedly that everything has changed since September 11. This is true in the sense that combating international terrorism has now become the central focus of American foreign policy after a post-Cold War decade in which no grand organizing principle similar to containment and deterrence emerged. During those years, many argued that the very definition of "security" had changed and expanded. No longer could it be viewed in strictly military or national terms. The economy, the environment, culture and especially "human" security were dominating international strategic relations. But combined with other trends in the 1990s, the impact of the "attack on America" has brought back aspects of the Cold War world and catapulted traditional concerns about national security to the top of the agenda in order to deal with a non-traditional challenge. Indeed, Secretary of Defense Rumsfeld has cautioned the American people that this is to be a protracted struggle comparable to the Cold War. Here again, we have Washington asserting indispensable leadership against a threat which, because it is directed primarily against America, endangers the entire Western world. Once more, the NATO alliance is called upon to lend its material support — and more importantly, legitimacy and unity — to American efforts. Once more, the United States is supplying military assistance and advisers, as in the Philippines, to help Third World governments deal with insurgencies. Indeed, Daniel Pipes has compared the campaign against Islamic terrorist extremists with the Cold War efforts to confront and contain communism wherever it existed.[44] We hear another presidential address to the American people asking like-minded nations in all parts of the world to join the United States in what will no doubt be yet another "long twilight struggle."

It was the global character of America's Cold War policies and its continued global activism in the post-Cold War era that made naval interoperability a priority for Canada. Sharing Washington's appreciation of the pervasiveness of the threat and sharing a continent, it was essential that Ottawa also become concerned with its own national security and how its policies would be coordinated and adjusted to accommodate and support its closest ally.

After September 11, Canada has once again been drawn into a global effort, one within even greater and more complex ramifications for bilateral security relations. Its response was to do what it has done in the past: join in the campaign alongside America and its Western allies. Along with NATO partners, it invoked article 5 of the North Atlantic Treaty, declaring the attack on America to be an attack on all alliance members. Ottawa dispatched forces overseas and turned anew to efforts to help secure the North American continent. Just as Canadians were glued to their television sets in early September 2001, so too have these events emphasized and reinforced the bonds of common strategic interests, concepts of world order and shared values that continue to underlay bilateral security relations. These will keep Canada firmly affixed to its seat at the table of the American-led and dominated Western alliance, including its maritime dimensions.

This means that interoperability with the USN, grounded in over half a century of multilateral and bilateral collaboration at sea, and given new emphasis in the post-Cold War era, will become even more important as the two countries confront the threat of terrorism in distant seas as well as their home waters. But as in music, where many instruments go into making a concert but all musicians are expected to play from the same sheet and follow one conductor, naval interoperability requires some sacrifice of autonomy. Thus sailing in concert with the USN will entail implicit recognition of the dominant position that the United States holds...and seeks to maintain in international relations.

But these are familiar waters for Canada. And while it may not mean smooth sailing, especially in these unusually troubled times, Ottawa has much experience in navigating them in a manner that protects and furthers Canadian interests. The world-class capabilities of the Canadian Navy, combined with careful and continual political oversight can make interoperability with the United States Navy a mainstay of Canada's national security policy.

Notes

1. Gause (2001, p. 1). The author would like to thank Mr. Gause for his permission to quote from this paper.

2. Department of National Defence (2001, pp. 111-2).

3. Brodie (1965, p. 226).

4. Till (1982, p. 56).

5. Huntington (1954, p. 490). For a discussion of Huntington's views in the light of the present situation, see Breemer (1994). See also Sokolsky (1998, ch. 2).

6. For example, see Maloney (1995) and Sokolsky (1991*a*).

7. Gray (1993, p. 2).

8. Elders (unpublished paper, 1997).

9. United States, USN (1992, p. 2).

10. United States, White House (1994).

11. United States, USN (1994*b*, p. 2).

12. Boorda (1994, pp. 1-2).

13. Breemer (1994, pp. 44-45).

14. United Kingdom (1995, p. 79).

15. Mastunduno (1997, p. 51).

16. Bell (1999, p. 60). Emphasis in original.

17. Gause (2001, p. 3).

18. Gause (2001, p. 4).

19. United States, Department of Defense (2001, p. 20). Emphasis in original.

20. Gause (2001, p. 3). Emphasis in original.

21. Gause (2001, p. 7).

22. Gause (2001, pp. 9-10).

23. Department of National Defence (2001, p. 54). Sokolsky (1991*b*, p. 38).

24. Department of National Defence (1999, p. 10).

25. Department of National Defence (2001, pp. 64-6, 128-9).

26. Record (1999-2000, p. 21).

27. Department of National Defence (2001, pp. 168-70).

28. Department of National Defence (2001, p. 112).

29. Eayrs (1965, p. 183).

30. As quoted in Jockel (1991, p. 1).

31. United States, Office of Naval Intelligence (1996, p. 24).

32. Kolisnek (1997, p. 9). The Virginian-Pilot (1996, p. A6)

33. McGrath (1996, p. 4); and Tummers (1995, pp. 37-44).

34. Department of National Defence (1994, pp. 23, 25).

35. Betts (1998, pp. 28, 41).

36. Rielly (1999, pp. 97, 99).

37. United States, Commission on National Security/21st Century (1999, p. 14).

38. *Marine Corps Gazette* (1999, p. 8).

39. National Guard (1999, p. 12).

40. United States, Department of Defense (2001, pp. 18-19).

41. Wimbish and Rodney (2001).

42. Rumsfeld (2002).

43. Axworthy (2002, p. A11); and Hellyer (2002, p. A11).

44. See Pipes (2002).

References

Axworthy, Lloyd. 2002. "Watch Your Step, Mr. Chrétien." *The Globe and Mail*, February 4, p. A11.

Bell, Coral. 1999. "American Ascendancy and the Pretense of Concert." *The National Interest* 57 (Fall): 55-63.

Betts, Richard. 1998. "The New Threat of Mass Destruction." *Foreign Affairs* 77 (January/February): 26-41.

Boorda, J.M. 1994 . "The Art of the Long View: Enduring Naval Principles." *Vital Speeches* 61, December 15 (Information Access R Reprint): 1-2.

Breemer, Jan S. 1994. "The End of Naval Strategy: Revolutionary Change and the Future of American Naval Power." *Strategic Review* 22 (spring).

Brodie, Bernard. 1965. *A Guide to Naval Strategy*. New York: Praeger.

Department of National Defence, Directorate of Maritime Strategy. 2001. *Leadmark: The Navy's Strategy for 2020*. Ottawa: Department of National Defence.

——. 1999. *Shaping the Future of Canadian Defence: A Strategy for 2020*. Ottawa: Department of National Defence, June.

——. 1994. *1994 Defence White Paper*. Ottawa: Minister of Supply and Services.

Eayrs, James. 1965. *In Defence of Canada*, vol. 2. Toronto: University of Toronto Press.

Elders, T. 1997. "The Navy's New Focus: From the 'Maritime Strategy to Forward from the Sea.'" Unpublished paper.

Gause, Kenneth. 2001. "US Interoperability with its High-End Allies." Center for Strategic Studies, Center for Naval Analyses, Arlington, VA.

Gray, Colin S. 1993. "Influence From the Sea: Naval Power and World Order." Address before the SACLANT Maritime Seminar, *The Role of NATO Maritime Forces in the 1990s*, June 17-18.

Hellyer, Paul. 2002. "The Americans Will Dictate Our Military Roles." *The Globe and Mail*, February 4, p. A11.

Huntington, Samuel P. 1954. "National Policy and the Transoceanic Navy." *United States Naval Institute Proceedings* 80.

Jockel, Joseph T. 1991. *Security to the North: Canada-US Defense Relations in the 1990s*. East Lansing, MI: Michigan State University Press.

Kolisnek, George. 1997. "The Russian Navy: Still an Enigma after All These Years." *Maritime Affairs* (April): 9.

Maloney, Sean M. 1995. *Securing Command of the Sea: NATO Naval Planning, 1948-1954*. Annapolis, MD: Naval Institute Press.

"USA COM Redesignated to U.S. Joint Forces Command." 1999. *Marine Corps Gazette* 83 (November).

Mastunduno, Michael. 1997. "Preserving the Unipolar Moment: Realist Theories and U.S. Grand Strategy after the Cold War." *International Security* 21, no. 4 (spring): 49-78.

McGrath, Dede. 1996. "Trinity's US Navy Staff Draws Curious Stares with Ceremony Abroad Sackville." *Trident* 30 (September).

"Guard General to Command Joint Civil Support Task Force." 1999. *National Guard* 53 (November).

Pipes, Daniel. 2002. "Who Is the Enemy?" January 31. Available at www.danielpipes.org/articles/article.php?id=103

Record, Jeffrey. 1999-2000. "Operation Allied Force: Yet Another Wake-Up Call for the Army?" *Parameters* 29 (winter).

Rielly, John E. 1999. "Americans and the World: A Survey at Century's End." *Foreign Policy* 114 (spring): 97-113.

Rumsfeld, Donald. 2002. "21st Century Transformation." Delivered January 31 at National Defence University, Washington, DC. Available at http://www.defenselink.mil/ speeches/2002/s20020131-secdef.html

Sokolsky, Joel J. 1998. *Projecting Stability: NATO and Multilateral Naval Cooperation in the Post Cold-War Era*. Halifax: Centre for Foreign Policy Studies, Dalhousie University.

——. 1991*a*. *Sea Power in the Nuclear Age: The United States Navy and NATO*. London: Routledge.

——. 1991*b*. *The Fraternity of the Blue Uniform: Admiral Richard G. Colbert, USN and Allied Naval Cooperation*. Annapolis: U.S. Naval Academy, War College Press.

Till, Geoffrey. 1982. *Maritime Strategy and the Nuclear Age*. New York: St. Martin's Press.

Tummers, Ed. 1995. "Undersea Surveillance Systems: Part of a Balanced Force." *Maritime Security Working Papers*, June 2: 37-44.

United Kingdom, Directorate of Naval Staff Duties. 1995. *The Fundamentals of British Maritime Doctrine*. London: Her Majesty's Stationery Office.

United States, Commission 21st Century. 1999. *New World Coming: American Security in the 21st Century*. Arlington, VA: Commission on National Security.

United States, Department of Defense. 2001. *Quadrennial Defense Review Report*. Washington, DC: Department of Defense, September 30.

United States, Office of Naval Intelligence. 1996. *World Submarine Challenges*. Washington, DC: USN.

United States, United States Navy. 1992. *From the Sea: Preparing the Naval Service for the 21st Century*. Washington, DC: United States Navy.

——. 1994*b*. *Forward... from the Sea: Preparing the Naval Service for the 21st Century*. Washington, DC: United States Navy.

United States, White House. 1994. *A National Security Strategy of Engagement and Enlargement*. Washington, DC: Government Printing Office.

"US Expects Russians to Resume its Bold Submarine Operations." 1996. *The Virginian-Pilot,* April 14, A6.

Wimbish, Michael, and Norinda Rodney. 2001. USJFCOM Public Affairs, "Homeland Security Mission Tops List of New Priorities for 2002." Available at http://137.246.33.101/NewsLink/StoryArchive/2001/pa122701.htm#

8

Multinational Land Force Interoperability: Meeting the Challenge of Different Cultural Backgrounds in Chapter 6 Peace Support Operations

Ann M. Fitz-Gerald

Introduction

Multinational military interventions that promote sustainable and enduring peace building measures have become increasingly challenged due to the complex environments and the many different players that are brought into these theatres. The reality of contemporary conflict environments is such that members of the local population, who themselves remain the only agent for sustainable and peaceful change, are now able to view the behavioural conduct and operational effectiveness of the peacekeeping forces due to the close proximity in which they operate. A common method used by warlords, nonstate actors and paramilitary regimes in garnering the support of local communities is to offer security guarantees in exchange for their support. As a result, the main task for the international community in responding to these conflicts involves determining the basis for local support and seeking to redirect the population's allegiance toward the interventionist forces by demonstrating the provision of credible security. Research has indicated that disparate national approaches observed in recent multinational peace support operations have had an adverse effect on the way in which the international military forces are perceived, due to a failure to build sufficient confidence measures at the grassroots level of society.

These situations test the professionalism of the military ground forces and their application of doctrine. However, troops serving in a multinational theatre of operations are often deployed under the strategic requirements and tactical procedures of regional organizations such as the North Atlantic Treaty Organization (NATO) and the Organization for Security and Co-operation in Europe (OSCE), as well as under the more global auspices of the United Nations (UN), the world's largest intergovernmental multilateral organization. The "organizational" approach to predeployment planning, therefore, does not always correspond directly with national doctrine. Additionally, where there is a direct correlation between the organizational and the national approach, the interpretation of the same set of rules and procedures can be completely different. This results in a disunity of effort on the ground, an increasingly blurred local perception toward the international effort, and a prolonged conflict.

Such factors are now considered of primary importance as calls from Canada and her allies increase for enhanced military interoperability. Much of the discussion on interoperability has focused on Canada-US defence relations. In the past decade, however, Canadian Forces (CF) have operated with a number of other countries in various peace support operations, especially in the former Yugoslavia. Canadian defence policy still maintains that the CF must be prepared for such operations in the future.

This paper will discuss the relationship between local populations and multinational military forces and examine the importance of this relationship in, what could be described as, "third generational conflict theatres."[1] It will then discuss some national disparities observed in Bosnia and Haiti and seek explanations behind these differences at the international, national and in-theatre levels. Lastly, the paper will discuss recent initiatives aimed at minimizing the differences and the impact this should have on national defence policy and military leadership at all levels.

The Importance of Local Dynamics

Much has been written on the "new" security environment and the growing complexity of conflicts and humanitarian emergencies. While these emergencies are not a new phenomenon, they have become an important subject area for academics and practitioners due to the increased tendency of the international community to respond. Military forces are no longer deployed along borders or demarcation zones that separate antagonistic states. Instead, they are expected to perform activities in the midst of these animosities, such as patrolling dangerous areas, observing human rights violations, protecting military compounds and international headquarters, rebuilding local infrastructures and assisting in the delivery and administration of humanitarian aid.

In the recent past, international military forces have been labelled as being "passive spectators" and accused of "turning their backs" on the atrocities and human rights violations committed by the warring factions in these environments. The charges have often evolved due not to the fault of the individual soldier but to the national and international political forces that control their behaviour in the field. In order to deploy troops to these regions, and search for a peaceful settlement to the conflict, the UN normally develops a mandate that falls within the UN's Chapter 6 operations. Chapter 6 refers to the section of the UN Charter that endorses military deployments that uphold the principles of consent, impartiality and the nonuse of force except in the case of self-defence.[2]

The issue of consent is fairly straightforward. In March 1995, the reinstalled Haitian government gave the UN permission to station international military forces in various areas on the Caribbean island to help maintain security and stability. The intervention force was called the UN Mission in Haiti (UNMIH) and was authorized by the UN Security Council under a Chapter 6 peacekeeping mandate. The troops would be impartial to all individuals and groups and pledged to maintain a stable and secure environment in which the democratic Haitian government could be reinstalled. Lastly, the troops were permitted to use life-threatening force only in self-defence, and only in a graduated and measured way that ensured minimum collateral damage.

These principles and procedures have applied to many other interventions, as far back as the 1956 UN Emergency Force (UNEF) deployed to the Sinai; the UN Force in Cyprus (UNFCYP), which is still stationed there today; the 1992 UN Protection Force (UNPROFOR) in Bosnia; the 1993 UN Preventative Deployment (UNPREDEP) in Macedonia; and the ongoing UN Mission in Sierra Leone (UNAMSIL). These types of operations are most appropriate for the issues discussed in this paper due to the restrictions imposed on the troops and the tasks they are expected to perform.

Chapter 6 mandates are most common during the earlier and latter stages of a conflict. If a conflict or humanitarian emergency deteriorates to the extent that more robust military action is required, a new mandate is often issued under Chapter 7 of the UN Charter, which authorizes the use of force.[3] The ratification process behind approving the more robust UN mandate has proven difficult in the past, particularly if it triggers sensitivities for certain UN Security Council members who have the ability to exercise a veto. Such a scenario prevailed during early talks on the deployment of military troops to Kosovo, and the subsequent decision for the OSCE to lead due to the dual veto exercised by both China and Russia.[4] Alternatively, the entire operation can be taken over by a "coalition of the willing" or a unilateral single-nation intervention. The American- and British-led "coalition of the willing" in the 1990 Gulf War and the 1994 US-led Operation Restore Democracy in Haiti serve as respective examples of these arrangements. Thus it is possible to categorize contemporary conflict interventions into the following three types: a UN-sanctioned/UN-led operation, a UN-sanctioned operation led by a regional organization, or a UN-sanctioned intervention led by an "executive agent" or a small "coalition of the willing." The American-led coalition, which launched airstrikes on Taliban military strongholds in Afghanistan on October 7, 2001, underlines more recent utility of "coalition warfare." This action drew on American-led rules of engagement and would be classified as the employment of air power to deter an aggressive threat under Chapter 7 of the UN Charter.[5]

When security and stability returns and humanitarian activity resumes, a new UN mandate that upholds the same Chapter 6 principles underwrites the new phase of operations. International troops are expected to perform a more integral role within the local society and assist in peace-rebuilding programs, security sector reform and democratic development. The forceful and more robust approaches featured in the earlier Chapter 7 mandates are no longer used, except in the case of self-defence. In some cases, the military forces retreat to carrying out a support function only, in an effort to give primacy to a newly developed security force and to re-empower local civil authority. Such was the case for the UN Support Mission in Haiti (UNSMIH), which succeeded UNMIH in 1996. A similar "single nation" parallel can be drawn with the British Army's ongoing intervention in Northern Ireland, where soldiers provide background support to the Royal Ulster Constabulary (RUC).

It is during these transitional stages in multinational peacekeeping environments that the local population either redirects its allegiance to the international community (represented by international military forces and humanitarian groups) or remains loyal to the individuals and groups behind the initial demise of the country. The latter occurs when conflict populations are not convinced that the international forces can provide them with security at the individual level. This encourages them to turn to their factional group leaders (or their own stakeholder group during the conflict) for reassurance. Local warlords, nonstate actors and paramilitary groups are aware of this phenomenon and have used it to garner support and entice young recruits and child soldiers.

On the other hand, the international forces can work hard to build the trust and confidence necessary for the local residents to believe in their efforts and support their programs. Over time, the loosening of ties with the paramilitaries and factional leaders encourages steps toward reconciliation and helps remove the bitterness. Achieving such an environmental transformation and changed mindset is necessary before international funds are spent on infrastructural reconstruction and societal rebuilding. Convincing each individual person that his or her own security is no longer at risk is paramount to a long-term solution.

There are many practical initiatives that can help to foster trust and credibility and remove the deeply rooted fear that helped sustain the status quo ante.

While it is beyond the scope of this paper to acknowledge all possible initiatives, recommendations that address this problem from a military doctrinal and leadership perspective will be explored. The problem is examined in the context of research carried out in Haiti during the third UN mandate, and during the 1996 Stabilization Force (SFOR) deployment in Bosnia. Both cases involve the participation of the CF, as well as many NATO and UN allies with whom Canada will continue to serve in the future. The choice of cases also reflects the contemporary nature of conflict, which includes a spectrum of activities ranging from low-intensity warfighting to more tranquil peace-building tasks.

Research Methodology

The research used to support the arguments in this paper formed part of a doctoral dissertation that examined the disparities of multinational land forces in peace support operations. The study covered the multinational interventions in Haiti (1994-96), Bosnia (1996, 1998 and 1999), Somalia (1997) and Northern Ireland (1999), the latter of which was used to explore whether or not parallels existed between the multinational research findings and a single-nation intervention. Research was undertaken in each country using rural and urban sample sets to increase the reliability of the data. For example, in Haiti, the urban area of Port-au-Prince and the rural areas in and around Cap Haitian were used. Similarly, in Northern Ireland, the urban area of West Belfast and the rural area of South Armagh were used. The case of Bosnia proved more challenging as, to maintain reliability and achieve accurate analyses, representation from many different rural areas was used to complement the research findings from the multi-ethnic region in and around Sarajevo. The same was true in the urban region of Mogadishu, Somalia, and the various tribal clans represented in the rural districts.

The data gathering, simulation and interpretation phases of this work employed a combination of both qualitative and quantitative research methods. Due to the nature of the problem, and the extreme shortage of statistics in an already underdeveloped area, a qualitative model based on interpretive quality perspectives dominated.[6] The mainstay of the data gathering involved numerous interviews with local representatives, all of whom were asked the same questions but who were also given room to elaborate on points which they clearly felt strongly about and to make observations that were directly related to some of the incidents that the author was investigating. Each national sample set was made up of the following proportions of respondents: at least 35 percent from individuals between the ages of 18-35; at least 40 percent from individuals between the ages of 35-50; and at least 20 percent from people in the 50+ age bracket. These proportions were applied equally to each of the different ethnic factions represented in the areas researched. The total number of people in the different sample sets varied, but remained consistent in relation to the population density of each area. The author also augmented the research with some quantitative research methods in order to monitor reliability[7] and ward off claims that the research relied too heavily on anecdotal comment.

Following the analysis of the research, the author returned to each region for validation and evaluation purposes. Using smaller, but wholly representative sample sets, a validation questionnaire[8] was used, which summarized the interim research findings. Individuals completing the questionnaires were asked to indicate on a scale of five gradients whether they strongly agreed or strongly disagreed with the findings. The questionnaires left room for additional comments, which were also incorporated into the research to enhance clarity.

The research methodology used was effective in penetrating the mindset of the conflict communities and discovering how the behaviour and performance of different national militaries affected their overall impressions of the multinational force and the role it would play in determining their future. These issues are important within the wider interoperability debate for the following reasons:

- Interoperability extends beyond weapon procurement programs and defence spending, which look at interoperability at the strategic level and not at the level of implementation.
- Efforts to achieve interoperability have encouraged mutual understanding and the development of collective defence doctrines, however, controls must also be put in place to monitor the different interpretations of the common doctrine to limit adverse effects on the ground.

The next few sections will explore some of the background detail and research findings from the Haiti and Bosnia case studies. While the involvement of the CF will remain the focus of this paper, it is important to remember that the broader study encompassed an in-depth view of twelve other nations, both NATO and non-NATO.

Haiti and Bosnia

Haiti

Haiti has suffered from civil unrest, government corruption and oppressive leadership for years. The circumstances of its birth as an independent state bestowed four potentially destructive legacies on Haitian society: the passing of power to the local Creole aristocracy; the precipitous and brutal changes of leadership that became a model for future Haitian governments, the violent tactics of Haiti's founding leaders and the use of voodoo; and, finally, the protracted wars of independence that destroyed the island's flourishing plantation economy.

The oppression of the Creole population by the Mulatto elite was challenged many times by renowned leaders like Toussaint Louverture, François Duvalier and his son who succeeded him, Jean-Claude Duvalier. However, this only created a black elitist regime, which continued to subject the majority of Haitian people to the same oppressive, impoverished and difficult life to which they had become accustomed. Each regime, along with their extended "families," controlled the few legitimate and illicit sources of economic wealth, political control and a powerful security apparatus that shared the riches.[9]

This kleptocratic nature of governance survived until the end of the Duvalier legacy in 1986, and through a series of similar regimes and bloody coups that lasted for four years.[10] In December 1990, the Roman Catholic priest Jean-Bertrand Aristide was sworn in as president by free and fair elections.[11] *A coup d'état* led by senior military officials and the capital city's chief of police removed him from power six months later.[12] Following US-led efforts to broker an agreement for the return of President Aristide, and the military regime's refusal to implement the terms of the agreement, a UN-sanctioned, US-led force mandated under Chapter 7 of the UN Charter was sent in to restore peace.

The mandate of the operation authorized the US force to use whatever means necessary to return President Aristide in accordance with the Governor's Island Agreement.[13] On March 31, 1995, the force was replaced by the UN Mission in Haiti, a multinational peacekeeping force acting under Chapter 6 of the UN Charter. The force was tasked with maintaining a secure and stable environment, assisting in the training of a new national police force, and facilitating a free and fair electoral process.[14] In 1996, the force was downsized and renamed the UN Support Mission in Haiti (UNSMIH) and was supported primarily by Canadian and Pakistani peacekeeping battalions, a French gendarmerie contingent and others participating in a UN Civilian Police Force (UNCIVPOL). A small group of US Army logisticians also provided support and was stationed at an airport compound. Its task was to assist in the professionalization of the national police force and in the maintenance of a secure and stable environment.

Bosnia

The events which preceded the deployment of the Stabilization Force (SFOR) in Bosnia had slight similarities to the international arrangements in Haiti, but came with complexities that were reminiscent of a region made up of about seven different ethnic factions, some of whom were fighting against themselves, and with conflict raging throughout the country (for different reasons depending on where one lived). In addition, the ambiguous political system that continued to change as the Yugoslav federation fell apart posed huge legal challenges to members of the international community who were poised to intervene.

Threatened Serb minority populations in Croatia and subsequent fighting along Croatia's border with Bosnia resulted in the deployment of a Chapter 6 peacekeeping force tasked with monitoring the designated UN-protected areas that contained those populations.[15]

The independent recognition of both Croatia and Slovenia encouraged more violence. By this time, similar problems had spread in and around the Bosnian capital of Sarajevo where all three ethnic groups were co-located. In the spring of 1992, the UN responded to a plea from Bosnian President Alija Izetbegovic and sent in a UN Protection Force (UNPROFOR) to facilitate the delivery and distribution of humanitarian aid.[16] Once again, the UN force served under a Chapter 6 traditional peacekeeping mandate and was expected to uphold certain rules of engagement and principles associated with the mandate.

As the fighting in the former Yugoslavia spread, the situation deteriorated, and several attempts at brokering a diplomatic solution failed, measures were increased to bring in NATO involvement and with it, a more robust mandate.[17] The deployment of the NATO Implementation Force (IFOR) in the fall of 1995 seriously weakened the position of the factional fighters,[18] which encouraged the factional leaders to seek diplomatic dialogue. They were eventually brought to the negotiating table where they signed on to the US-brokered Dayton Peace Accords.[19]

NATO, under IFOR, lasted for one year and was replaced by the Stabilization Force (SFOR), which is still there today. There are three area commands, all of which answer to a central command in Sarajevo.[20] While SFOR represents a UN-sanctioned NATO-led force subject to the authority of the NATO commanders, it is still deployed in a peacekeeping/peace-building capacity and is therefore expected to carry out the rebuilding and reintegration role inherent in postconflict operations.

The local response in Haiti

Research was carried out during the UNSMIH deployment to examine whether or not different national military conduct and behaviour affected the local population's impression of the UN Force. Interviews were conducted on the streets, in residential neighbourhoods, restaurants and cafes, prisons, municipal offices at the local police stations, in the more rural areas, and in the aid agency and military compounds. Views were gathered from the local inhabitants and international personnel assisting in all phases of the operation. Although this information was collected during the UNSMIH deployment in the spring of 1996, feedback on national military troops also included those who participated in earlier phases of the intervention.

Feedback on the American troops was divided according to time periods: during the earlier Chapter 7 operation that authorized the use of force, and the later support role the American forces contributed to the UN force. People generally felt that the American military was the right force to bring in during the earlier days of the conflict; a lightly-armed peacekeeping force would not have deterred the violence, crime and political unrest. This was particularly the case for the people interviewed in Port-au-Prince where the worst violence was erupting.

In the northern city of Cap Haitian, the American response to a particularly violent firefight with the paramilitary group Force Armée d'Haiti (FAD'H) resulted in increased support for the international force. The incident saw American warning shots to deter a gang member from shooting a pro-Aristide demonstrator outside the Cap-Hatian police station answered with direct fire toward the American troops. In response, the Americans shot and killed 10 of the paramilitaries. The response to the incident strengthened support for the Americans in particular, for at least two reasons.

First, it demonstrated to young potential paramilitary recruits that similar behaviour would not be tolerated and showed the disincentives of subscribing to the cause.

Second, residents of Cap Haitian commented on the renewed confidence instilled by the American action, which led to the reopening of local businesses that had been continually looted by the paramilitaries. For the majority of people in Haiti, any extra income besides state allowances was usually made from market stalls in the city and town centres; thus the American performance had brought hope that the markets could function once again. One former mayor even suggested that the robust, resolute approach proved to many people that the American's current involvement in Haiti was different from the nation-building tactics used between 1915 and 1934 that had generated so much resentment toward the US.[21]

After the arrival of other national troops contributors and the official handover of command from the US to the UN, local groups were able to observe the behaviour of several different military forces and remark on the way they were managing the transformed peacekeeping/peace-building environment. Local Haitians living around Port-au-Prince grew to resent the American military forces' insistence on using dedicated military vehicles (and not the open-sided UN trucks used by the other national battalions). Moreover, the locals questioned the need for the tall, heavily-manned guard towers that the American forces had constructed at each of their sites, and the requirement to travel in groups of no less than eight with heavy military vehicles wherever they went. This approach during a more peaceful environment had a compelling psychological impact on the Haitian population and enhanced the understanding of the UN presence.

The Pakistani battalion, which had been deployed since the transition to the UN force in 1994, had seemingly developed a good rapport with the local groups.

Many of those interviewed commented on the Pakistanis' determined look, the positioning of their guns and their attentiveness during patrolling activities, which made the Haitians believe that the Pakistanis were very much aware and in control of the situation. Their ability to combine this structured approach with constant interaction with people, whether it was helping someone push a wheelbarrow down the street or building a soccer field for the children in a bad neighbourhood, built tremendous support for the Pakistani battalion in Haiti. People acknowledged that this more than made up for their inability to communicate in the local language. If any violence broke out, the majority of people interviewed were convinced that the Pakistanis would resist any aggression and protect the population.

At a conference held in 1996, one British academic and former war defence correspondent described the Canadian approach to peace support operations as "enormously generous" but reliant more on the use of "soft tactics" shaped by peacekeeping policy approaches similar to those of the Dutch and Scandinavians.[22] Recent debacles in Rwanda, Somalia and internal problems within the Canadian National Defence Headquarters have, in the recent past, put enormous pressure on the individual serving soldiers and subjected them to rigid procedures that have, according to Canadian soldiers and their allied partners, restricted their operational freedom. The continuous need for the Canadian Forces to be seen as militarily "clean" and "politically correct" due to these past experiences has caused the average soldier to feel more limited in using traditional robust approaches. Thus, for the sake of satisfying a public and government back home that seem ill-informed about current operational requirements, the reputation of the Canadian land forces as credible security providers has been compromised. The point here is that this is not due to the individual level of competence or professionalism brought into theatre by each soldier, but to the discouraging ethos that has been built by Canadian politicians who repeatedly fail to see the role these individual men and women play in defending Canadian national interests — at home and abroad — and support them appropriately.

In Haiti, shortly after the transition to the UN-led operation in 1995, Canada was forced to modify its interpretation of the UN rules of engagement (ROEs) in order to protect a group of Canadian hydro workers who had been commercially contracted to restore electricity to the capital city of Port-au-Prince.[23] When a warehouse they were working in came under paramilitary fire, Canadian troops had to request permission from the highest authorities to use force to deter the attack. The existing Canadian ROEs only permitted the troops to use force "in the case of self-defence" due to Canada's insistence on the removal of "and in defence of property," from the same clause prior to deployment of troops. Had Canada agreed to the original UN text, the ROEs during this incident would have been clear and the Canadian troops would have had the freedom to deter the attack from wherever they stood. The only counterattack that their initial ROEs would have permitted would have been for the troops to position themselves in between the paramilitaries and the hydro compound in order to use force "in the case of self defence" — clearly not an option in these violent circumstances. The ROEs were later modified to include the use of force in the case of self defence "and also in defence of the mandate," which could justify the protection of the Canadian civilians. Most countries already use this text despite the Canadian belief that ambiguity in determining what would and would not threaten the mandate might result in unnecessary violence.[24]

The restrictions on the Canadians were obvious even to the local inhabitants. When people described the Canadian's approach to patrolling and escorting, they remarked that they did not appear to be in as much control as the Pakistani troops. In addition, feedback suggested that the troops tended not to hold their guns at the ready position like their Pakistani colleagues. There was also a perception that the Canadian soldiers tended to avoid the more turbulent areas of the city where needy reconstruction projects required their assistance. Reports on their failure to defuse and control several street riots and student protests also highlighted the constraints under which they had to operate.

Almost all of the respondents appreciated the kindness shown by the Canadian military forces and the ease of communication through French cultural and linguistic affinities. Many respondents also commented on the friendliness of the Canadians and their consistency in smiling and waving to the local inhabitants. Nevertheless, in the case of heightened security measures, only 12 percent of the 147 local respondents interviewed in the city of Port-au-Prince stated that the confidence they had in the Canadians as credible security providers was quite low.[25]

The local response in Bosnia

Similar themes were noted from the research findings in Bosnia. Local Muslim, Serb and Croat residents were interviewed in and around Sarajevo, the Bihac region of southwest Bosnia, Mostar, Banja Luka, Prijedor and in the Central Bosnian corridor of Drvar, Jajce, Gornji Vakuf and Bogojno. Canadians had served in many of these regions during different phases of the UN and NATO interventions in the former Yugoslav Republic.

Several Canadian soldiers recalled a well-known incident that occurred on April 1998 in the central Bosnian town of Drvar. During the repatriation of Serb groups to the now Croat-dominated town, the Croat residents of the area rebelled and ignited riots on the streets. The reactions of many Canadians were described as being "confused" and "fearful." This apparently became more evident when many of the troops jumped back into their SFOR trucks in the hope that the problems would die down. Perhaps these reactions also reflected the fact that, at that time, Canadian land force doctrine did not include crowd or riot control. British troops arrived shortly after in armoured vehicles and secured the area. Many Croats and a significant number of the repatriated Serbs said that, at the time, they were very happy to see the arrival of the British troops.

A number of Canadians interviewed concluded, fairly or otherwise, that, although it would have been in the ROEs to shoot, the political ramifications back home in Canada made them resist.[26] Other more senior onlookers also acknowledged that the ROEs were such that the troops could have fired, but that the response would not have reflected the Canadian approach to these incidents.[27] The recommendations of a more senior officer were to always threaten areas with future military presence as opposed to immediate robust reactions. As such, immediately following that incident, local authorities were instructed that any further incidents would result in an indefinite deployment of SFOR troops in the town. Apparently, after that statement was issued, no further problems in the area were reported. However, feedback from the local residents indicated that the belated but robust intervention of the British soldiers had served to deter any further incidents.

Serbs in most areas did not warm to the American ground troops due to several reasons. When diplomatic efforts reached an impasse during the UN deployment in 1993, the American support for the "lift and strike" option caused some degree of resentment. Moreover, in various press releases and official statements, visiting US officials rarely acknowledged the problems caused by the Muslim and Croat populations in Bosnia. This was particularly the case when the Serbs received strong condemnation by the US in the February 1994 mortar incident in the Sarajevo market despite the fact that incident reports analyzing the projection and impact of the firing weakened the argument that the Serbs bore any responsibility.[28]

Most military personnel living and serving in Bosnia were aware of the "heavy" approach used by the Americans when serving on the ground. In the context of the more recent American intervention in Kosovo, Professor Lawrence Freedman offers the following thoughts on the American obsession with force protection:

> As a US Army brigade moved into Kosovo as part of the force that intended to bring calm into the country after the war, its mission statement listed as its first priority "self-protection" with the "peacekeeping tasks" secondary. While the troops of US allies intermingled with the local population, US troops stayed in a guarded and well appointed compound, separated from the society that they were supposed to help calm.[29]

In Bosnia, inquiries into the travel plans of American troops at checkpoint stops were often met with soldiers jumping out of heavily armed military vehicles in order to guard the delegated spokesperson while he or she dealt with the factional representatives. This top-heavy approach, particularly during the SFOR mandate that focused on peace building and reconciliation, was viewed as unnecessary and served only to raise anxieties among the ill-informed and aggravate others. As in Haiti, the troops never travelled in groups of less than eight people with a minimum of two armoured personnel carriers.

Local residents believed that the Americans had little interest in speaking and interacting with them. They believed that the troops felt indifferent toward understanding the local situation and lacked the capacity to think laterally and beyond their own cultural beliefs. Following the August 1997 shooting of the Bosnian Serb paramilitary Simo Drljacha by British Special Forces,[30] demonstrations were mounted in front of the SFOR Civil-Military Centre (CIMIC) in Prijedor which, at that time, was being manned by American officers. The CIMIC had been set up earlier that year to serve as an information centre for locals and a place where they could speak to ground troops about various SFOR programs and initiatives. After this incident, the Americans on duty at the CIMIC refused to speak to the locals or make efforts to defuse the situation, which generated even more hostility. Soon after the incident, they were replaced by a group of Czech officers.

Bosnian Muslims living in the Bihac region recalled an incident that further underscored the Americans' reluctance toward understanding local dynamics. A number of American soldiers had been tasked with distributing IFOR newsletters (a peace-building tool used to improve communications and understanding) in the area of Bos Krupa. The newsletters were translated in slightly different dialects and emphasized slightly different issues, depending on whether the target audience was Serb, Muslim or Croat. Both the interpreters and a substantial number of local residents realized that little care was being taken with the distribution of the material. This indifferent and detached attitude, which was exhibited on many occasions, undermined ethnic sensitivities and did not help the Americans gain respect and credibility in a very ethnically mixed region.

Another incident in 1995 in the Bosnian town of Doboj also illustrates the same top-heavy American approach that tainted the Haitians' view of the American forces during the peacekeeping and peace-building stages of that conflict. A group of Muslims had been given permission by Danish and British troops to cross a bridge to visit a cemetery (in the newly proclaimed "Serb" side of town) where relatives had been buried. Riots broke out and Serbs began chasing Muslims, throwing stones and physically beating them. As the British and Danish troops fired warning shots into the air, their efforts were overshadowed by the sudden appearance of American gunships, with blades tilted downward in order to spray stones and objects into the air in an "overkill" effort to move back the crowd. Several Danish and British officers who witnessed the incident suggested that alternative, more graduated levels of force could have been used to prevent further threats and violence.[31] Many of the local residents who were interviewed felt that the use of gunships had sent a very powerful message to the factional militant groups in terms of consolidating more resources and heavier equipment. Other individuals felt that the Americans were trying to use "scare tactics" to increase compliance in the area.

Due to the broad range of national troops contributors that served in Bosnia between 1995-99, the local populations identified several other national tendencies that affected their perception of the international effort. The behaviour of the Bangladeshi and Malaysian soldiers toward the Bosnian women and the significant time they spent in local bars and restaurants had affected IFOR's reputation in both Sarajevo and Bihac. In the more southern area of Mostar, while the locals warmed to the Italian and Spanish troops, they worried about their ability to offer sufficient protection in the cases where fighting reignited between the Croat and Muslim factions in the area. Residents in and around Sarajevo and Mount Igman felt very strongly toward the French Foreign Legion's "over the top" approach to "spraying bullets in response to a tree branch breaking in the wind."[32]

The results proved that certain categories of military "professionalism," perceived from the local perspective, were observed in Bosnia. On one end of the scale were troops described as "net users," as opposed to "net contributors," of security. These included the Malaysian, Jordanian, Bangladeshi and Ukrainian troops. To a certain extent, the Russians were also grouped in this category, despite their potential to take strategic advantage of their historical alliance with the Serb factions. The national contingents described above were considered a liability to the multinational effort, particularly in many instances where impartiality, credibility and professionalism were often compromised. This suggests that differences in approaches to ROEs, and thus effectiveness, are more the result of which countries the forces come from rather than a lack of common training. While common training programs would, without question, help bridge the gaps, the cultural ethos that shapes a national military's interpretation and leadership approaches toward a common doctrine and ROEs are central to achieving a more unified response.

Grouped in the next category were troops like the Spanish, the Dutch and the Canadians. These groups were known to practise softer and less robust soldiering, which, to the locals, would not be effective during periods of heavy violence. In addition, the local perception was that these troops lacked the operational freedom necessary to offer adequate protection if tensions between the ethnic factions resurfaced. On a more socio-behavioural point, the locals generally felt that these troops kept a certain "distance" from the local groups and did not engage themselves in the local environment as much as they could have. Adjectives such as "withdrawn" and "detached" were used several times to describe the approach of these groups to patrolling.

However, it should be noted that recent discussions with personnel from the Office of the High Representative (OHR) and the office of the United Nations High Commissioner for Refugees (UNHCR)[33] suggested that the Canadian contribution in the Canton 10 region of central Bosnia has made notable progress in the past few years in terms of interfacing with the local community, coordinating with other civilian agencies and assisting in the peace-building strategies. Admittedly, the threat in Bosnia has diminished and social and economic rebuilding in multiethnic areas is now the priority. Perhaps a move away from the ambiguity inherent in the volatile Chapter 61/2 theatres has instilled the confidence required for the CF to have a more magnified presence.

The French and Czech forces generated fairly positive feedback on their military conduct and professionalism. Locals from all ethnic backgrounds felt safe in the company of Czech troops and applauded their way of handling tense situations. These observations were quite complementary to the Czech battalions, considering the fact that they were still adjusting to western military practices and only just being folded into the NATO Alliance's integrated military structure.[34]

The Americans were criticized for their "top heavy" approach, particularly during times when it was not necessary. They were also described as being insular and noncommittal toward understanding and interacting at the local level. Generally speaking, local residents from all sides felt that the Americans should not be used in such sensitive environments, where relationship building and positive encouragement were considered priorities. This has implications for a country whose technological superiority and projection of military power excels at the highest of diplomatic and operational levels. Perhaps there is significant merit in John Hillen's argument that "superpowers don't do windows," which recognizes that NATO best serves its many different security roles by playing to the core competencies of its members,[35] which, for the United States, should perhaps not include Chapter 6 peacekeeping in complex environments.

The response to the British presence was more encouraging. Local residents applauded their firm approach to and impatience with obstructionism. However, they also felt that the British troops demonstrated a firm commitment to understanding local circumstances in each community and made efforts to remain informed at all times. Individuals and groups seemed very aware of the help that the British troops provided to other forces and were cognizant of the leadership roles they assumed in different multinational situations. The one criticism that surfaced in Bosnia was the tendency for the British troops to refrain from concealing their weapons, or carrying them into public places during sensitive events aimed at rebuilding the social fabric of country, such as local elections and multiethnic council meetings, even during the post-Dayton phase when peacebuilding strategies were progressing quite well.

These disparities in national approaches have also been observed in multinational interventions elsewhere. The 1993 slaughter of 25 Somali civilians carried out by the Pakistani military contingent in response to the shooting of two Pakistani soldiers, and the subsequent exoneration of those who orchestrated the slaughter, had interesting ramifications.[36] It not only upset any hope for an interim peace agreement to be followed but also illustrated the differences between how these incidents are handled by national governments and, therefore, what does and does not serve as a deterrent for such actions.

Of recent interest has been the overwhelming support for British intervention in Sierra Leone and the criticism expressed toward the multinational UNAMSIL force led by a Nigerian general, who allegedly answered to Abuja rather than New York.[37] The strategy of the British troops to offer better protection for residents of Freetown, to take control of several northern paramilitary strongholds and to return refugees fleeing for Guinea, while incurring only one fatality, represented a remarkably positive turn of events in a country controlled by rebel forces.

There are obviously many other incidents that may be investigated to assess the collective impact of different national military approaches on a conflict population. However, such preliminary observations recognize that inconsistent and incongruent national interpretations of multinational military procedures, conduct and leadership required to fulfill a mandate can have a negative impact on the overall effort. For example, the current war on terrorism will carry interesting implications for the recently debated Turkish-led UN peacekeeping force proposed by Algerian Ambassador Lakdar Brahimi to succeed the interim British-led force in Kabul.

Nor can different national contributors be heavily faulted for fine-tuning rules of engagement according to their own national law. Notwithstanding the fact that multinational forces often operate under the operational control of a regional organization or the UN, donor nations will never be willing to have their forces governed (and held legally accountable) to a standard that is not in accordance with the donating country's domestic law.[38] But when it is known that national legal constraints and political pressure will disturb the multinational unity to the extent that it has a negative impact on the peace process, restrictions should be communicated and core competencies should be more clearly defined. Arguably, the reality of coalition warfare has already advanced to the point where national exceptions to the force ROEs are accommodated; however, it is questionable whether or not the UN's very political and ad hoc approach to organizing Chapter 6 (and Chapter 61/2) peacekeeping forces is familiar with this template.

Implications for interoperability and multinational training

Current multinational training programs neither address this gap in developing standard approaches and uniform interpretation, nor provide adequate resources to tackle the problem in the near future. The UN has recently been stripped of its "gratis program" under the aegis of the UN Department of Peacekeeping Operations (UNDPKO), whereby an increasing number of military staff officers from the armies of the member states were assigned on loan, or "gratis," to UNDPKO. While nearly one-quarter of the department's 400-plus staff was in fact made up of national armed forces personnel, being "gratis" meant that they worked for the UN and not their home country. These gratis officers contributed invaluably to contingency planning and operational training prior to multinational deployments to areas such as Rwanda, Somalia and the former Yugoslavia.[39]

Due to internal bureaucratic problems and discontent fuelled by the financial limitations imposed on the less-industrialized nations to fund officers employed in New York City, the UN General Assembly passed a resolution in September 1997 that called for an expeditious phasing out of all gratis personnel. This left the training unit of UNDPKO with a staff complement that had been reduced from 27 to 4. The UNDPKO's training unit had previously used the input of the gratis officers to develop the curricula for all UN military training programs and to represent the department as course trainers. The diversity of backgrounds and military experience added value to the training programs and ensured that training was multinational in nature.

However, even with this program in place, most of the gratis officers came from NATO countries or US allies, which did not accurately represent national contributions in the most recent peacekeeping operations (i.e., Bangladesh, Pakistan and Nigeria). As a result, there was a tendency to duplicate NATO procedures and other templates that had been offered in the past by leading NATO contributors. The lack of proportional representation in peacekeeping training forums, compounded by the phased-out gratis program, poses major challenges to multinational military training for future UN deployments.

At the moment, the very small staff complement in the UNDPKO Training Unit continues to run their biannual program called "Train the Trainers," whereby those responsible for peacekeeping training at the national level gather in Turin, Italy, for courses in multinational peacekeeping training. They are then expected to return to their national training centres and incorporate the instruction and content into courses run at national training centres.

Discussions with many leading military nations indicated that only a limited number of member states sent designated trainers to these courses. Certain national representatives felt that their own training modules were more up-to-date and reflected current operational requirements for multinational interventions much better than the UN-sponsored course. Some commented that the research behind most of the UN course content was very outmoded and reflected the old Sinai-based peacekeeping principles that lacked a great deal of contemporary relevance.[40] Others did not send any of their officers because they did not consider the training to be "half as advanced" as their own national training programs.[41]

Another problem, and perhaps something that could be developed and facilitated more easily than an expanded training program, was a lack of quality control mechanisms in place to ensure that the national interpretation of UN ROEs and operating procedures was reunited under the multinational mandate once troops arrived in theatre. After the UN mandate and ROEs are agreed to at the highest levels and further endorsed by the UN Security Council (UNSC), contributing nations send these UN templates back to their own national authorities for further analysis. The documents are sent to national legal experts who ensure that the UN's legal interpretation of the mandate is compatible with the contributing nation's own legal interpretation. In addition, the Ministry of Defence assesses the procedures and requirements according to its own national doctrine and codes of conduct to ensure compatibility at the military level. After any requested changes or modifications are dealt with, final directives are issued and operational planning for the multinational deployment begins.

Depending on whether time permits, predeployment training may take place at national training centres. However, the speed at which troops deploy, particularly during the first rotation into theatre, becomes critical and therefore minimizes the chances of providing training tailored to the specific mandate and geographic region. Therefore, the responsibility for ensuring uniformity of response ultimately falls into the hands of the UN force commander once all troops are deployed in theatre. Faced with a multitude of other more demanding responsibilities, such as holding meetings with political factions, negotiating peace agreements and liaising with heads of state, the UN secretary-general and the Security Council, the force commander cannot be expected to monitor how different nations interpret the same mandate.

Concepts and doctrine that embrace the more contemporary principles of peacekeeping are articulated in the NATO doctrine for peace support operations. The growing pre-eminence of this doctrine, and its influence in Europe and the US, is also encouraging a degree of apathy toward UN peacekeeping concepts and training. As the declared "custodian" for NATO doctrine for peace support operations, the UK has been hugely influential in shaping recent military training programs in all of NATO's member states. It recognizes the complex evolution of peace support operations and how different contingencies affect the response requirements. The doctrine also acknowledges the vast number of civilian agents and the continuum along which transitional management and lead-agent responsibilities become exceptionally important.

There is now a significant divide between countries that subscribe to the NATO doctrine and those who remain loyal to the UN's outdated approach. Defence analysts might argue that NATO's lead on military training is more appropriate, considering its recent involvement as lead agency in peace support operations in the Balkans. Nonetheless, it is imperative that the international community decide which organization should take the lead in training future multinational forces and recognize that softer, more traditional peacekeeping principles cannot be definitely separated from more robust postures.

A few recent initiatives have attempted to bridge the gulf between UN and NATO approaches to the development of contemporary peacekeeping practices. NATO's Euro-Atlantic Disaster Response Coordination Centre (EADRCC) at NATO Headquarters in Brussels is now reviewing the Oslo Guidelines, first drafted by the UN's Office for the Coordination of Humanitarian Affairs (OCHA). The guidelines advise UN and NATO members on how their assets can become immediately deployable to help respond to natural disasters within NATO's area of operations. More recent discussions have advanced the debate to also include out-of-area "complex humanitarian emergencies" in light of NATO's precedent-setting non-article 5 interventions in Bosnia and Kosovo. This process has called for the inclusion of UN OCHA staff within the EADRCC, which may drive a more prosperous and transparent relationship between the two organizations.

Other initiatives have arisen from ambitious recommendations published in the Brahimi Report, the most important of which include the development of meaningful criteria and techniques to identify and respond to conflicts in a more robust fashion; an increased role for regional organizations in peacekeeping; and greater coherence and coordination among security, trade and development institutions in conflict prevention. However, these efforts are being stymied by the insistence of developing countries that equal organizational focus and resources be directed toward development issues such as the elimination of poverty and HIV/AIDS.[42] The office of the UN secretary-general has acknowledged these concerns and, as a result, has requested that priority be placed on developing strategies for conflict prevention, something that each side of the debate regards as a halfway house between sustainable development and conflict intervention.[43] The change of focus has also posed difficulties in generating consensus among member states to endorse further job appointments in the DPKO necessary for the full implementation of many of Brahimi's recommendations.

National Solutions

The results of the research hardly seem encouraging in the wake of increased calls for better multinational interoperability. From a Canadian perspective, phrases such as "achieving close strategic partnerships" and "interoperability with Canada's principal ally" feature significantly in the Canadian chief of defence's *Strategy 2020*, as well as in the most current literature on Canada's strategic defence priorities. These statements were in reference not only to the United States, but also to "other like-minded states" with which the CF would find themselves deployed in the future.[44] This type of interoperable cooperation requires clearly defined, commonly accepted and attainable objectives, all of which should create the conditions to achieve a unity of effort in theatre. This is particularly the case when peace support operations and humanitarian missions still remain high on Canada's agenda and reflect a closer cooperation with other government departments, such as the Canadian International Development Agency (CIDA), in keeping with "joined-up" government.

While it is beyond the scope of this paper to elaborate on the mammoth subject of interoperability with the United States — a relationship which then-Canadian defence minister Art Eggleton described as the "closest defence relationship in the world"[45] — it is important to delineate between levels of interoperability and what Canada should be moving to achieve. The technological advancements and defence trade and partnering at the strategic level must be complemented with the correct training and knowledge-sharing at the operational and tactical levels. What good is the most advanced shared weapons system if it is not operated in harmony and cohesion by like-minded allies? The same argument holds true for the need to unite a headquarters' vision with the same vision of the soldiers representing that headquarters on the ground. If multinational interoperability is to be achieved and progress, it must be taken right down to the unit level. The coalition and allied warfare of the future will see armoured infantry companies of one nation serving under the battalion leadership of another. Such was already the case when the Canadian infantry provided a reconnaissance support company under the British battalion and brigade commands in 1998 in Kosovo.[46] Building the weapons defence and intelligence systems to deter attacks such as September 11, 2001, is only one part of multinational interoperability. Postattack overseas deployments, ceasefire monitoring, ground force operations and postconflict peace building are also essential elements of the broader equation. This extends the more parochial view of interoperability and builds it into a broader, more strategic context. Such a vision may make it possible for Canada to achieve its five-year target objective of "managing our interoperability relationship with the US and other allies to permit seamless operational integration at short notice."[47]

But is it in fact Chapter 6 peacekeeping to which the CF will aspire in the future? Evidence from their recent intervention in Afghanistan suggests otherwise. While carrying out combat operations and conducting joint training with the Americans in Kandahar, there is still no interim peace in Afghanistan, implying that the Canadians are in fact operating in a war zone characterized by the legalities embedded in Chapter 7 of the UN Charter which underwrites "peace enforcement." This raises the question of whether or not the CF, through interoperability, are implicitly taking themselves away from peacekeeping (even the more robust kind) and moving toward peace enforcement. Given the limitations on Canadian military resources, as well as the need to be selective about future operations, does this mean that Canada will not be able to do both? The fact that Canada placed itself under the operational control of the American forces and not the British-led peacekeeping force could open up arguments that make this debate even more complex — there now may be a NATO way of doing things, a UN way of doing things, and a new Canada-US approach to peace support.

Whatever the case may be, it is important to remember that Canada prides itself on being an advanced liberal democracy with a quality of life that is second to none. Much of this profile derives from a responsible government that is committed to policy based on popular vote. It is perfectly acceptable that the Canadian government wishes to be selective about its future involvement in other peoples' wars — in fact, the government should be able to decide whatever it wants for their armed forces because it reflects the choice of the Canadian people. However, for the tasks that are given to the CF, it is essential to give them the tools that they need, including public opinion and political resolve.

As for training and doctrine, a focus on national leadership and policy may be the best step forward. Both doctrine and training are important functions of ethical military leadership and, where troops serve in multinational theatres of operations, commanders must realize that different national interpretations can result in disparities within an agreed series of legal responses and procedures. As such, the following recommendations are proposed to improve Canada's position as an interoperable land force partner in future peace support operations:

- National Defence headquarters should undertake research that explores the "positioning" of different national troops contributors in terms of manpower, fighting power, capability and deployability, along with the cultural mindset and public and political support each nation brings to a theatre. Such a study could further identify nations in broader categories, perhaps labelled as "front-line," "second-line" or "support-line" peace interventionists. As each country finds its position and establishes the goals it must pursue to achieve multinational interoperability with its allies, benchmarking the successful approaches and conduct of "like" countries may help them to reach their objectives and, more importantly, improve unity of effort in difficult and challenging circumstances

- Multinational land force training must also underscore the fact that the different interpretation of ROEs by one national military contingent can seriously undermine the progress achieved by another. "Lessons learned" and "after-action reports" must be carefully monitored and fed back into the national and multinational training curricula, supported by explanations of why other multinational partners handled an incident in a particular way and the short- to medium-term impact of such disparities. This is currently covered at the Canadian Land Forces Command and Staff College, but with more emphasis on warfighting rather than peace support operations.
- The seven-day course held at the Canadian Forces Peace Support Training Centre in Kingston, Ontario, should include an element of "partner culture" within the current module on cross-cultural awareness. This part of the course focuses mainly on the cultural norms in the peace support theatres into which the delegates are about to be deployed, as well as including some basic language training. Combining some background knowledge on the other national military forces with which the Canadians would be serving would be an added benefit.
- Canada should continue its active participation and current level of commitment within the Partnership for Peace (PfP) framework at NATO. The CF play a significant role in many of the exercises that fall under the Partnership Work Program (PWP), a "menu" from which NATO partners can choose to pursue goals and objectives that increase their chances for full accession to the alliance. Should Canada continue to play a peace support role in Afghanistan, benefits will accrue from PfP experience with countries like Uzbekistan, and from being familiar with its doctrine and equipment.
- Modules offered to Canadian and foreign civilian and military representatives at the Lester B. Pearson International Peacekeeping Training Centre should include lectures on the disparities of ROE interpretation and behavioural conduct and the impact these disparities can have, on not only the peace process but also the ongoing development programs. The Pearson Centre could provide a forum for rich dialogue between these partners, particularly as it attracts delegates from both UN and NATO countries. Parallel efforts should be made to publish the results of these syndicate discussions, presentations or group assignments in order to build on the debate in a much broader and more representative way.
- Theatre war games and tactical exercises integrated into staff college modules should consider the implications of joining certain national military

partners serving under the same multinational flag (although, quite understandably, theatre war-gaming exercises are geared more toward warfighting rather than peacekeeping). The conclusion drawn by the research in this paper suggesting that UN member states remain divided between subscribing to the NATO as opposed to the UNDPKO approach to doctrine and training, should be conveyed to both junior and senior staff courses. Efforts should also be made to inform Canadian military representatives serving in foreign delegations and on the international military staffs of intergovernmental organizations such as NATO, the UN, the European Union and the OSCE, of the division between UN and NATO approaches and the impact this has on conflict communities.

- The Canadian government should review the relationships between the National Defence Headquarters, the Department of Foreign Affairs and International Trade and the Canadian International Development Agency in terms of "joint" policy and planning functions. In today's peace support operations, the spectrum of activities in which ground forces find themselves engaged requires close cooperation and transparency between those responsible for diplomacy, defence and sustainable development. This closer interface will be required for planning at the strategic and operational levels.
- The National Defence Headquarters should consider developing a Combined and Joint Doctrine and Concepts Centre dedicated to the continuous development of Canadian Defence Doctrine, including doctrine for peace support operations. This should be viewed as a dynamic process which remains in line with national interests and in keeping with a "joined up" government policy on defending Canadian interests abroad.
- Lastly, in terms of civilian education, university modules that teach undergraduate and postgraduate students about the principles of contemporary peacekeeping should discuss the need for multinational forces to combine a "human face" and a commitment to the local dynamics with a robust military posture. This requires a sound understanding of not only the cultural environment and the wider spectrum of activities ranging from peace enforcement to postconflict peacebuilding, but also effective transitional management approaches necessary to support these changes. Emotional wounds take a long time to heal and fractious and divided communities can easily regress if they perceive that the international civil and military communities cannot guarantee their basic safety and security needs for stability. Much of the Canadian literature on peacekeeping principles has been based on the founding principles upheld by the Pearson Peacekeeping Centre,[48] principles that are much less absolute in contemporary Chapter 61/2 conflict zones.

Conclusion

This paper has examined the realities and uncertainties of current conflict landscapes and the challenges posed to military interventionist forces deployed to these areas. It has also emphasized that, despite claims of a changing future strategic threat, most interventions will almost always include the deployment of ground forces, either before, during or after an actual targeted attack. For this reason, it is essential that future policies on multinational and bilateral interoperability consider the more operational and tactical requirements that are essential for improving cohesion and unity of effort on the ground.

Empirical research undertaken in Bosnia and Haiti underlined the significance of the local populations in these areas and how international military forces must view these groups as the main stakeholders of conflicts and the only key to a sustainable peace. If, through inconsistent behaviour and disparate approaches of multinational troops, the local groups re-evaluate their allegiance to the international effort, they may easily fall vulnerable once again to the forces that ignited the conflict in the first place. Once this happens, the prospect of a sustainable peace becomes more uncertain and the task of the interventionists becomes increasingly difficult.

Calls for increased efforts for Canada to achieve multinational interoperability with her allies has resulted in a plethora of documentation on the subject, primarily with regard to fostering the country's defence relationship with the United States. However, while much of the literature stresses the importance of trade partnerships, the shared defence industrial base, and advanced weapons systems, complementarity at the tactical and operational levels must also be included in the interoperability equation. These objectives will be met only through more multinational training, interagency coordination and joint doctrine development, as well as enhanced clarity between Canada's political masters and the soldiers on the ground.

Notes

1. For more information on the concept of "second generational peacekeeping," which looks at the less absolute nature of concepts like impartiality and consent, see Mackinlay (1994, pp. 149-73), and Mackinlay and Chopra (1993).

2. See United Nations (1994, p. 19).

3. United Nations (1994, p. 22).

4. Evans (1998a).

5. United Nations (1994, p. 25).

6. See Rubin and Rubin (1995).

7. Throughout the study, the reliability figure of 77 percent was maintained.

8. The validation questionnaire used a semantic differential scale to summarize feedback.

9. The demise of French domination in Haiti left the French slave owners and their black concubines to produce a new class, the mulattos, also known as the *gens de couleur* or *affranchis*. This new class's social status rested between those of whites and blacks. In spite of the institutional discrimination against them, many mulattos became wealthy landowners, establishing a viable class unto themselves.

10. For an excellent overview of the Duvalier legacy, see Abbott (1988).

11. President Aristide led the *la fanmi Lavalas* political party. In Creole, the word *Lavalas* means "avalanche," which reflected the approach Aristide promised to take to clean up the mess that prevented the "seeds" of development from growing in the country's "gardens." Integral to this strategy was Aristide's promise to prosecute the criminals of the past and rid Haiti of any remaining *ton ton macoutes*.

12. Ironically, the leaders of the coup were the same officers Aristide had promoted in hopes of reforming the military.

13. Under the Governor's Island Agreement, President Aristide would be immediately reinstated, appoint a new commander-in-chief of the armed forces, and engage in political dialogue with the OAS and the UN. The dialogue would pave ways to creating political and social conditions to ensure a peaceful transition that would enable the Haitian Parliament to resume its normal functioning as quickly as possible. Other provisions dealt with issues of amnesty, the creation of a new police force and international cooperation. Lastly, the agreement specifically requested the presence of UN personnel in Haiti to assist in modernizing the armed forces and establishing the new police force.

14. United Nations Security Council Resolution 867.

15. It should be noted that although the Serbian Yugoslav President, Slobodan Milosevic, contested that Serb minorities were under threat due to Croatian secessionist aspirations, federal military reinforcements sent to the region by the government were responsible for many of the Muslim and Croat atrocities.

16. Traynor (1992).

17. Evans (1993a) and Evans (1993b).

18. See Bellamy (1996).

19. For an excellent overview of the Dayton Peace Accords, see Bass (1998, pp. 95-109).

20. Multinational Division Southwest (MND SW) was moved in 1997 from Gorni Vakuf in Central Bosnia (in the Croat-Muslim Federation) to Banja Luka in the Bosanska Krajina area (in the Republika Srbska). The other two commands, MND North and MND Southeast, are located in Tuzla and Mostar, respectively.

21. Moore (1996).

22. Bellamy (1996).

23. Canadian Permanent Mission to the United Nations (1996).

24. Canadian Permanent Mission to the United Nations (1996).

25. This number excludes input from the humanitarian and UNSMIH representatives interviewed and represents a local opinion only.

26. Based on discussions with soldiers and officers of the Royal Canadian Regiment stationed in Bihac, Bosnia, and interviews with other Canadian staff officers at MND SW, Banja Luka, Republika Srbska.

27. MacLean (1999).

28. For an interesting account of this incident, see Rose (1998).

29. Record (2000, p. 5); Freedman (2001-02, p. 72).

30. Simo Drljacha was shot dead at a lake just outside of Prijedor, Republika Srbska.

31. Some Danish officers remarked on their use of "non-lethal weapons" during this incident.

32. Based on discussions with Muslim groups living on the slopes of Mount Igman, November 1996.

33. Based on discussions with UNHCR and OHR representatives, Sarajevo, February 26, 2002.

34. The Czech Republic joined the Partnership for Peace program in January 1994, and officially became a full member of NATO at the Washington Summit on April 24, 1999.

35. Hillen (2001, p. 17).

36. Iqbal (1999).

37. Riley (November 2000-May 2001).

38. The Canadian Forces Staff College uses an excellent exercise that reinforces a clear understanding of how ROEs are developed for multinational forces. Lawyers are assigned to advise groups (acting as a Joint Staff) writing a set of ROEs that must be approved by a Joint Chief of Defence. (One example might be the restrictions on military engineers whose country has not yet signed up to the terms of the Ottawa Treaty.)

39. McClure (1999, p. 99).

40. The expression "Sinai-based" refers to peacekeeping principles characteristic of one of the UN's first peacekeeping operations, the UN Emergency Force (UNEF) that deployed to the Sinai in 1956 to monitor the ceasefire agreement between the Israeli and Egyptian forces. It is appropriate to mention here that the initial UNEF (UNEF 1) was under Canadian command and applied the traditional principles of peacekeeping that assumed a zone of separation between the warring factions, consent for the force to occupy a position in the disputed territory, the nonuse of force, and the requirement for the troops to remain impartial. More recent circumstances have demonstrated that these conditions, and therefore the principles that apply to them, are no longer characteristic of contemporary peace support operations.

41. These statements were taken by the author in a series of meetings with national military advisers in New York between September 4 and 6, 1999.

42. Coleman (2001).

43. Khan (2001).

44. Department of National Defence (1999).

45. Eggleton (2001).

46. Smith (2002).

47. Department of National Defence (1999, p. 4).

48. The Lester B. Pearson International Peacekeeping Training Centre gains its name from a former Canadian prime minister whose contribution to the first formalized UN peacekeeping policy was invaluable. While there is still much utility in the three basic principles (consent, nonuse of force and impartiality) underlining Pearson's original idea of UN peacekeeping, these tenets are much less absolute and the conflict theatres in which they are applied are much more complex. The Pearson Centre has modelled its ideas on a "New Peacekeeping Partnership" paradigm that includes the media, the police, the military, NGOs and election monitors. The author would contend that this model is more relevant to postconflict peace building and has limited utility in volatile theatres where contemporary peacekeeping still involves robust soldiering.

References

Abbott, Elizabeth. 1988. *Haiti: The Duvaliers and Their Legacy*. London: Robert Hale.

Bass, Warren. 1998. "The Triage of Dayton." *Foreign Affairs* 77, no. 5 (Sept/Oct): 95-109.

Bellamy, Christopher. 1996. "Knights in White Armour." Presented at "Defence-Development-Disaster: Security Imperatives in the New Millennium," Prague, November 7-8.

——. 1995. "NATO Takes Step into the Unknown." *Times* (London), December 21, p. 8.

Canadian Permanent Mission to the United Nations. 1996. Discussions with the author, New York, May 7.

Coleman, Christopher. 2001. Discussion with the author, New York. May 16.

Department of National Defence. 1999. *Shaping the Future of Canadian Defence: A Strategy for 2020*. Ottawa: Ministry of Public Works and Government Services.

Eggleton, Art. 2001. "Speaking Notes for the Honourable Art Eggleton, Minister of National Defence To the Toronto Board of Trade." Ottawa: Department of National Defence. Available: http://www.dnd.ca/eng/archive/speeches/2001/dec01/14dec01trade_s_e.htm

Evans, Michael. 1998. "Eyes in Sky Will Back Unarmed Peace Terms." *Times* (London), October 14, p. 13.

——. 1993a. "NATO Jets Take on a Reconnaissance Role Over Bosnia." *Times* (London), April 16, p. 1.

——. 1993b. "UN and NATO Plan Joint Attack Role in Bosnia." *Times* (London), August 7, p. 1.

Freedman, Lawrence. 2001-2002. "The Third World War?" *Survival* 43, no. 4 (winter): 61-88.

Hillen, John, and Bill Nash. 2001. "Can Soldiers be Peacekeepers and Warriors?" *NATO Review* 49, no. 2 (summer): 16-20.

Iqbal, J. 1999. Discussion with the author, New York. September 5.

Khan, Jehangir. Adviser, UN Department of Political Affairs. 2001. Discussion with the author, New York. May 17.

Mackinlay, John. 1994. "Improving Multinational Forces." *Survival* 36, no. 3 (fall): 149-73.

Mackinlay, John, and Jarat Chopra, eds. 1993. *A Draft Concept of Second Generational Multinational Operations*. Providence, RI: Thomas J. Watson Institute for International Studies, Brown University.

MacLean, Douglas. 1999. Discussion with the author, Bosnia. May 10.

McClure, Robert L., and Morton Orlov II. 1999. "Is the UN Peacekeeping Role in Eclipse?" *Parameters* 29, no. 3. (fall): 96-105.

Moore, Ti Don. 1996. Interview with the author. May 9.

Record, Jeffrey. 2000. "Force Protection Fetishism: Sources, Consequences and Solutions." *Aerospace Power Journal* 14, no. 2 (summer): 4-11.

Riley, Jon. 2000/2001. Discussion with the author. November/May.

Rose, Michael. 1998. *Fighting for Peace.* London: Harvill Publishers.

Rubin, H. J., and I. S. Rubin. 1995. *Qualitative Interviewing: The Art of Hearing Data.* London: Sage Publications.

Smith, Nick. 2002. Discussion with the author, Osnabruck, Germany. February 1.

Traynor, Ian. 1992. "Muslims seek UN protection." *Guardian* (Manchester), March 28, p. 13.

United Nations. 1994. *Charter of the United Nations and Statute of the International Court of Justice.* New York: United Nations, Department of Public Information, Article 45, Reprint DPI/511-93243.

9

Missed Opportunity II: The Canadian Forces, Force Structuring and Operation Iraqi Freedom

Sean M. Maloney

Introduction

The Chrétien government decided in 2003 not to commit the Canadian Forces to combat operations against the Hussein regime in Iraq. Instead it committed a 1900-man, partially mechanized battalion group to serve with the International Security Assistance Force (ISAF) in Kabul, Afghanistan. These decisions have once again prompted some to argue that the Canadian Forces are really a peacekeeping force, that they are not, or should not be, capable of fighting in a mid- to high-intensity war.[1] Indeed, the ISAF deployment was announced as a peacekeeping operation, not a warfighting operation. These arguments echo the sentiments made during the debate process over the *1994 Defence White Paper* back in 1993, which drew on the nondeployment of Canadian ground combat forces to the first Gulf War (1990-91). But these sorts of simplistic arguments, made in the media and academic arenas to support policy advocacy, deliberately overlook critical factual and contextual details relating to force structuring, deployment capabilities and political will.

This study examines the military forces Canada has contributed to other wars in the 1990s, the principles underlying those commitments, the force structure as it existed in the fall of 2002 and spring of 2003, and what options existed for a possible deployment to Iraq. It illuminates critical contextual aspects of force generation and deployment as well as factors that need to be taken into account when considering any commitment of Canadian combat forces to crisis situations.

Definitions and Terminology

There are several existing and competing typologies of mission types which, in turn, inform the types of forces needed for their successful execution. This study uses the following typology, which was developed in 2000 by representatives of the Department of National Defence for a North Atlantic Treaty Organization (NATO) Small Scale Contingencies (SSC) working group.

Since the 1960s, Western analytical organizations and service doctrine shops have generated many different means of differentiating between military operations. In general, the need to define emergent types of low intensity conflict during that decade coupled with the rise in political science as a discipline produced this proliferation of descriptors.[2] Indeed, as we entered the 1990s, there was a rush to expand the term "peacekeeping" beyond its originally intended definition(s).[3]

The current Canadian model is inadequate. It posits three states: peace, conflict and war. Conflict is the largest band and the various types of war remain vague. It does not, for example, take into account the variety of operations conducted by naval forces, some of which are extremely nuanced.[4] A newer and more discriminate typology based on an older Canadian typology appears to provide a better means to achieve the aim.[5]

This project, therefore, differentiates "war" from "situations short of war." The war category first distinguishes between high- and mid-intensity war: high-intensity war is used to categorize conflicts in which there exists the possibility of the use of weapons of mass destruction (WMD) between identifiable nation-states either during conventional combat or independent of it. Here, WMD refers to nuclear, biological or chemical weapons, as opposed to electromagnetic-pulse and high-energy weapons or forms of computer warfare.

Mid-intensity warfare includes conflicts in which there is combat between identifiable conventional military forces operating as representatives of recognized nation-states operating in an environment where the possibility of WMD use does not exist.

Again, in all of these cases the emphasis is on traditional conflict but does not rule out nontraditional methods operating within the context of the larger conflict. Clearly, nontraditional methods can augment traditional methods but are not necessarily dominant.

"Situations short of war" may consist of traditional conflict or nontraditional conflict, or traditional means or nontraditional means. In addition, situations short of war may consist of governmental responses to traditional or non-traditional conflict. In these cases, the concept of "traditional" generally refers to nation-state interaction, while "nontraditional" encompasses conflict overlapping with and beyond it. Situations short of war include the following:

- *Asymmetric support operations*, which include military assistance and training, the use of special operations forces (SOF), and military and/or non-military support to an ally or third party
- *Interventions*, which may include international counterterrorism, non-combatant evacuation operations, nonpermissive humanitarian assistance, and Chapter 7 UN operations
- *Peace operations,* which include peace observation, sanctions enforcement, interpositional peacekeeping, nonlinear peacekeeping, support to peace observation and interpositionary operations, and support to non-governmental organizations (NGOs)
- *Humanitarian assistance*, which includes humanitarian and NGO support operations conducted within a permissive environment
- *Diplomatic manœuvre*, which is the use of military forces to apply pressure, send signals, bolster support and/or exert presence, all without resorting to actually employing the military forces for combat purposes
- *Domestic operations*, which are conducted by sovereign states within their jurisdictional boundaries. They may involve a variety of military activities that break down into "aid of the civil power" operations (where force is used) and "assistance to the civil authority" operations (where force is not used)

Both war and situations short of war use military forces for political means. The nonpolitical use of force or violence is purely criminal and a judicial matter; it is in the domestic operations category and is therefore not included here in detail.

As with any typology, new forms and novel combinations of types conspire to generate exceptions to the rule. For example, operations in Afghanistan conducted by Operation Enduring Freedom progressed from an asymmetric support operation against the Taliban regime to counterterrorism against Al Qaeda embedded within a mid-intensity war fought against forces of the Taliban regime. Since the original NATO SSC study, further clarification of mission types has produced the "stabilization operation," which lies between counter-insurgency and interpositional peacekeeping. In effect, *stabilization operations* has emerged and become doctrine. Fundamentally different from peacekeeping, stabilization operations are not based on policing a thin blue line; they are based on the willingness and ability to use military force (tanks, special forces, attack helicopters, psychological operations) to coerce well-armed belligerents to comply with political agreements and to cooperate with humanitarian organizations.

UN peacekeeping forces, outgunned by increasingly sophisticated belligerent forces and incapable of protecting themselves (with poor command, control and leadership, and a muddled mandate), could not keep the peace in the early 1990s.

Stabilization operations rectify those failings and are either NATO-led (in the case of the Implementtion Force (IFOR), Stabilization Force (SFOR) and Kosovo Force (KFOR) in the Balkans) or led by an ABCA (American-British-Canadian-Australian) member (in the case of INTEFET in East Timor, the initial Haiti force in 1994, or the Multinational Force (MNF) in Zaire in 1996). Stabilization operations are closer to 1960s-style counterinsurgency than 1950s-style peacekeeping, and they perhaps constitute a new form of warfare in which the threat of force and its careful application can shape the environment and produce stability.

The importance of term definition lies in the problem of force structure. Too many analysts and media commentators assume there is one particular force structure for a particular form of warfare. In some cases, this is correct. However, we have seen forces structured for high- and mid-intensity war adapted for use in situations short of war: the use of a mechanized brigade group to contain and defeat (without resorting to lethal violence) an armed insurgency outside of Montreal in 1990 was one example. We have also seen forces dedicated to situations short of war utilized in mid- and high-intensity war. The use of a force originally trained for domestic counterterrorism in Canada and then committed to Afghanistan against Al Qaeda within the context of Operation Enduring Freedom is one such example.

ISTAR stands for Intelligence, Surveillance and Target Acquisition. It is not only a command process that fuses information from a variety of sources (human as well as electro-optical and electronic) and provides it to a commander; it is also used to designate a physical capability, like the Coyote reconnaissance and surveillance vehicle or an unmanned aerial vehicle (UAV), and now even a unit type.[6] ERSTA ("air force-ese" for ISTAR), Electro-optical Reconnaissance Surveillance and Target Acquisition, refers to the physical capabilities, e.g., sensors attached to an aerial platform, which feed information into the ISTAR process.

PGMs are precision guided munitions, known colloquially as "smart bombs."

Canada and Wars in the 1990s

Canadians are ceaselessly inundated with the ideology that Canada doesn't fight wars, that Canadians are peacekeepers and that combat is something others do for crass economic motives. This ideology neatly taps into latent feelings of post-colonial insecurity and anti-Americanism lying dormant in the Canadian psyche. Peacekeeping is altruistic, peaceful and good. War is selfish, violent and bad.[7] This simplistic construct was operative in the media and in government throughout the fall of 2002 during the run-up to Operation Iraqi Freedom. It ignored the context of Canadian Forces employment not only historically (Boer War, First and Second World Wars) but also in the 1990s. From the end of the Cold War in 1990-

91, Canada used military forces to intervene six times. This number includes Canada's participation in three of the four major wars of the period involving ABCA and NATO powers: The Gulf War, Kosovo and Afghanistan. Canadian combat forces were also involved in no less than 12 stabilization operations. The bulk of what people know as peacekeeping, interpositional peacekeeping and peace observation missions, were mostly holdovers from Cold War-era operations in which Canada provided service support troops, or were short-term missions with limited manning. Indeed, on four occasions the Canadian Forces were involved in armed missions with combat arms forces on Canadian soil against armed insurgents. When combined, these statistics eclipse in magnitude and duration the number of occasions on which Canada conducted unarmed support or assistance. In no way can the use of Canada's military forces during this period be characterized as strictly UN peacekeeping or humanitarian assistance.[8]

What, exactly, has Canada sent overseas to fight since 1990 and how was it employed?

The Iraq War, Phase I, 1990-91

The first post-Cold War war was the Persian Gulf War of 1990-91. The Saddam Hussein regime invaded Kuwait, which threatened to destabilize the entire Persian Gulf region and with it the post-Cold War economy. The fact that the Hussein regime had weapons of mass destruction and the means to deliver them against Israel meant that the situation could escalate into Israel's use of nuclear weapons against Iraq, with dire consequences.

Initial coalition operations were designed to impose economic sanctions on Iraq. Canada sent a three-ship task group (Operation Friction) to participate in the maritime interdiction force: a helicopter-carrying destroyer, a Harpoon-equipped destroyer and an operational support ship also equipped with Sea King helicopters. As the situation deteriorated, 24 CF-18 fighters and an infantry company group to protect their base were deployed (Operation Scimitar). The initial role of the CF-18s was to provide air cover for the naval task group.[9]

A debate over how the CF-18s would be employed broke out in the Canadian political system, and in the media, during the conflict when Operation Desert Shield shifted into Operation Desert Storm. Some wanted the CF-18s to be used for defensive purposes only, while more realistic observers knew the distinction between defensive and offensive use of military forces was outdated, artificial and ridiculous in a war designed to liberate Kuwait. The main problem was that the CF-18s had a very limited air-to-ground bombing capability. This reflected a cultural problem in Air Command going back to the late 1950s: there were those who yearned for the "Snoopy versus Red Baron" age of dogfighting air-to-air combat, while there were others who understood the need for Air Command to conduct operations in support of ground manouvres (the so-called air-to-mud role). At this point, the Red Baron aficionados had ascendancy and the PGMs that NATO wanted Air Command to acquire in the early 1980s, so that the CF-18s based in Germany could contribute to the NATO Follow On Forces attack strategy, had not been purchased. In the Persian Gulf War, most of Iraq's air force retreated to Iran and was *hors de combat*, leaving virtually no effective role for the Canadian CF-18s, save the lobbing of an air-to-air missile at an Iraqi patrol boat.

A series of ground force options were explored. The predominant one, Operation Broadsword, revolved around the deployment of a complete mechanized brigade group (which consisted of a 59-tank regiment, three mechanized infantry battalions, a self-propelled artillery regiment of 24 SP guns, plus support units), based on 4 Canadian Mechanized Brigade Group located in Germany, to serve with either the British division or VII US Corps in Saudi Arabia. Operation Broadsword, however, was not mounted for a number of reasons:

> Several factors conspired to prevent [the deployment]. Most of these were related to the Canadian Forces' stagnant ability to plan and execute an expeditionary operation of greater than company or battalion size as opposed to specific political opposition to fighting on the ground. Conditioned as the Canadian Forces was to preparing for war against the Warsaw Pact in the NATO Central Region, there was only a nascent joint planning structure, little strategic lift, no mobilization capability for sustainment and high expectation for revitalization of several equipment programs cancelled after the 1989 budgetary reassessment by the Mulroney government. Scaremongering casualty predictions based on outdated estimate processes were used by a minority of bureaucrats within DND who opposed ground operations in the mistaken belief that it would compromise Canada's (mythical) peacekeeping image or tradition.[10]

Canada instead sent a field hospital (Operation Scalpel) which operated in forward areas. There was an infantry company group deployed to protect it, but absurd legalistic evolutions emanating from the judge advocate general's (JAG) office at National Defence headquarters in Ottawa interfered with the infantry company's ability to take heavy weapons with it to carry out its function. Antitank (TOW) missiles, mortars, heavy machine guns and other equipment were left in Canada; only basic small arms were taken.[11]

Haiti, 1993

The elected leader of Haiti, Jean Bertrand Aristide, was deposed by a junta led by Raoul Cedras. Canadian and American efforts, through the UN, resulted in negotiations between the parties, which in turn agreed to permit a Canadian-American construction force to land in Haiti to improve local infrastructure (Operation Cauldron). The situation was tense. As a result, a maritime command (MARCOM) task group consisting of an operational support ship, two helicopter-carrying destroyers and a destroyer escort sailed for the Caribbean to participate in "exercises" at the same time the UN Security Council resolution allowing the deployment of the infrastructure team was adopted in September 1993.

The Canadian task group was designated Operation Dialogue, which also included an alert of 1st Commando of the Canadian Airborne Regiment for deployment to Haiti.[12] The Dialogue task group remained outside Haitian territorial waters, prepared to intervene and extract Operation Cauldron personnel if necessary. The Canadian construction engineer team, aboard the landing ship USS *Harlan County*, was not permitted to land because of a demonstration conducted by hired Haitian goons. The *Harlan County* left and the Operation Dialogue force withdrew. This incident sparked the 1994 American-supported UN intervention, which toppled the Cedras junta.

Contingency plan COBRA, 1995

In the spring of 1995, the possibility that the United Nations Protection Force I and II missions in Croatia and Bosnia might collapse prompted the creation of a NATO contingency plan, Operation Determined Effort, to extract UNPROFOR if necessary. This plan involved several American, British and French divisions and a Canadian contribution. It also involved securing various ports on the Adriatic and covering the withdrawal of UN forces while under fire. Other versions of the contingency plan involved fighting into Bosnia to add some steel to the UN force and imposing peace on all belligerent forces through firepower.

Contingency Plan Cobra, the Canadian component of Determined Effort, advanced to the point where units were identified, some equipment upgrades implemented, and some training initiated for the mission. Air Command was to provide 12 C-130 Hercules transports, 2 KC-130 tankers and 18 CF-18 Hornets. Maritime Command was to deploy 2 operational support ships, 2 helicopter-carrying frigates and a helicopter-carrying destroyer, plus the 7 embarked Sea King helicopters. The army commitment shifted during planning. Initially, a mechanized brigade group was envisioned, but this was reduced to the following:[13]

- a brigade group headquarters
- a mechanized battle group of three companies and a tank squadron (Leopard C-1)
- a reconnaissance squadron

- an artillery battery (155 mm M-109 SP guns)
- a combat engineer regiment
- an enhanced field ambulance and
- a service battalion.

Planning assumed that two other units would come under command of the brigade headquarters: possibly the two Canadian Army units already committed to UNPROFOR, perhaps units provided by other NATO members.

Contingency Plan Cobra was not activated, in part due to the successful diplomacy of the Dayton Accords that produced the NATO-led Implementation Force (IFOR). The force structure for IFOR was generally based on NATO contingency planning and essentially converted an intervention force into a stabilization force. The planned Cobra force was never deployed, though Canada contributed to IFOR with 1047 personnel in a composite organization that consisted of a recce (pronounced *rek-ee,* for reconnoitre or reconnaissance) squadron, a mechanized infantry company, an engineer squadron, and a National Support Element (logistics) and National Command Element (headquarters and signals). This was called Operation Alliance.

The Iraq War, Phase II, 1991-2002

The second phase of the Iraq War lasted for 10 years. For the most part, it consisted of a series of feints and counterfeints that resembled Cold War posturing. On occasion, however, the Hussein regime made dangerous moves that threatened to escalate into war. Throughout this period, the Canadian military contributions included combat engineers, as part of the UN shield force in Kuwait, a frigate or destroyer with the maritime interception operation, intelligence support to the United Nations Special Commission (UNSCOM), medical support to the stabilization effort in the Kurdish regions on the border, and crews for airborne warning and control system (AWACS) aircraft monitoring the no-fly zones.

The most dangerous period was throughout 1998 when the frigate HMCS *Toronto* and two KC-130 tanker aircraft deployed to the Persian Gulf, added to coalition forces conducting a diplomatic manœuvre to compel the Hussein regime to back down (Operation Determination). The Canadian Forces deployed at this time were prepared to fight and indeed expected to do so. In the end, Operation Desert Fox, an air campaign, was implemented by British and American forces in the region. Canadian AWACS personnel contributed to Desert Fox.[14]

Kosovo, 1999-2000

The deterioration of the political situation in the Serbian province of Kosovo throughout 1998 brought forward concerns of further lethal ethnic violence in the Balkans. Attempts to deploy international observers to ameliorate the Milosevic regime's military and paramilitary response to an ethnic Albanian guerilla movement produced the Kosovo Verification Mission (KVM). When the Milosevic regime continued to use disproportionate levels of force against Kosovar Albanians, a major refugee crisis emerged as nearly one million Kosovar Albanians flooded into Macedonia and Albania, which in turn threatened to destabilize the region and wreck the peace in Bosnia.[15]

Operation Allied Force was implemented by NATO powers to coerce the Milosevic regime, while a series of ground options were prepared. The 78-day air campaign involved 18 Canadian CF-18 fighter-bombers (Operation Echo), many with the ability to deliver precision-guided missiles (PGMs). Unlike the Gulf War operations, Canadian air operations in Kosovo were significant and effective contributors to Operation Allied Force.[16] The primary limitation on CF-18 operations revolved around the small number of Nitehawk targeting pods needed for all-weather PGM strike operations, as not all 18 aircraft could be equipped at once. Other deficiencies lay with the inability of the CF-18s to communicate with allied forces operating the HAVE QUICK radio system and the comparatively small initial holdings of Canadian PGM stocks.[17]

Initially, there was a belief that the KVM would succeed and that an SFOR-like NATO stabilization force would conduct a permissive entry into Kosovo. This force, called KFOR, was British-led (a request was put in for a Canadian contribution, Operation Kinetic). After negotiations, Canada decided to deploy a Coyote armoured recce and surveillance squadron (18 Coyotes from the Lord Strathcona's Horse Recce Squadron) and a Griffon tactical utility helicopter unit (eight Griffons from 408 Tactical Helicopter Squadron) which was also equipped with a surveillance capability. The intention was that the Coyote and Griffon squadrons would be part of and commanded by a British brigade group, while a National Command Element and a National Support Element for Canadian-specific materials would also be deployed.[18]

When the Milosevic regime reneged on its agreements, KFOR planning shifted from a stabilization mission into a combat mission. The Canadian elements were retained: the Coyotes and Griffons were integrated into the forced entry plans. A further request was made for Canada to contribute a mechanized battle group to include tanks (Operation Kinetic+). Doctrinally, a battle group should have a squadron (16) of Leopard tanks, three companies of mechanized infantry, TOW under armour (TUA) antitank vehicles, mortars, and support units, some 1200 men in total.

A variety of bureaucratic problems and mistakes reduced the planned battle group to 600+ personnel, so the battle group for Operation Kinetic+ consisted of two mechanized infantry companies, five Leopard tanks, six TUA antitank vehicles, plus support units. It was based on 1st Battalion, Princess Patricia's Canadian Light Infantry (PPCLI) and the Lord Strathcona's Horse (Royal Canadians). By this point, negotiations and coercion produced a permissive entry into Kosovo: the recce squadron, the helicopter unit and the battle group deployed as part of a British-led multinational brigade group. There were serious problems in strategic lift: National Defence headquarters had to rent a RO-RO ship and 747 air transport to deploy the Operation Kinetic force.[19]

The Operation Kinetic mission in Kosovo was not a pure stabilization mission. There were serious concerns throughout the first year that the Milosevic regime might conduct a mechanized incursion into Kosovo to seize the resource-rich northern region and create an ethnically Serbian-only zone. The Canadian units, working alongside NATO allies, participated in Cold War Central Region-like deterrent operations on the Kosovo-Serbian provincial boundary and participated in what amounted to a scaled-down version of NATO Central Region ground defensive operations, planning to repel any incursion. Despite their small numbers, the Canadian tanks, antitank, and combat engineering resources were deemed important enough by KFOR to cover an important sector. The unmatched surveillance capability provided by the Coyotes and Griffons was an essential component of the operation as well. All of this was conducted simultaneously with stabilization operations designed to suppress a nascent Kosovar Albanian insurgency and prevent an outbreak of Serb-Albanian violence in the province.[20] The follow-on rotation in December 1999 came from the 1st Battalion, RCR; the Royal Canadian Dragoons Recce Squadron; and 430 Escadron Tactique de Helicopter.

Afghanistan, 2001-02

The radical Islamist Al Qaeda organization declared war on the Western way of life in the mid-90s. After a series of what appeared to be unconnected attacks and thwarted operations, Al Qaeda operatives mounted extremely destructive attacks on the Pentagon and World Trade Center in September 2001. Al Qaeda training, command, communications and logistics sites were located in Afghanistan, hosted and shielded by the radical Islamist Taliban regime led by Mullah Omar. An American-led coalition, Operation Enduring Freedom (OEF) was established to take the fight not only to Afghanistan but also to wherever Al Qaeda entities operated from.

The initial Canadian contribution to OEF, Operation Apollo, consisted of six ships (four frigates, a destroyer and a support ship) incrementally deployed to the Indian Ocean over a four-month period, three C-130 transport aircraft and a pair of Aurora maritime patrol aircraft. These forces, operating alongside coalition forces, were responsible for the conduct of leadership interdiction operations in the Indian Ocean to prevent Al Qaeda and Taliban leaders from escaping by sea. At the same time, OEF mounted an ambitious campaign to take down the Taliban regime and uproot those portions of the Al Qaeda network based in Afghanistan.[21]

The initial phase of the operation involved the deployment of coalition special operations forces supported by air power to conduct a series of asymmetric support operations against the Taliban, using Afghan warlords who could be bought, coerced or otherwise induced to support OEF efforts. As these operations expanded in October-November 2001, Canada contributed a portion of the JTF-2 special operations unit that was trained for counterterrorism operations. JTF-2 operators, working alongside coalition special operations forces, were employed in critical direct action and sensitive site-exploitation missions.[22]

As the campaign evolved and the Taliban started to lose control over Afghanistan, the decision was made to deploy a light infantry formation to secure a base area and an airhead from which OEF special operations forces could operate. Canada was asked to contribute to a light infantry brigade. This brigade would take over Kandahar International Airport and secure the immediate area. The Canadian contribution was a light infantry battalion (3 PPCLI) of three rifle companies and a Coyote recce troop, plus the Lord Strathcona's Horse Recce Squadron. An electronic warfare troop was deployed to conduct signals intelligence gathering. Support units were also deployed. In all, the battalion group had some 1100 personnel deployed.

The Operation Apollo battalion group rotated missions with the other units of the American brigade group: 3 PPCLI conducted stabilization operations in the communities around the airport to enhance the security of the facility; it also participated in Operation Anaconda, a sweep mission designed to catch and kill a substantial number of Taliban supporters hiding in a mountain sanctuary. Throughout early 2002, JTF-2 special operations personnel continued with direct action missions against Al Qaeda high-value targets.[23]

In terms of troop contributors exclusive of special operations forces, Canada was the only country other than the UK to commit conventional ground fighting forces to Afghanistan at this time (early to mid-2002).

The Matters of Saliency and Effectiveness

It is not enough to merely commit Canadian Forces in an alliance setting to a coalition. Some diplomats and analysts suggest that it is enough that "Canada is there," that some form of low-risk or minimalist participation in operations should be pursued as policy. This is fallacious reasoning and ignores the realities of alliance politics and national interests. Canada's strategic traditions include the principle of saliency.

Compared to the nations that will lead alliance or coalition operations, Canada cannot commit large numbers of personnel to military endeavours. Instead, the forces that Canada can contribute must be able to provide a unique capability that no other nation can bring to the table, or use the forces committed in an unorthodox or unusual way, or be prepared to accept missions that no other nation in the coalition would accept. Indeed, the Canadian commitment must be militarily significant and effective in the schema of the operation. If it is not, it will be ignored. There will be no operational influence exerted within the alliance or coalition command structure, which in turn can lead to the misuse of the Canadian contribution by other members.

Without saliency, Canadian diplomatic personnel in the coalition or alliance will be unable to claim that Canada is making an effective contribution, and therefore their influence in policy or strategic realms far from the battlefield may be impaired.[24]

The previous cases are illustrative. The forces deployed to the first phase of the war with Iraq were not salient. There were plenty of other coalition aircraft that were multirole and not hindered by the absurd "defensive versus offensive" debate. The naval component was able to develop some effectiveness through the persistence of local Canadian commanders, but the main effort of the war was on the ground. With no effective Canadian participation, there was no Canadian influence. Canadian interests in the region were overshadowed by the allies who helped liberate Kuwait.

The Kosovo war, however, encompassed the exact opposite state of affairs. Canadian air and ground forces fulfilled every principle of saliency: the air component was relevant, and it was used vigorously. On the ground, no other nation had a capability like the Coyote vehicle. The British lacked utility transport helicopters, and the Griffon unit, which also had a "hip pocket" surveillance capability, provided unique capabilities to the KFOR effort.

In Afghanistan, the pressing need for always scarce special operations forces meant that the JTF-2's contribution was salient, while the Coyotes in the 3 PPCLI battalion group once again provided a capability that no one else possessed. The infantry unit, already trained and equipped for light operations, backfilled stretched American resources and was employed in ways that highlighted Canadian capabilities. The Canadian electronic warfare unit even wound up running the coalition EW effort in the Kandahar region because its equipment was superior to the American equipment.

The framework

It is clear from the examples that since 1990 the leadership of the Canadian Forces has learned that it is capable of contributing salient and effective forces to alliance and coalition war efforts in geographical areas as diverse as the Balkans and Southwest Asia. Could Canada have done so for the third phase of the Iraq War, Operation Iraqi Freedom? A detailed answer to that question rests on the answers to several more:

- What was the nature of Phase III of the Iraqi war expected to be, and what was it actually?
- What forces did the allies send to achieve their objectives?
- What Canadian Forces capabilities existed at the time?
- What CF capabilities were available at the time for use in Iraq?
- What options did Canada have?
- Would those capabilities have been salient and effective?

Phase III of the Iraq war, like Phase I in 1990-91, was expected to be a high-intensity war; i.e., one conducted using all available conventional means supported by unconventional means and fought against the backdrop of the possible use of weapons of mass destruction by one or both sides. It was expected from the outset that mechanized and air-mobile forces would manœuvre after preliminary special operations (to include various forms of information operations as well as SOF direct action and special reconnaissance), and air operations had been initiated to shape the battlefield. The terrain would include desert as well as complex built-up areas in the south, with more mountainous complex terrain in the north. Coalition operations were presumed to take place under an umbrella of air dominance. The rapid seizure and protection of vital resources, particularly petroleum production facilities, was critical. It was equally evident to any trained observer that the collapse of main force Iraqi resistance and the damage done to the infrastructure would necessitate the immediate shift to some form of stabilization operation in preparation for reconstruction and reconstitution of the country.[25]

The ground forces allocated by the United States, the United Kingdom and Australia to carry out Operation Iraqi Freedom (OIF) have been incorrectly categorized as purely mechanized, armour-heavy forces. Actually, the OIF coalition mix was diverse.[26] In the south, operating from Kuwait were the following:

- Three division-sized formations: 1st Marine Division, consisting of three mechanized infantry regiments equipped with LAV-25 and LVTP-7 vehicles, an artillery regiment and two M-1A1 tank battalions. Task Force Tarawa had a mechanized infantry regiment, two Marine Expeditionary Units (battalion-sized light infantry) and two M-1A1 tank companies. III MEF also had the 1st (UK) Armoured Division. A composite formation, 1st (UK) Armoured Division had three brigades: 7th Armoured Brigade with

Challenger tanks and Warrior MICV's; 16 Air Assault Brigade, an air-mobile formation which also had AH-64 attack helicopters; and 3rd Commando, Royal Marines, which had two British light infantry battalions and the 15th Marine Expeditionary Unit, a light infantry battalion, attached as its third manœuvre unit. III MEF also had the 3rd Marine Air Wing, equipped with Harrier AV-8B fighter bombers, Cobra helicopter gunships and F/A-18 fighter bombers

- V (US) Corps: V Corps had two divisions and a brigade. 3rd (US) Infantry Division was a standard American mechanized division with three brigades, 270 M-1A1 tanks, 200 M-2 and M-3 MICVs and two regiments (six battalions) of AH-64 attack helicopters. 101st Airborne Division (air-mobile) with its three light infantry brigades and organic helicopter units equipped with AH-64, UH-60 and CH-47 machines. A brigade from the 82nd Airborne Division, with three battalions of light infantry trained for parachute and air-mobile operations. 4th (US) Infantry Division, scheduled to deploy to northern Iraq, was shifted to Kuwait late in the planning. It resembled 3rd (US) Infantry Division in structure, but was unavailable for initial combat operations
- Combined Joint Special Operations Task Force 20 (CJSOTF-20): the British Special Air Service Regiment, the Australian Special Air Service Regiment and the 5th Special Forces Group (three battalions). Two ranger battalions and an independent airborne battalion from the 82nd Airborne were also included. Elements of the Joint Special Operations Command (JSOC) (made up of NAVDEVGROUP, the Combat Applications Group, and their associated USAF and US Army air support elements) are assumed to have been embedded into CJSOTF-20 for the apprehension of high-value targets and other direct action missions

Operating from northern Iraq alongside Kurdish forces were the

- 172nd Airborne Brigade: three light infantry battalions trained in parachute operations
- 26th Marine Expeditionary Unit: a light infantry battalion
- 10th Special Forces Group (three battalions)

All coalition formations were supported by the following air resources from the Royal Air Force, Royal Australian Air Force, US Air Force, US Navy and US Marine Corps[27]:

Close support
18 X Harrier GR.7 (UK)

4 X Jaguar (UK)
70 X AV-8B Harrier
30 X A-10 Thunderbolt II
8 X AC-130 Spectre
Strike
12 X F-117 Nighthawk
30 X Tornado GR.4 (UK)
Fighter
236 X F/A-18 Hornet (USN/USMC)
14 X F/A-18 Hornet (RAAF)
56 X F-14 Tomcat
42 X F-15C Eagle
48 X F-15E Eagle
60 X F-16 Falcon
71 X F-16CJ Falcon
14 X Tornado F.3 (UK)
Bomber
11 X B-1B Lancer
4 X B-2 Spirit
28 X B-52 Stratofortress

Canadian Capabilities and Options

Canadian Forces capabilities as they existed in late 2002 and early 2003 and their applicability to Operation Iraqi Freedom can be broken down by element.

Maritime forces

The CF has 18 Aurora maritime patrol aircraft; 3 Arcturus unarmed patrol aircraft; 4 helicopter-carrying destroyers (DDH), each capable of carrying 2 Sea King helicopters; 12 helicopter-carrying frigates (FFH) each capable of carrying 1 Sea King helicopter; 2 operational support ships (AOR), and 12 maritime coastal defence vessels.[28] Between 3 and 6 of the destroyers and frigates, 1 AOR and 2 of the Auroras were engaged in Operation Enduring Freedom in the Indian Ocean. The MCDVs and Arcturus patrol aircraft are really sovereignty tools and unsuited for operations overseas.

Manning problems ensured that at least one-third of the existing ships were tied up in port, unable to deploy. Maritime forces would have had no saliency in Operation Iraqi Freedom, though their effectiveness elsewhere is without question, as other operations like the Multinational Interception Force (MIF) in the Indian Ocean and Operation Enduring Freedom missions demonstrate; but maritime forces alone cannot be decisive against a land power. The only system that could have been used over land was the Aurora, perhaps in a reconnaissance role, but there were faster, less vulnerable resources available.[29] If, however, there was some form of asymmetric threat to the sea lines of communications to the Iraqi theatre of operations, then Canada's maritime forces could have played an important role. As far as can be determined, there were no contingency plans to do so.

If Canadian ships had been capable of firing Tomahawk cruise missiles, in the same way British submarines HMS *Splendid* and HMS *Turbulent* contributed to supporting Operation Iraqi Freedom,[30] the saliency level might have been raised somewhat. It still would not have been as salient as the commitment of ground forces and, as in the 1990-91 phase of the Iraqi war, it most likely would have been seen by other coalition members as a "coat-holding" exercise.

Air forces

Potentially deployable Air Command forces at end 2002 consisted of CF-18 Hornet fighter and fighter-bomber aircraft and CH-146 Griffon utility tactical helicopters. There were four CF-18 squadrons (425, 433, 416 and 441),[31] each with an estimated UE (Unit Equipment) of 18 aircraft per squadron, for a total of 72. At least two of these squadrons have North American Aerospace Defence (NORAD) taskings and must remain in Canada, leaving two which could be deployed. Though they can be equipped with the Nitehawk pod for PGM delivery, it is probable that not all aircraft in the deployable inventory have the capability. Similarly, not all CF-18 pilots had attended the Air-to-Ground School, or were trained to use the Nitehawk pods themselves, or trained in night vision goggle work or qualified for air-to-air refueling. It is possible that one squadron of 18 Hornets could have been made available for service in Operation Iraqi Freedom, assuming that the top third of each squadron's pilot pool had the required skills and experience for PGM delivery.

The saliency aspects of CF-18 Hornet employment would have revolved around how aggressive the Air Command leadership would have been in pushing for Canadian participation in high-profile PGM delivery missions in the context of the Operation Iraqi Freedom air support campaign. As Operation Allied Force in Kosovo demonstrated, a high operational tempo might have made up for the smaller numbers of Canadian aircraft deployed relative to American and British numbers. Indeed, a Canadian Hornet commitment might have resembled the Royal Australian Air Force commitment in numbers.

As for tactical helicopters, there are five Griffon tactical helicopter squadrons which support land operations (400, 408, 427, 430 and 438 Squadrons), each with an estimated UE of 12 aircraft. In addition, there are four combat support squadrons equipped with Griffons (444, 439, 414 and 417).[32] The combat support squadrons have a lower UE and tend to support localized search-and-rescue and base activities. Like the Hornets, some Griffons can be temporarily modified to augment their primary role. The ERSTA program is designed to enhance the Griffon's long-range all-weather surveillance, target and acquisition capability. How many Griffons could have been modified with ERSTA in time for Iraqi Freedom is unclear, but the capability was successfully employed during Operation Kinetic in Kosovo where the Kosovo Rotary Wing Aerial Unit operated closely with the army's Coyote surveillance vehicles.

The comparatively limited capabilities of the Griffon vis-à-vis the UH-60 Blackhawk and AH-64 Apache would have, perhaps, forced Canada to consider deploying to support the US Marine Corps (which operates a similar machine) or to support British forces as it did in Kosovo. The Griffon's ERSTA enhancements would have provided more "eyes" on the battlefield, something there is never enough of, and creative couplings with ground-based ISTAR units might have provided a salient capability. Additional tactical helicopter transports might have been welcome in the American air-mobile units, but the lift limitations of the Griffon would probably have militated against that mission.

Land forces

The Canadian Army regular component in late 2002 consisted of three brigade groups and several independent units. Each brigade group had approximately one tank squadron with Leopard C-2 medium tanks; one Coyote surveillance squadron and one Coyote cavalry squadron (Coyote minus the surveillance package); two LAV-III equipped mechanized infantry battalions, each with a platoon of TUA antitank vehicles; one light infantry battalion; and an artillery regiment, some of which were equipped with the 105 mm Light Gun, other with M-109A1 155 mm self-propelled guns or a mix of batteries. There are, of course, support units: one combat engineer regiment, one field ambulance, one MP platoon, one service battalion, and one command and signals squadron per brigade group.[33]

Practical totals for combat arms units

1 X tank regiment
1 X Coyote recce and surveillance regiment
1 X Coyote cavalry regiment
6 X mechanized battalions
3 X light infantry battalions
3 X artillery regiments (mixed batteries M-109 and LG-1)

Independent units include 2 (electronic warfare) Squadron; a test bed nuclear, biological and chemical recce detection unit with plans to equip it with modified Coyotes; an air defence regiment; and an engineer support regiment.[34]

The intractable problems with determining where, exactly, Canada's reserve forces fit into the land force structure and concept of operations outside of providing individual soldiers to augment regular force units had not been solved by late 2002 and remain to be solved, if they can be. Consequently, the deployment of formed militia units or subunits would not have been seriously considered for Operation Iraqi Freedom. Obviously, if this Gordian knot had been dealt with as mandated, it is quite possible that substantial and perhaps unique capabilities could have been deployed to Iraq.

Despite the seeming formidability of the Canadian land force structure, not all of the ground units would have been able to deploy in spring 2003. Existing ongoing deployments, for example, would have eaten into the numbers. Every six months, a new battle group based on either a mechanized or light infantry battalion must rotate through the NATO-led Stabilization Force (SFOR) commitment in Bosnia. Each rotation usually takes a Coyote troop or sometimes a squadron. Canada has also committed a flyover troop of LG-1 105 mm guns and a regimental headquarters to a multinational artillery support unit committed to SFOR's Multinational Division Southwest. On the Golan Heights, Canada sends 200 logistics and support troops to the United Nations Disengagement Observer Force. This composite unit is formed from individual units across Canada.

Furthermore, it is Canadian Forces' policy to designate soldiers as nondeployable for one year after any six-month deployment. This policy was established because of the mounting human cost of the high operational tempo established by the Canadian government in the 1990s, while the same government reduced the size of the army. Though the exact figures are not publicly available, large numbers of army personnel might not have been deployable in the spring of 2003 unless they had been given waivers by the deputy chief of defence staff. If we assume that three infantry battalions were tied up with the SFOR commitment (one in place, one preparing to deploy, one resting from having been deployed) and that 3 PPCLI was resting after participating in Operation Enduring Freedom in Afghanistan, this would leave five available infantry battalions, two light and three mechanized. One of the Coyote squadrons, perhaps two, would also have been in rest mode, leaving four available squadrons. The tank regiment and at least two of the three artillery regiments (albeit with mixed batteries) would have been available.

However, there was another commitment that could have tied up ground force units. The Canadian Army had not collectively trained at the brigade group level for the better part of a decade. The operational tempo and the emphasis on peacekeeping and stabilization training, in the main, prevented this in the early

1990s and forced training down to the battle group/battalion level. The Land Staff was preparing a brigade training event (BTE) for the summer of 2003, and it had taken years to staff and arrange resources for it. To what extent could the BTE be disrupted for a deployment to Iraq?

Canadian policy-makers would have needed to take a decision on one of three courses of action very early on. There would have been a combat operations phase, followed by a stabilization operations phase to the war. Canada could either contribute forces for both, to the combat phase only, or to the stabilization phase only. For saliency purposes, Canada would have needed to contribute saliently and effectively to the combat phase in some fashion. The mainstream options developed by Canadian planners for the combat phase of Operation Iraqi Freedom probably consisted of the following:

- A mechanized brigade group of one tank regiment, a recce squadron, two or three mechanized battalions, and an artillery regiment
- A mechanized battle group of one mechanized infantry battalion plus a tank squadron
- A brigade group headquarters, a recce squadron plus a battle group, hoping that a nation or nations would provide two other manœuvre units
- A Coyote recce squadron
- A light infantry battalion

The problem with the first option was the inability of Canada to sustain it longer than six months, probably less if the formation started taking casualties. The lack of a battle casualty replacement system or plan, and the lack of ear-marked reserve units or equipment for them, would have posed serious problems. Option two was more realistic. However, there is a question of saliency. What would have been unique about another mechanized battle group? How would it have been employed in ways different from other allied mechanized battle groups? It would in all likelihood have been submerged in the mass of other units. Option three would have had more saliency and would also have permitted Canada to have staff representation at the divisional level above the brigade, thus allowing for operational influence. It was also more sustainable than the first option. Let us, however, examine some less mainstream and perhaps more creative force options, based on what other allies provided and which might have afforded opportunities to Canada.

1st (UK) Armoured Division

The composite structure of this British division could have included Canadian Forces in at least four areas. First, a mechanized battle group with tanks could have put additional steel into 3rd Commando Brigade, which only had two light infantry battalions. Second, one or two Canadian light infantry battalions plus Griffon helicopters could have joined 16 Air Assault Brigade. Third, a mechanized battle group could have augmented 7th Armoured Brigade. Fourth, some form of composite ISTAR unit consisting of a Coyote recce and surveillance squadron, an ERSTA-equipped Griffon squadron and an electronic warfare squadron could have contributed significantly in many ways to divisional operations. Canada had an established track record working with British formations in Bosnia and Kosovo, but at the time of Operation Iraqi Freedom the relationship between the two armies was at a low point because of the perception in Canadian planning circles that Britain and the Europeans roughly spurned Canadian participation in the International Security Assistance Force in Kabul.[35]

III Marine Expeditionary Force

A composite ISTAR regiment, consisting of two or more Coyote squadrons (surveillance equipped and cavalry variant equipped), plus ERSTA Griffons, and an EW squadron could have been used as a corps-level resource. Canada, however, has no experience in dealing doctrinally or operationally with the US Marine Corps, an organization that has a tendency to remain as self-contained as possible. Additionally, the marines conducted essentially a flank guard operation and were dependent to a certain extent on the US army logistics system.

Task Force Tarawa

Canada could have provided the third mechanized battalion, LAV-III equipped, in the second regiment, plus a Coyote recce squadron. The same limitations as option 2 still apply.

The airborne and air-mobile forces

The brigade of the 82nd Airborne deployed to Iraq could have accepted a Canadian light infantry battalion, a light gun battery and Griffon helicopters. Similarly, the 101st Airborne Division, a brigade of which had already incorporated 3 PPCLI battalion group into it during Operation Enduring Freedom, could have taken in one or two light infantry battalions, a light gun battery and Griffon helicopters. Indeed, the 101st would have welcomed working with a Canadian light infantry battalion during Operation Iraqi Freedom.[36]

Northern option

The American forces operating in northern Iraq were spread very thin and could have used reinforcement. A mechanized battalion with LAV-IIIs, plus a Coyote squadron, or a light infantry battalion plus a light gun battery or regiment could have worked with the 173rd Airborne Brigade. Alternatively, a Canadian brigade group headquarters with a recce squadron, a light gun battery or regiment, a light or a mechanized infantry battalion or both could have taken the 26th Marine Expeditionary Unit under command. Having the 173rd Airborne Brigade and a Canadian brigade headquarters would have justified the deployment of a division headquarters, which Canada would have had some significant representation in. On the downside, there would have been problems sustaining a northern option.

V (US) Corps, specifically 1st (US) Armored Division

A composite ISTAR regiment, consisting of two or more Coyote squadrons (surveillance equipped and cavalry variant equipped), plus ERSTA Griffons, and an EW squadron could have been used as a corps-level resource. A Canadian contribution would have been welcomed by the US Army, which has a close doctrinal and working relationship with the Canadian Army, one far closer than between the Canadian Army and the US Marine Corps. This would probably have been the best option for Canadian participation.

It should be recognized that the ability of the Canadian Forces to support these options would have rested on having access to certain allied service support structures. Specifically, Canada would have been hard pressed to provide in-depth medical capabilities. Commonality of ammunition would also have been a factor: for example, the 25 mm ammunition used by LAV-IIIs and Coyotes would have been similar to that used by US Marine Corps LAV-25 vehicles.

Which of the above options would have provided Canada with visibility within the coalition, while at the same time contributing in a militarily effective manner? Generally, the northern option or any of the options that involve contributions to some higher formation's ability to collect intelligence and targeting information would have achieved both criteria. We must not ignore the probability that forces engaged in executing the other less visible options would have acquitted themselves well on the battlefield and may have generated saliency through their sheer valour.

What of the stabilization phase? As demonstrated by Operation Kinetic in Kosovo and Operation Apollo in Afghanistan, all Canadian units are flexible and experienced enough to downshift from combat operations and conduct stabilization operations. Indeed, all Canadian equipment is compatible for such a downshift, including, as we have seen in Kosovo, tanks. In addition to the existing options for the combat phase of the war, certain capabilities would have to be brought in from Canada for the stabilization phase for all units and formations concerned. Four capabilities come to mind. First, some form of civil-military cooperation (CIMIC) capability would have to be brought in to augment the ad hoc measures that the deployed units would have already been implementing.

Second, counterintelligence and human intelligence gathering detachments would have been critical to feed the ISTAR system established to monitor the defined area of operations. Third, additional electronic warfare resources would have to be brought in. Fourth, a military information team would have to deploy to handle the psychological operations (PSYOPS) tasks critical to any stabilization effort.

We must, of course, address the problem of strategic lift. Canada's clapped-out C-130 Hercules fleet plus the five Polaris airliners would not have been enough. Alternate service delivery (ASD) aircraft would have had to be rented. Furthermore, Canada has no strategic sealift capability, other than the infamously seized (and rusting) GTS *Katie* RO-RO vessel. Again, ASD shipping would have had to be found. This leads us to the matter of when to commit. As Operation Broadsword in 1990 demonstrated, and Operation Kinetic in 1999 confirmed, if Canada does not commit early enough, Canada will be outbid in several critical areas, which severely complicates or even prevents deployment.

Other coalition members will buy up Antonov heavy airlift and RO-RO shipping: these are scarce resources. Other nations will have made paper commitments to get personnel onto coalition staffs early so they can influence planning. Prime sites for support and logistics facilities with good access will be quickly gobbled up by others. Air ramp space, always at a premium in the Gulf region, will be occupied by those who commit early. Access to critical intelligence will be given to those who commit early. Indeed, those who seek to bureaucratically thwart a Canadian deployment because they disagree with it in principle have many means at their disposal to do so and can therefore influence a decision before it is even defined.

Equipment effectiveness

The question of equipment effectiveness is a morass influenced by sensitive issues related to the acquisition process, domestic political factors in production, and what limitations are accepted on the equipment by the bureaucracy and how that relates to risk. The following equipment types were available for deployment to Iraq (these are very broad characterizations as to how Canadian equipment would have rated in the coalition):

- LAV-III Kodiak infantry fighting vehicle: as effective as infantry fighting vehicles deployed by the US Marine Corps; less protection than the Bradley Fighting Vehicles employed by the US Army
- Coyote reconnaissance and surveillance vehicle: no equivalent US or UK capability
- CH-146 Griffon helicopter: less capable than American helicopter types flown by the US Army; as capable as some utility aircraft flown by the US

Marine Corps; as capable as most equivalent aircraft flown by the UK, when equipped with ERSTA package, with the exception of TOW-armed Lynx (note that Canada has no equivalent to AH-64 or AH-1 attack helicopters or heavy lift machines like the CH-47)

- Leopard C-2 medium tank: not as effective as the US M-1A1 Abrams or the UK Challenger II, but would have been able to defeat most Iraqi armoured vehicles; not as "survivable" as US or UK vehicles, but more mobile than Challenger II
- TOW under armour: as capable as equivalent US and UK vehicles
- M-109A1 self-propelled gun: less effective than upgraded US M-109A6 and UK AS-90 vehicles (note that Canada lacks any equivalent to the US and UK MLRS rocket systems, which complements tube artillery)
- LG-1 towed gun: as effective as equivalent US and UK systems
- Electronic warfare capability: better than US and UK systems
- Unmanned aerial vehicles: Canada lacks an equivalent capability to US Predator or UK Phoenix systems[37]
- CF-18 Hornet: with PGM capability and upgraded communications, Canadian fighter bombers would have been as capable and effective as equivalent US systems

The special question of special operations forces

Possessing special operations forces is something relatively new for the Canadian Forces. Special operations forces, due to their elite nature, have unique qualifications for the types of personnel required, have access to specialized equipment, have higher levels of readiness and are smaller in number. They are also more expensive. Special operations forces are in a category of their own when it comes to saliency. Special operators are always in short supply. The fact that Canada's JTF-2 has emerged as a respected Tier I unit[38] and has seen considerable operational activity in Afghanistan and elsewhere would have made a Canadian special operations commitment to Iraq extremely significant.

There are some analysts, planners, and commentators who mistakenly believe that JTF-2 is some kind of reincarnated Canadian Airborne Regiment, or view it as a passing fad and seek to downgrade the potential that exists in this special unit. Such thinking is simplistic and nearsighted. JTF-2, had it been deployed to Iraq, would have been the "ace in a royal flush" of effective and salient Canadian capabilities. With its integral air support capability and a proven track record in direct action missions alongside American JSOC units, there would have been unlimited scope for JTF-2 employment in Operation Iraqi Freedom, particularly against high-value targets.

Conclusions

There can be little doubt that Canada had the military means to contribute in an effective and salient fashion to Operation Iraqi Freedom in 2003. The force structure currently deployed to Operation Athena in Kabul, Afghanistan, as part of the NATO-led International Security Assistance Force and the previous deployments to Operation Enduring Freedom in Afghanistan and Operation Kinetic in Kosovo demonstrate that the capacity and capability to deploy battle group- or battalion group-sized forces exists (with some creative thought and jury-rigging) and that Canadian soldiers are well trained. Operation Allied Force demonstrates that Air Command has learned from its previous experiences in the Persian Gulf in 1990-91 and would also have been capable of contributing air resources for combat in Iraq. The main problems of long-term sustainment remain, as do the ongoing problems of reliable strategic lift.

The questions of political will and strategic direction, however, will continue to be the deciding factors as to whether the Canadian government continues to present ad hoc solutions to ongoing problems or develops a more effective and more rational approach to force generation and deployment in pursuit of Canada's global interests.

Notes

1. Thus the original title of this paper, "Are We Really Just Peacekeepers?" The author has rapidly become the expert on Canadian ground nondeployments to fight in Iraq. See "Missed Opportunity: Operation BROADSWORD, 4 Canadian Mechanized Brigade and the Gulf War, 1990-1991" (updated version) in the *Army Doctrine and Training Bulletin* 5, No. 1 (spring 2002), 24-31. This was originally an annex for the history of 4 Canadian Mechanized Brigade, but the lack of space resulted in its omission. It was subsequently published as a separate article.

2. See Frank Kitson, *Low Intensity Operations.* (Harrisburg: Stackpole Books, 1971).

3. See Christopher Bellamy, *Knights in White Armour* (London: Hutchison Publishing, 1996) and John Hillen, *Blue Helmets* (Washington: Brassey's Publishing, 1998).

4. Land Staff, *BGL 300-000/FP-000 Canada's Army: We Stand on Guard for Thee* (Ottawa: DND, 1998).

5. CFHQ, *CFP 165: Conduct of Land Operations* (Ottawa: DND, 1967).

6. Operation ATHENA, the Canadian force deployed to Kabul, Afghanistan, includes an ISTAR squadron which is based on a Coyote squadron.

7. Sean M. Maloney, "We're Not Peacekeepers," *Maclean's,* October 21, 2002.

8. As quoted in Douglas Bland and Sean Maloney's *Campaigns for International Security* (McGill-Queen's University Press, 2004).

9. See Sean M. Maloney, *War with Iraq: Canada's Strategy in the Persian Gulf 1990-2002* (Kingston: Queen's Centre for International Relations, 2002).

10. Sean M. Maloney, *War with Iraq: Canada's Strategy in the Persian Gulf 1990-2002*, p. 8.

11. Interview with Lieutenant Colonel Michael Blanchette, Kingston, October 2001. Lt. Col. Blanchette commanded the hospital protection company.

12. DND Access To Information (ATI). 1994, "Speech Module: DND and the Situation in Haiti," (August 4, 1994), memo from the CDS to the Minister, "Operation LANCE and Operation PASSAGE: Canadian Airborne Regiment Participation."

13. ATI DND, (April 12, 1995), letter, CDS to SACEUR.

14. Sean M. Maloney, *War with Iraq: Canada's Strategy in the Persian Gulf 1990-2002.*

15. Sean M. Maloney, *Operation KINETIC: The Canadians in Kosovo 1999-2000* (forthcoming: Vanwell, 2004).

16. David Kinsman, "Canada's Air Operations in the Balkans," *The Defence Associations National Network,* Vol. 6, no. 2, winter 1999.

17. Bob Bergan brought these deficiencies to light when he presented his paper, "Balkan Rats and Balkan Bats" at the CDAI Students Conference held at RMC in Kingston, Ontario on October 25, 2003.

18. Sean M. Maloney, *Operation KINETIC: The Canadians in Kosovo 1999-2000.*

19. Sean M. Maloney, *Operation KINETIC: The Canadians in Kosovo 1999-2000.*

20. Sean M. Maloney, *Operation KINETIC: The Canadians in Kosovo 1999-2000.*

21. *http://www.forces.gc.ca/site/Newsroom/view_news_e.asp?id=490 Backgrounder*, "The Canadian Forces' contribution to the International Campaign Against Terrorism."

22. http://www.cbc.ca/news/indepth/background/wtc_jtf2.html "JTF2: Canada's super-secret commandos."

23. Presentation given by Lieutenant Colonel Pat Stogran to the author's WS 500 class, RMC Kingston, January 2003.

24. Sean M. Maloney, "The Canadian Tao of Conflict," in Bernd Horn (ed.) *Forging a Nation: Perspectives on the Canadian Military Experience* (St. Catharine's: Vanwell Publishing, 2002), pp. 271-86.

25. See Williamson Murray and Robert H. Scales, *The Iraq War: A Military History* (Cambridge: Harvard University Press, 2003) and Bing West and Ray L. Smith, *The March Up: Taking Baghdad with the 1st Marine Division* (New York: Bantam Books, 2003).

26. The ground forces order of battle depicted here is drawn from Murray and Scales, *The Iraq War*, pp. 62-70.

27. This is an abbreviated list taken from Tim Ripley, "Iraq War: By the Numbers," *Air Forces Monthly* (August 2003) pp. 26-29, and it in turn is based on an initial US Air Force "hot wash up" document that was made publicly available. I have not included all of the support aircraft and have instead focused on aircraft that shoot.

28. http://www.navy.forces.gc.ca/mspa_fleet/fleet_home_e.asp "The Fleet."

29. Note that Canadian Auroras have been used in overland recce missions, specifically over Montreal during the 1990 Oka Standoff (Operation SALON). American P-3s also conducted recce missions over Zaire during Operation ASSURANCE in support of the Canadian-led MNF. In both cases, these missions were conducted in an unopposed environment with no IADS threat. Canadian Auroras were asked to fly a recce mission over Serbian bases in Montenegro during Operation SHARP GUARD, but the on-scene commander refused since the aircraft could not protect itself.

30. "Hammer Blows from Tomahawks Precede a Daring Invasion of Al-Faw," *Warships*, May 2003, p.6.

31. *http://www.airforce.forces.gc.ca/organization3_e.htm* "Air Force Wings Across Canada" and "Canadian Forces Aircraft."

32. *http://www.airforce.forces.gc.ca/organization3_e.htm*, "Air Force Wings Across Canada" and "Canadian Forces Aircraft."

33. *http://www.army.forces.gc.ca/lf/English/7_0_1.asp* "Army Units."

34. Note that the NBCD recce unit is still in the formative stages but could have been made available if prioritized.

35. Confidential interviews.

36. Confidential interview.

37. Note that Canada is renting a UAV system for Operation ATHENA in Kabul.

38. Special operations forces are generally grouped into tiers depending on their experience and capability. A Tier I unit is capable of highly specialized counterterrorism tasks involving direct action. Other tiers include special reconnaissance or other tasks like foreign internal defence.

10

Military and Postconflict Security: Implications for American, British and Other Allied Force Planning and for Postconflict Iraq

Ann M. Fitz-Gerald

Introduction

Security sector reform (SSR) is a subject that has garnered significant attention from the development community and has crystallized into a debate that has been tackled holistically by the national governments, as well as by many multilateral and nongovernment actors.[1] While efforts to generate and build on the theory and concepts have encouraged the development of wider approaches and mindsets, the way in which these concepts have been translated into practice remains unclear at worst, and disparate at best.

It is necessary to acknowledge the wide spectrum of activities and actors involved in SSR issues, and the institutional cooperation this implies both at the international and local levels. For example, it is impossible to initiate a comprehensive military reform program without addressing legal and constitutional frameworks that ensure an acceptable degree of accountability from, and transparency of, the armed forces. Similarly, a country's defence sector cannot undergo a completely successful transformation without equal efforts being extended to the reform of the internal security forces and their civilian oversight mechanisms, such as the police and judicial systems. The plans to overthrow Saddam Hussein's regime in Iraq will result in an extensive agenda for the international community to take on, particularly with regard to Iraq's security structures — a potential vacuum which, if left unaddressed, will have a further fractious effect on the region.

While a plethora of additional humanitarian and civil protection responsibilities (which have come in the early stages of many recent military interventions) has triggered cries of "mission creep" from national military contributors, today's interventionists must be cognizant of the wider security needs of the transitioning societies to which they are deployed. This notion bridges the gap between the emergency response phases of a conflict and the wider development agenda — a gap which, if left, can have a enormous destabilizing effect on a society, and reignite the roots of the conflict quite quickly.

This paper examines security in its broader context and identifies the key actors and activities of the wider security sector. It gives an overview of the main program and policy areas for SSR and discusses the potential contribution of international military forces to each of these areas. Lastly, this paper looks at the current state of doctrine for peace support operations, and the operational training that supports contemporary approaches to peace operations, and makes recommendations as to how SSR concepts can be blended into strategic planning and policy to account for the evolving developmentalization of security.

Security Sector Reform

The agenda for international military forces has always included elements of broader security concerns, many of which are addressed in the later stages of a conflict intervention, also known as the "postconflict development" or "postconflict reconstruction" phases. For example, international military forces serving under the NATO Stabilisation Force (SFOR) mandate in Bosnia now find themselves responsible for operations such as minority returns, the settlement of internally displaced persons (IDPs), the support of mine action programs and the collection of small arms and light weapons (SALW) caches from an array of different areas in the country. More recently, in Afghanistan, British forces serving in Kabul have focused on garnering local support through the implementation of "quick impact programs," which have included community-rebuilding and local integration programs.[2]

Even in different types of interventions, such as NATO's Kosovo Force (KFOR) and the British Army's ongoing operations in Northern Ireland, a focus on the need for civil protection for minority groups dominates the agenda. Both cases have parallels with experiences in Haiti, El Salvador and other internal crises where internal policing became the priority, and the international military forces strove to reach an endstate where the police could regain primacy. This will be important in Iraq. Although events in Northern Ireland have progressed to a level where the Royal Ulster Constabulary has held primacy since the early 1990s, the military still plays a support role to boost the effectiveness of the overall security sector and works toward achieving wider development goals.

Due to the complex and multidisciplinary nature of the subject, SSR has recently been identified as one of the most important challenges facing a range of government ministries and agencies. Chris Smith, of King's College, London University, suggests that the academic history behind the evolution of SSR is one of fits and starts, lacking continuity and lucidity.[3] It has emerged with the wider examination of military institutions in the Third World and with the study of defence diplomacy or the development of democratic market-based defence forces. Important linkages have also been drawn between defence expenditures and economic development in these countries, which is often seen in bloated defence budgets and the absence of any civilian oversight governing the activity of the military.

The subject further evolved when the relationship between defence forces and internal security forces was examined, and it was recognized that in many countries there was little difference between the military, what constituted internal police forces, and systems underpinning justice and the rule of law. A wider security community became identified, which — in addition to the police, the armed forces and the judiciary — included border guards, civil defence forces, intelligence services and paramilitaries. Professor Robin Luckham, from the Institute for Development Studies, Sussex University, describes SSR as

> the quintessential governance issue. This is so both in the sense that there is enormous potential for the misallocation of resources and also because a security sector out of control can have an enormous impact on governance — indeed, be a source of malgovernance.[4]

Other academics and practitioners typologize the actors involved in the wider security sector, which include those statutory and nonstatutory security services authorized to use force, civil society actors and oversight mechanisms, and nonstatutory forces that are not authorized to use force on any occasion. Nicole Ball describes the "wider security family" as including

- the security forces (armed forces, police, paramilitary and intelligence services)
- the relevant ministries and offices within the executive branch charged with managing and monitoring the security forces (such as the ministries of defence, finance, internal and foreign affairs, national security councils, and budget and audit offices)
- informal security forces

- the judiciary and the correction system
- parliamentary oversight committees
- private security firms
- civil and political society[5]

Because this paper will focus on the contribution of the military to SSR activities, a framework will be developed that is slightly more focused than Ball's list of actors and draws a distinction between the players — the enablers/disablers and the activities/programs — which, collectively, draw on most of Ball's descriptors listed above.

Figure 1 outlines a framework that describes the central players within a country's security sector along the horizontal axis. These include the intelligence services, the armed forces, the police, the judiciary, paramilitaries, and border and customs officials. A number of policy imperatives, which can serve as "enablers" or "disablers," are listed along the vertical axis, all of which taken together should underpin the activities of each of the security-related agencies listed below. These include accountable and transparent legislative frameworks; civilian oversight; capacity, structure and capability; and policy and budgetary planning. The application of professional competencies to some of these combined areas results in bilateral and multilateral policy being developed to address the more prominent problems inherent in some of these categories. Examples of this may include the development of programs and policies on United Nations (UN) peacekeeping, disarmament, demobilization and reintegration (DDR), small arms and light weapons (SALW), democratic governance and civilian policing.

It is important to note that these program areas can respond to the needs of a number of security agencies and their respective enablers/disablers,

Figure 1

Combined Areas for Program and Policy Development[6]

	Security sector actors				
Program and policy areas	**Military**	**Police & judiciary**	**Paramilitaries**	**Intelligence**	**Border guards/customs**
Legislative framework					
Civilian oversight					
Budgetary planning					
Capacity, structure & capability					
Policy and planning					

simultaneously, depending on the program and the way in which it is implemented. It is these program areas that this paper will focus on, and not the elements characterizing the vertical and horizontal axes of the model. Moreover, the analyses will draw linkages between each program area and study the implications for international military forces.

Security Sector Reform and Peace Support Operations

The realization that UN peacekeeping spanned a number of undefined modes of military intervention — some of which did not necessarily require a specific "peace" agreement to be in place prior to the deployment of international military forces — brought about a number of new definitions for what constituted peacekeeping in its broader sense. While the United States popularized the term "military operations other than war" (MOOTW),[7] the UK crafted the term "peace support operations" (PSO),[8] which would serve to embrace all the UN-bespoke "tools of peace" such as peacekeeping, peacemaking, peace enforcement and peace building (the details for each can be found in the UN's 1995 *Supplement to the Agenda for Peace*).[9]

Over the last decade, the UK Armed Forces have further developed doctrine and concepts dealing with the broader notion of PSO. As a result, the UK has earned itself the position of custodian for the development of NATO doctrine on PSO, and has served as an influential lead nation on these issues at UN headquarters in New York. Moreover, it has projected these ideas further afield and has played a role in doctrine development and training in African, Latin American and Southeast Asian countries.

There are essentially three types of military responses to regional war, intrastate war or internal civil strife: (1) a UN-led, UN-endorsed intervention, such as the 1992 UN Protection Force (UNPROFOR) in Bosnia and the UN Mission in Haiti (UNMIH); (2) a regional-organization-led, UN-endorsed intervention, such as NATO's Implementation Force (IFOR) in Bosnia or the Organization for Security and Co-operation in Europe's initial intervention in Kosovo; or (3) an "executive agent" or coalition-of-the-willing-led, UN-endorsed intervention (the type of intervention that the international community is seeing more of due to bureaucratic constraints to rapid reaction, mobility and the high readiness of multinational troops).

The common element to all of these interventions is the endorsement of the UN Security Council; or, in the case where the Council has, in the past, been gridlocked by the veto of any of the five permanent members, resolutions have still been passed in the UN General Assembly to sanction the efforts of others. Thus, the evolution of programs and policy on United Nations PSO requires continuous support. This will undoubtedly be facilitated by the UN's commitment to implementing recommendations for improving UN peacekeeping as articulated in the Brahimi report.[10]

Recent debate over the situation in Iraq has also prompted the UK government as well as most other European heads of state and international diplomats to reiterate the importance of international responses to international crises, which implies the need for the response to Iraq to be supported by a UN resolution if possible. Security Council failure to respond only complicates the post-conflict peace support issues. However, strengthening the expediency of political decision-making mechanisms must run parallel to a more robust capacity to respond militarily, even if this capacity is outsourced.

The concept of PSO continues to be defined in terms of level of consent and degree of impartiality of the intervening forces. When both these factors are low, the environment tends to be quite volatile, with fighting continuing and little prospect of peace in sight. This is normally characterized as a *peace enforcement* intervention, in which troops are sent into theatre under Chapter VII of the UN Charter, where consent and sometimes even a stable government are nonexistent. As the levels of consent and impartiality increase, international military forces often find themselves in a peacekeeping environment, which is typically supported by some sort of interim peace agreement. Figure 2 illustrates this argument.

However, there is often a blurred distinction about what constitutes a pure peacekeeping environment, as peace agreements can remain quite fragile and often exclude many nonstate actors that perpetuate the violence in the first place. Lastly, a state of *peacebuilding* comes when the peace agreement has gained wide acceptance, when the international community has secured a degree of local trust and confidence, and when the international military forces take a backseat to the primacy of local actors who develop new democratic roles in an effort to sustain peace.

Not surprisingly, the primary problem with this military model is that it presumes that these states exist on a continuum and that one always precedes the other. All too often we have witnessed the case where this continuum works backwards and peace processes disintegrate overnight, as was the case in Sierra Leone in May 2000 when the Revolutionary United Front (RUF) attacked Freetown and caused havoc to ongoing development activities and reform programs. The continuum-based approach therefore does not give the attention it should to backward transitional management, or forward, as the state of the security environment regresses. As SSR has traditionally been associated with the last phase on this continuum, and one that is often not reached due to the backward swings characteristic of a regressive peace or flawed intervention, a lack of attention to the wider security issues often serves to impede progress.

It is essential that doctrine and concepts on PSO further evolve to reflect the wider security challenges that can be encountered in a theatre of operations, challenges that should not necessarily be left to the peacebuilding, or postconflict reconstruction phase of the operation. Intervening military forces should be fully aware of how their activities may impact the development of a new national police force or the rebuilding of a national justice system. Similarly, knowledge acquired on small arms, light weapons, paramilitary groups and excombatants might usefully feed into the information repositories developed for Disarmament, Demobilization and Reintegration (DDR) and SALW programs, which may be more formalized following an interim peace agreement. If a government already exists, and the level of consent for the international intervention is high, the role of the peacekeepers should also contribute to strengthening civil society and educating parliamentarians, journalists and NGOs on the democratic role of the military, something which still undermines levels of public confidence in Belgrade, where a relatively stable government has now been in

Figure 2

The Peace Support Operations Continuum

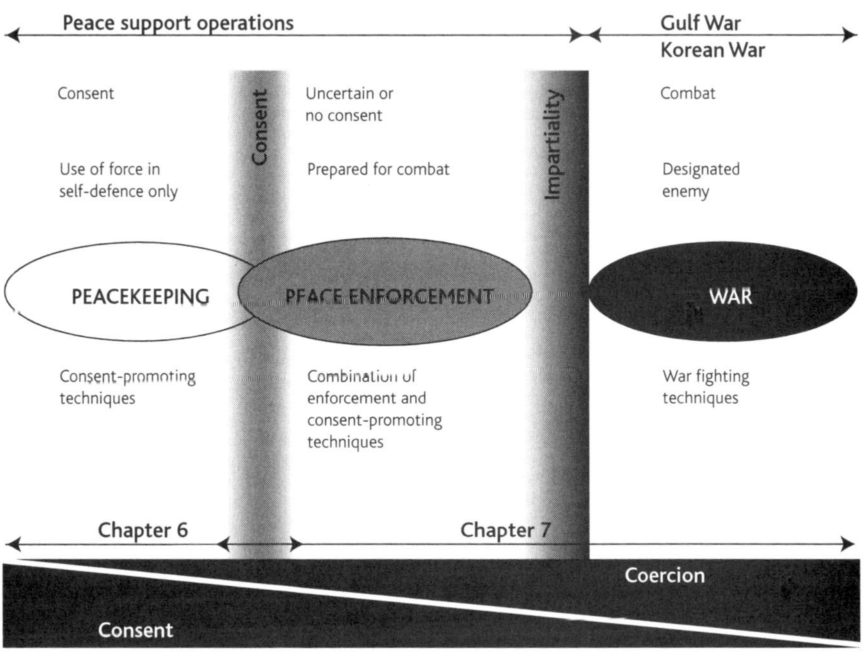

Source: United Kingdom, *Peace Support Operations*, Joint Warfare Publication 3-50, Joint Doctrine and Concept Centre, Ministry of Defence, 1998.

place for some time. These local actors play an important role in feeding government policy, advocacy, agenda-setting, educating young opinion-formers and in communicating these issues to the greater public.

Efforts to build SSR into PSO doctrine and training should extend to both NATO and UN peacekeeping countries, due to the influence both organizations exert on the policy and procedures of troops deployed under their respective organizational mandates. At the moment, large disparities exist in the approaches of different countries in implementing effective peace support activities.[11] Different levels of commitment to local dynamics become apparent, particularly in the postconflict phase of an intervention when the more professional national military contributors, normally earmarked for the more intense phases of military activity, are replaced by forces from countries that are more suited to contributing once some degree of peace and stability has been achieved and more of a monitoring and confidence-building role is required by the international military forces. All too often, many peacekeeping contributors in this "second tranche" are quite removed from the current debate on PSO doctrine and training and, as a result, are less committed during this critical stage of SSR programs. The Nigerian-led UN Assistance Mission in Sierra Leone (UNAMSIL) serves as a good example of where a lead military contributor was ill-prepared for the deterioration of a conflict and failed to respond to the wider security needs that could have prevented the resurgence of violence. For example, despite the initiation of a large-scale DDR program called for in the framework of the 1999 Lomé peace agreement,[12] Nigerian troops were unsuccessful in deterring further attacks by the RUF on Freetown, a series of which culminated in the hostage-taking of 500 UN peacekeepers. The UN and the British government immediately called for the deployment of troops to solve the hostage crisis, secure the airport and assist the Sierra Leone Army in more credibly deterring further rebel attacks.

Once again, this underlines the danger in using the continuum model, which is still imbued in the mindset of many national militaries and aid organizations, and which comes across quite vividly in the illustrations of doctrine on peace support operations. SSR has been misplaced as an activity in the postconflict-reconstruction or peace-building phases of a conflict, and is often left to the authority of development agencies and international peacekeepers, which play only a marginal, and often disinterested, support role. Military and civilian interventionists who ignore requirements to reform security sectors in the earlier phases of intervention do so at their peril, and they fail to provide more comprehensive solutions to wider security needs. Broadening their stakeholder base and awareness of interrelated activities does not suggest a shift toward "mission creep" — security is the core business of the military and SSR issues must be incorporated into their mindset, mandate and operational planning.

The following sections will examine some of the SSR program areas and the potential contribution that international military forces could make to these activities. While it is beyond the scope of this paper to address all program areas, the linkages between PSO and police reform, DDR, SALW and civil society will be discussed, albeit each in modest detail only.

Linkages between Peace Support Operations and Policing

As described in earlier sections, the transition from military to civilian policing has been a common element of many complex emergencies and military interventions, particularly in countries where there is a breakdown in the rule of law and civil protection. Generally speaking, there are three circumstances in which there is need for police reform:

- where there is a political transition and a need to transform the police forces from an oppressive to a democratic institution (e.g., South Africa and Russia)
- where there was no civil war, but where the military was the primary source of security and the police forces stood as the poor relatives with no resources (e.g., Nigeria, Uganda)
- in a postconflict situation where the police forces were either nonexistent or needed rebuilding by an external third party such as UN Civilian Police (UNCIVPOL) (e.g., Sierra Leone, Mozambique, Afghanistan)

Police reform normally involves a tripartite process, whereby international military forces are initially responsible for securing and stabilizing an environment before handing over to civil authorities. An international civilian police force, such as UNCIVPOL,[13] then assumes police primacy as a newly developed, or newly reformed, force trains under international authorities and prepares for an eventual handover of the civil protection function. Figure 3 illustrates this need for supporting mechanisms to underpin each end of the "military/police primacy spectrum." If overlooked, the resulting vacuums (shown in black) must be filled for a cooperative and amicable relationship to develop in the future.

During this process, it is absolutely critical that the initial military mandate includes an element of civil protection and does not convey the impression that the presence of international peacekeeping forces is merely to offer protection to politicians and international personnel. As a result, peacekeepers must be perceived to be offering credible security guarantees to the majority of the civilian population, or individuals and groups will go in search of security by other means. Paramilitary groups and warlords who do not have an interest in disarming and

Figure 3

The Military-Police Primacy Spectrum

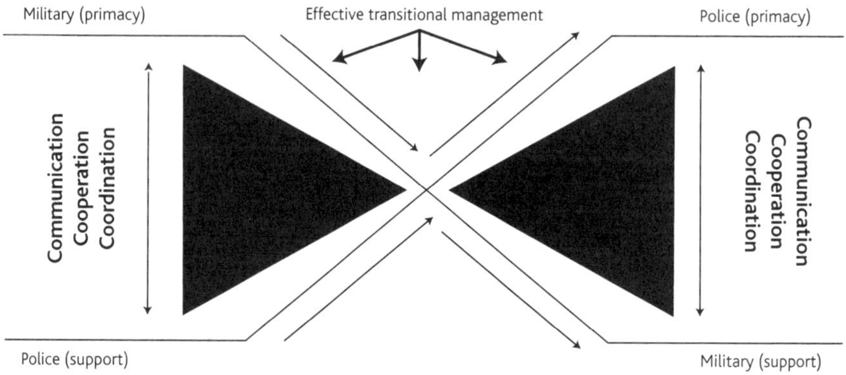

demobilizing garner the support of these local groups, who then serve as a prime target market that justifies their continued existence. This subsequently makes the job of the international police forces even more difficult, as they lack the local confidence necessary to rebuild a national police force in an environment where local acceptance provides a more promising foundation for sustainability.

As outlined in figure 3, in many cases there is a limited distinction between a country's military force and police force, and the former is usually the dominant ruler of both internal and external security. As a result, the separation of these powers becomes hugely challenging, let alone the efforts to encourage forces to work in concert with each other, as opposed to in competition with each other. Thus, it is difficult to directly draw on the templates used in Northern Ireland and Bosnia for police intervention in Africa, Southeast Asia and Iraq.[14] The former cases assume a degree of transparency between the two forces and the willingness to offer mutual support regardless of which party maintains primacy. Other entry points or, perhaps, supporting strategies, would have to be developed for different global regions where issues such as primacy and "aid to the civil power" remain unfamiliar terms.

It is therefore essential that issues concerning the rule of law and civil protection be factored into a robust international peacekeeping mandate at the earliest stages of a military intervention. This does not mean that the military should take on civilian policing activities, but it should acknowledge the end-state required for both a successful withdrawal of military forces and a well-managed transition to civilian protection for internal security. As a result, a more comprehensive approach to rebuilding the security sector can be undertaken by both internal and external actors. This underlines the importance of understanding the full security development spectrum at the earliest stages of international involvement.

Linkages between Peace Support Operations and Disarmament, Demobilization and Reintegration

As civil wars and regional conflicts continue to be fuelled by the proliferation of rebel forces, paramilitaries and breakaway armies, the need for effective and comprehensive Disarmament, Demobilization and Reintegration (DDR) programs remains critical for postconflict peace. DDR programs come with enormous social, economic and political implications and, as a result, must be integrated into the wider institutional development strategy.

Questions have been asked as to whether or not a peace agreement must be in place in order to initiate a successful DDR program.[15] In most cases, this would be a necessary prerequisite for the wider buy-in of the national government and the necessary support for local structures needed to sustain an effective DDR program, such as a national commission for DDR. Arguably, much of the relative success of the DDR program in Mozambique could be attributed to the fact that it was strongly supported and subsidized by the Mozambican government, as well as supported regionally by the South African government. As the successful transition of postapartheid events in South Africa preceded the disarming of RENAMO rebels in Mozambique, border security interests represented one of the many mutual interests between the two countries in stabilizing the region. This encouraged South Africa to provide a proactive contribution to Mozambique's DDR process.

The case of the DDR program in Mozambique also opens the debate for more regional solutions to DDR. Small arms and excombatants do not become obsolete easily, are not easily traceable and can continue to operate in vigilant and criminal networks beyond the immediate borders of their countries. For example, despite the relative success enjoyed by the National Commission for DDR in Sierra Leone, many RUF fighters are continuing their aggressive action in Liberia, Guinea and the Ivory Coast. Thus, national DDR programs must engage the many existing regional and subregional structures, such as the Economic Community for West African States (ECOWAS), the G-8 countries, the Organization for African Unity (OAU), and the New Partnership for Africa's Development (NEPAD), and encourage them to be instrumental to the wider SSR agenda for the region.

For example, the 1998 deployment of the Nigerian-led ECOMOG force (ECOWAS set up the armed monitoring group, ECOMOG) and the subsequent contribution the Nigerian government made to the UNAMSIL force serve as examples of regional solutions to regional problems. Bilateral discussions between the governments of Sierra Leone and Guinea prepared Guinean forces to meet rebel groups attempting to penetrate the Guinean border, which sent a powerful message to the RUF concerning the support to the international community offered by Sierra Leone's neighbours. Similarly, immediate efforts to implement UN sanctions on Liberia also sent out an important message and served as an effort to break ties between Liberian leader Charles Taylor and the RUF. The creation of an International Contact Group (ICG) on Liberia (which included France, Monrovia, Senegal, Britain, Nigeria, the UN and ECOWAS) also sought to create stability in the subregion and find a resolution between Liberia's government forces and the rebel groups.

Beyond the regional level were the consistent efforts at the international level to keep the conflict in Sierra Leone high on the agenda of the UN Security Council (UNSC). The problem with diamonds providing an incentive for violence, paying for weapons and fuelling the war was also mitigated by the US-government-led Kimberly Process, which sought to combat the conflict-diamond trade through the implementation of a global rough diamond certification system.

A more recent trend that has surfaced from ongoing DDR programs is the tendency to concentrate on the disarmament and demobilization parts of DDR, but not as much on the reintegration aspect. Due to the chronological nature of this equation, donor fatigue tends to have an eroding effect on the DDR programs, creating a reluctance to accept the wider community needs for successful reintegration.[16] These difficulties arise from the tendency to disaggregate the terms in DDR when, ironically, the emphasis should be on reintegration, with disarmament and demobilization seen as necessary prerequisites. Besides a loss of interest and lack of resources, funds are often borrowed from the reintegration budget for stop-gap measures during disarmament and demobilization. Lastly, DDR straddles parts of both the emergency relief and the development domains. As bilateral and multilateral donors and operational agencies increasingly pursue more coherent strategies that place them in only one of these camps, more divisions are created between what may be considered as the more emergency-related phases of the DDR program and the longer-term development activities.

Furthermore, the international community often underestimates the economic challenges for the successful reintegration of excombatants when poverty levels characteristic of postwar economies present limited opportunities for civil society. Because of the public disdain for and lack of confidence toward armed forces, there is a reluctance to consider the possibility of reintegrating excombatants into a democratic armed force as a viable source of employment. The postwar emphasis on downsizing often rules out areas of employment most compatible with the training and core competencies of the excombatants. Should the US-led military intervention in Iraq result in provisional governance by an international administration, as in Kosovo, the restructuring of the security forces, complemented by a well-orchestrated public information campaign, will be quintessential elements for consideration.

Beyond this, however, is the need to focus on the reintegration of the wider community, and not just the excombatants. More effective research, and demographic and market analyses, must underpin DDR programs in order to produce skill sets to support a country's wider development needs. For example, the infrastructural rebuilding of roads and bridges to stimulate internal economic activity and improve logistics to attract foreign investors requires engineers and labourers accustomed to working in team environments. Institutional rebuilding requirements could also produce skill set profiles that could directly influence national DDR objectives.

The wider community may also be affected by large groups of refugees who have either been internally displaced or who have no home to return to. Reintegration strategies must consider these other groups that do not form part of the pool of excombatants but face similar problems with regard to community reintegration. Families of excombatants will also require support, as will former female soldiers who often represent a significant part of a rebel force, as was the case in Eritrea and El Salvador.

Most importantly, DDR programs must be integral to the national plan and link with economic, social and political imperatives. This has important implications for strategic planning with regard to other strands of the wider security sector and the linkages between programs such as DDR and the development and training of democratic armed forces and a transparent and accountable internal security system. From the highest foreign policy objectives — which are a function of national interests and core values — should be drawn the broader implications for security planning. It is important to note that the national interests of many countries will differ from most Western templates, as other societies are built upon different sets of values. Consequently, different objectives to support a national security strategy will be developed.

A national security strategy should therefore embrace all strands of the wider security sector and should clearly articulate objectives and priorities for each strand. At the operational level that follows, implementation imperatives for each area should be specified, such as details on national DDR programs, which may then be linked back to the highest levels of a country's national strategic plan. The same is true for other SSR program areas, such as the requirement for a strategic defence review, a review of national defence expenditure, boosting the degree of civilian oversight for security agencies, and perhaps preventive measures like weapons stockpile management and legislation on arms exports. This encourages wider buy-in at all levels of government, as well as codes of transparency and accountability, which underpin how these objectives should be achieved.

The above merely touches on the extent to which PSO, strategic planning for a country's defence sector and DDR programs are so tightly linked to more comprehensive SSR strategies. The linkages create opportunities for military intervention at the strategic level as well as the operational level. For example, force commanders negotiating with local governments must be aware of how the international military intervention will support other strategic priorities in the national plan, particularly if the international military force is still the dominant actor in the transitioning society.

At the operational level, international militaries can contribute information on excombatants to local government authorities as well as the appropriate international and local commissions tasked with addressing DDR requirements. In addition, they can advise on the type of weapons the commissions could expect to collect prior to encampment and what may be required before rebel soldiers are able to renounce their combatant status. Lastly, the international military force could contribute to the security of the demobilization camps and advise on the skill sets of excombatants and their applicability in civilian employment.

Linkages between Peace Support Operations and Small Arms and Light Weapons

The linkages between SSR and small arms and light weapons (SALW) relate to much of what has already been discussed in earlier sections. It is important to acknowledge the relationship between the demand for arms and the perception of a threat and, therefore, the associations between SALW and areas such as policing, international peacekeeping, judicial reform and the role of civil society.

Once again, the common denominator for all of these areas is the place of SALW in the national strategic plan. Implementation objectives and priority areas for security planning across the sectors have wide-reaching implications for SALW programs. For example, the need for legislation on central procurement can serve to specify procurement needs for national defence and internal security purposes. Similarly, requirements for weapons stockpiling and inventory management should also underpin the objectives of all the security service portfolios and be a priority for external actors tasked with assisting in military training and DDR programs.

International military forces can also contribute to the planning of SALW programs. As mentioned in earlier sections, UN peacekeepers are often on the ground before any thought has been given to SSR. With a broader understanding of comprehensive SSR requirements and the activities scheduled to ensue following the departure of these forces, peacekeeping troops should be able to provide critical knowledge to other incoming external actors, particularly on excombatants and armaments. One often assumes that lists documenting the location and details of arms, land mines and rebel forces are readily available. While this has been the case in a very small number of postwar societies, it is certainly not the case for most countries.

Due to the earlier interventions of NATO forces in Bosnia, as well as improved relations between the Bosnian entity armed forces and the international military forces, Stabilisation Force (SFOR) now plays a vital role in mine action programs. The same logic could be used to increase the contribution international peacekeeping forces could make to SSR programs, simply by sharing information on rebel and paramilitary forces, as well as the location of arms caches, types of weapons used, weapons trafficking channels, etc.

In their role as postconflict military training advisers (as with the International Military Assistance Training Teams, which operate in different African and Central and Eastern European countries), knowledge of the military's contribution to SSR and wider development strategies could be communicated to the newly trained forces. This information should also be channelled toward civil society groups in order to boost confidence in the newly formed military force.

Linkages between Peace Support Operations and Civil Society

As described in the framework outlined in figure 1, as well as in Nicole Ball's list of members of the wider security family, civil society plays an enormously important role in the development of democratic security forces and SSR. Following the end of a civil conflict, particularly when it involves serious human rights abuses by the country's military and internal security forces, the public perception of security forces in general is quite negative. The long-lived nature of these forces in many countries prevents civil society from understanding or accepting any other use for the military.

In societies where remnants of elitist and oppressive military regimes still haunt postwar ministries of defence and headquarters of general staffs, parliamentary oversight becomes key to ensuring an acceptable degree of transparency and democracy within military politics. Such is currently the case in Serbia, where civil servants and former generals of Slobodan Milosevic's regime still hold prominent positions within the Federal Republic of Yugoslavia's Ministry of Defence. It is essential that these government representatives be presented with well-informed debates by parliamentarians to ensure political accountability to the wider electorate. Unfortunately, newly elected members of Parliament are often ill informed on the essentials of democratic security forces and lack the ability to pose challenging debates. In Belgrade, the fact that the perceived status of a member of Parliament is directly proportional to one's role in the corporate world precludes any commitment to current debates, particularly in more remote portfolios such as defence and security.[17] Similarly, news reporters and editors formally controlled by information monopolies of the oppressive regimes lack the skills required to conduct investigative, as opposed to sensationalized, journalism.

These problems open opportunities for international nongovernmental agencies (NGOs) to provide support to local actors to strengthen their knowledge and skills and provide a more informed level of oversight to the elected officials. While the role for international military forces here is more marginal than in other SSR-related program areas, the forces can certainly help to inform NGOs and media groups about the activities of newly trained forces, newsworthy information, and areas requiring further research and empirical studies that could be fed into parliamentary questioning. Furthermore, they could engage their own defence ministries to fund capacity-building programs in-country and advise on the democratic control of armed forces and the management of defence resources in market-based environments.

The Need for Regional Solutions

Earlier sections highlighted the need to engage existing regional mechanisms to provide regional solutions for postconflict and transitioning states. Also identified was the need for local constituents to develop national plans, in order to secure wider local buy-in to the solutions proposed.

It is essential to underpin both of these imperatives with clear direction from the international community, or the particular bilateral donors who wish to see themselves engaged in a specific area. More explicitly, a country supporting the postconflict rebuilding of another country must be able to state, categorically, why it has a national interest to do so. All too often, national militaries and development agencies are deployed overseas with no clear idea of the strategic interest supporting their mandate. It is just as important for the private soldier on the ground as it is for a Western nation's head of state to fully understand why their country is assisting in the effort. If it is simply to fill a widespread humanitarian/security vacuum, intervention agents should still understand the longer-term objectives and wider development agenda.

This theory must also be applied to supplementary support given to these countries, which often falls under the well-known label of "outreach programs."

For a long time now, many different bilateral and multilateral programs have come to the assistance of postconflict and transitioning countries by offering foreign military training and courses on peacekeeping. One must sometimes question whether or not this is indeed what these countries are really in need of, and whether or not courses and training on the professionalization of armed forces might better serve their requirements. More blatantly, what some countries might really desire is training on how to be a professional platoon or company commander, or on how to run an accountable and transparent ministry of defence. After all, what good is the most effectively trained army if it cannot be controlled?

Conclusions and Recommendations

International and national military forces intervening in civil wars, collapsed states and transitioning societies must consider the wider security agenda in their overall mandate, campaign planning, doctrine and training curriculum. SSR requires comprehensive solutions and the joined-up government[18] efforts of bilateral donors that respond to these problems. In addition, what was traditionally an issue for the postconflict reconstruction group of players must now be seen as an issue for the conflict prevention and conflict phases of an intervention. For example, international development agencies and embassy staff should concentrate on preventive mechanisms such as better arms export control, legislation on customs and excise, the state of a country's armed forces and other vulnerabilities that could trigger the collapse of the security sector.

During the conflict phases of an intervention, when international military forces occupy the most prominent position on the ground, measures must be put in place to consider existing security structures and information gathered on the wider security sector, to ensure both better transitional management from military to civilian primacy and a sustainable development agenda. This paper has taken a cursory look at certain SSR program areas such as police reform, DDR, SALW and civil society, and discussed how the work of international military forces can usefully contribute to these program areas.

As we live in the shadow of an American-led war on Iraq, the requirement for postconflict peace and support to the security sector remains large, and looming questions must be addressed and married with any proposed strategy to install an interim government or an international custodial administration. Even in Kosovo, which seems now to have become a long forgotten land for which no one is interested in taking on responsibility (including Serbia), internal security and civil protection of minority groups, as well as democratic governance, still remain the biggest obstacles to a sustainable peace. Weighed against the postconflict security challenges following a military engagement in Iraq, Kosovo's problems seem negligible.

The following section will outline specific recommendations to enhance the military's contribution to SSR at the multilateral and bilateral levels.

Recommendations

1. It is essential that formal mechanisms be developed within states (particularly donor countries) for joined-up decision-making between the relevant ministries that have an interest in SSR issues. This would clearly involve the participation of the ministries of defence, international development, foreign affairs, the interior, and the intelligence services or intelligence and critical infrastructure. SSR steering committees also should be created across these ministries, as should a common pool of funding to which each of the ministries contributes. The funding should be used to support policy development on wider security issues and capacity-building initiatives like in-country training and education in regions where the government has a national interest.
2. The development of in-depth assessment tools and methodologies. To improve information and knowledge management throughout the relevant ministries, in-depth assessment/analytical tools and methodologies should be produced for country/regional surveys and scoping studies. Other subject experts such as historians and social anthropologists should be brought in to add to these studies to ensure that there is a well-developed information repository for all areas to which the country may be asked to respond.
3. National defence ministries should be encouraged to build SSR concepts into their respective doctrine on peace support operations to recognize the wider spectrum of security-related activities and the linkages to international military mandates. The concepts should also be factored into campaign planning and into identifying lines of activity that in the past have included issues like humanitarian assistance and support for refugee programs. Defence ministries should build SSR concepts into their operational training courses and modules, particularly those that prepare new units for rotational deployments to different operational theatres. These courses should include briefings on the relationships and linkages between PSO and other SSR program activities, as well as the notion of "security development" beyond military endstates. They should ensure that the concepts of transitional management and change management become integral parts of management modules at the armed forces command and staff colleges. Module managers should spend time applying these concepts to a more comprehensive understanding of SSR imperatives in PSO.

4. UN military advisers to permanent diplomatic missions at the UN headquarters in New York should help stimulate the awareness of the relationship between SSR concepts and UN peacekeeping doctrines, particularly within the UN Department of Peacekeeping Operations.
5. More detailed scoping studies and research should be undertaken, as a follow-up to this paper, on the role of international peacekeeping forces in SSR programs and the contribution these forces can make to DDR and SALW programs. As well, research should be undertaken that looks at the development of national strategic planning tools that work to the benefit of the intervention agencies and donor governments, as well as to the benefit of the local government. This will encourage the wider buy-in of local governments to SSR programs and articulate separate and interrelated objectives for each of the security portfolios. For external actors, a national strategy will encourage their efforts to work toward a wider development agenda that has already garnered the support of local constituencies.

Notes

1. The extent to which a country's security sector requires reform depends on whether or not the country is undergoing significant political, social or economic transitions or whether or not it is considered a "postconflict" state. In the latter case, SSR is a much better reference, due to need for widespread reform. In the former case, Security Sector Transformation (SST) is increasingly being used. For the sake of consistency, this paper will use "SSR" to refer to all contingencies.

2. Based on discussions with Captain Ian Richardson (RN), Assistant Director, Joint Doctrine and Concepts Centre, UK MOD, July 2002.

3. Smith (2001, pp. 5-17).

4. Robin Luckham, IDS, University of Sussex, May 1998.

5. Ball (2001, 2001).

6. An SSR planning tool developed in conjunction with Colonel Witek Nowosielski, UK Defence Attaché, Belgrade.

7. See *US Army Field Manual* (2001, sec. 1-30).

8. See UK Joint Warfare Publication 3-50, *Peace Support Operations,* United Kingdom Ministry of Defence Joint Doctrine and Concept Centre, p. 4-2, 1998.

9. See Boutros-Ghali (1995).

10. In response to calls put forward by several UN members for a major revision of United Nations peacekeeping operations, to be submitted to the organization's Millennium Summit in New York, Algerian foreign minister Lakhdar Brahimi was commissioned to draft what has become known as "The Brahimi Report." The report details recommendations on how the UN's capacity to undertake more effective peacekeeping operations can be strengthened in the future.

11. See Fitz-Gerald (2002).

12. The UN's Special Representative in Sierra Leone continued diplomatic efforts to eventually bring all parties to sign an agreement at Lomé, Togo under which rebels would receive posts in government and assurances that they would not be prosecuted for war crimes. The agreement also called for a more expanded role for the UN Observer Mission in Sierra Leone (UNOMSIL). This resulted in the deployment of the UN Mission in Sierra Leone (UNAMSIL), which was mandated to implement the Lomé peace agreement and assist with the DDR plan. In the 10 months that followed, efforts to keep the peace and rebuild the country's infrastructure were interrupted periodically by further rebel activity and unrest.

13. The involvement of the Royal Canadian Mounted Police in Bosnia and the Royal Ulster Constabulary in Kosovo provides another example of bilateral, as opposed to multilateral, international police intervention.

14. Based on discussions with Inspector-General Keith Biddle, Sierra Leone Police Force, 30 January 2003.

15. Julia Taft (2002).

16. Based on discussions with Ambassador James Jonah, Senior Fellow at the Ralph Bunche Institute, New York, and former finance minister, Sierra Leone, 1998-2001, 16 December 2002, New York.

17. Based on discussions with Miroslav Filipovic, Head of the Defence and Security Committee, Federal Republic of Yugoslavia's Parliament, Belgrade, 26 November 2002.

18. "Joined-up" government relates to the Labour government's policy of promoting coordination of the various elements of government activities, in all their phases.

References

Ball, Nicole. 2001. "Transforming Security Sectors: The World Bank and IMF Approaches." *Journal of Conflict, Security and Development* 1, no. 1: 45-66.

Boutros-Ghali, Boutros. 1995. *Supplement to the Agenda for Peace.* New York: UN Publications.

Fitz-Gerald, Ann. M. 2002. "Multinational Land Force Interoperability: Meeting the Challenge of Different Cultural Backgrounds in Chapter VI Peace Support Operations." *Choices* 8, no. 3 (August).

Smith, Chris. 2001. "Security Sector Reform: Developmental Breakthrough of Institutional Re-engineering?" *Journal of Conflict, Security and Development* 1, no. 1: 5-17.

Taft, Julia. 2002. "From Promise to Practice: Strengthening UN Capacities for the Prevention of Violent Conflicts." IPA-UNDP workshop, *A Framework for Lasting Disarmament, Demobilisation and Reintegration of Ex-Combatants in Crisis Situations,* German House, UN, New York, 12-13 December.

United Kingdom. 1998. *Peace Support Operations,* Joint Warfare Publication 3-50. Joint Doctrine and Concept Centre, Ministry of Defence.

United States. 2001. *Operations,* US Army Field Manual, Department of Defense, publication FM 3-0.

11

On Being an Ally: Why Virtue Is Not Reward Enough

Thomas S. Axworthy

The American moralist Ralph Waldo Emerson advised that "virtue is its own reward," a sentiment heartily endorsed by many Canadians.[1] At its best, Canadian moralizing reflects a generous value system of sharing, tolerance, civility and diversity that we enjoy at home and wish others to have abroad. At its worst, it is the hectoring of others who believe in their own virtues and see no need to adopt ours. Dean Acheson, whose family came from Toronto and who worked summers in northern Ontario before becoming the greatest American secretary of state in the twentieth century, gave vent to this feeling when he once admonished Canada as the "stern daughter of the Voice of God."

My argument is that virtue is not reward enough. Feeling good about oneself is less important than doing well, and to do well in this difficult, complicated world you need capacity more than rhetoric. Canada has been underinvesting in the tools of international policy for a generation, and we are now in danger of becoming the Uriah Heep of the international community: promising great things but delivering little. Lest you think I am being unduly harsh, last year the Centre for Global Development ranked the 21 richest nations on how their trade, aid, environmental and peacekeeping policies helped or hurt poor nations — Canada was ranked 18th.

This self-delusion would not matter much except to our self respect, but it impinges greatly on the critical subject which is the focus of this conference: how best to deal with the United States. Canadians have had experience in dealing with Americans for over 400 years. For a time — between 1763 and 1776 — we were even loosely united with our American cousins in an early British Empire experiment of sovereignty-association. For proponents of sovereignty-association in general, or union with the United States in particular, this initial integration of North America may not be a happy precedent.

But from the first moments of Canadian history, as the Huron Nation worried about the Iroquois moving north or Champlain heard about Plymouth Rock or a young John Macdonald joined the militia to repel agitators intoxicated with the American invention of mass democracy, to the present-day topics of this conference — trade, migration and security — the United States has been a daily presence in the lives of Canadians. John Holmes, the Canadian diplomat and scholar who more than any other individual sparked my interest in foreign policy, put it best: "Coping with the fact of the USA," he wrote, "is and always has been an essential ingredient of being Canadian. It has formed us just as being an island formed Britain."

Therefore, one of the pre-eminent questions of Canadian public policy has always been how to peacefully and usefully share North America with the dynamic colossus to the south. As the smaller, weaker power, however, our overwhelming preoccupation with the United States has rarely been reciprocated. Margaret Atwood developed a wonderful image to capture this phenomenon: the 49th parallel, she said, is like a long, one-way mirror with Canadians anxiously pressed to its face watching the frenzy below while Americans careen about endlessly fascinated with each other and unconcerned with the world around them. Less politically, the historian J. Bartlett Brebner made the same point when he wrote, "Americans are benevolently ignorant about Canada, while Canadians are malevolently well informed about the United States."

As the smaller player, Canada has to be the proactive, informed, engaged partner in the relationship because it is simply a fact of life that, while the United States is central to us, we have always been peripheral to them. This is not because Americans are insensitive — though Adlai Stevenson, twice a presidential candidate, once said the technology Americans most needed was a hearing aid — it is because the United States has been a great power since the 1860s, a superpower since 1945, and a hyperpower since 1990; and we have never been a threat, only occasionally an irritant.

In short, in Canada's international policy there is the United States and then everybody else. This is so because a well-managed American relationship is central to the achievements of our basic national interests. National interests, unlike management fads or pop stars, rarely fade. In preparing these remarks, I thought back to when I first came to Ottawa in the mid-1960s, as a very junior research assistant to Walter Gordon, a former finance minister and hero of the progressive wing of the Liberal Party. Writing in *A Choice for Canada*, soon after the Cuban Missile Crisis, Gordon outlined our basic national interests:

- An expanding economy to provide good jobs for the young people who will be looking for them
- Contributing to security to avoid the threat of thermonuclear destruction
- A fair division of incomes to lift the disadvantaged from poverty
- To remain free and independent, economically and politically, "or as free as it is possible or desirable for any single nation to be in the shrinking and increasingly interdependent world in which we live"

Last year, J.L. Granatstein, one of Canada's most eminent historians and definitely not a hero to the progressive wing of the Liberal Party, outlined four goals that are remarkably similar to Gordon's:

- Canada must protect its territory, the security of its people and its unity.
- It must strive to protect and enhance its independence.
- It must promote the economic growth of the nation to support the prosperity and welfare of its people.
- It must work with like-minded states, in and outside international forums, for the protection and enhancement of democracy and freedom.

The critical point is that, in securing the Gordon-Granatstein list of national objectives, the United States is central. We need a constructive relationship with America, not to please them but to promote ourselves. And to have a constructive relationship with anyone, let alone the United States of America, you have to give as well as get. What exactly are we bringing to the party? To quote Mr. Gordon again: "Our influence depends primarily on the importance of our resources and power, both moral and material, and upon our willingness to contribute aid and peacekeeping assistance when called upon to do so." In other words, Canada needs capacity to contribute to the common good. Being an ally means that you are neither a sycophant nor a freeloader. You are, rather, a partner, and partnership means sharing the load.

Now being an ally or partner with the United States has its own special, high-maintenance characteristics. The United States has high ideals, all of which Canada shares, but we often differ on how best to promote them. Differences in means should never obscure agreement about ends. The United States is also huge and endlessly changing; and we have to understand her, not the other way around. So, Canadians have to work harder. Then there is the United States Congress. Promoters of the democratic renewal of Parliament be forewarned: Congress is the most independent and difficult legislature in the world.

We come from a parliamentary system where the American ambassador has to get to know, at most, 10 or 20 decision-makers who count. The Canadian ambassador in Washington has to get to know all 535 members of Congress because they all count. And each member of Congress has a score of assistants, and they all count too.

As an ally trying to influence these 50,000 Washington influentials, you must never threaten their definition of the national interest; you must have excellent intelligence to understand how an issue is developing before the decision gels; you can offer arguments or support for one side of the debate, if they are inclined to listen to you; and, if quiet diplomacy fails, you can try to persuade the American public, which is the ultimate influence on Congress and the administration.

To influence, you must have an audience willing to listen. Their willingness to listen depends on your goals, your capacity and your manners. Do you share the same objective, can you contribute to achieving it, and do you have the appropriate style? Mr. Pearson, for example, usually practised quiet diplomacy; he sometimes practised public diplomacy; but he never practised rude diplomacy.

My argument, therefore, is that Canada has basic national interests that have not changed much over time, that a constructive engagement with the United States is essential to achieving every one of them, and that the way to influence Washington is to have enough capacity or assets to make a meaningful contribution.

My final point is that we are not investing enough in the capacity needed to defend our national interests. Let me briefly refer to the 4 Ds: defence, development, diplomacy and democracy.

The primary responsibility of the state is to defend its citizens from harm. The privatization of war through malignant individuals gaining access to powerful weapons, plus the failed states that give terror a place to breed, is an immediate threat to the security of Canadians. Yet the Canadian armed forces have been so run down that a report from Queen's University, *Canada Without Armed Forces*, forecasts that in 5 to 10 years, as trained personal and technicians retire, Canada will be without effective military resources. If that occurs, the Government of Canada will be negligent in its first duty of protecting us from harm, and we will be reneging on the 1938 pledge of Mackenzie King that "we too have our obligations as a good friendly neighbour, and one of them is to see that…our country is made as immune from attack as possible…and that enemy forces should not be able to pursue their way, either by land, sea or air, to the United States."

Canada does not devote enough resources either to development, trade access or trade promotion. Aid is important, and here Canada's giving has fallen in half, from .54 percent of GNP in the 1970s to .25 percent today. But the best way to reduce world poverty is to give developing countries access to developed markets. This would also give Canadian consumers a break. Yet, we are as restrictive as many and more restrictive than some. We should be a G-20 leader in unilaterally reducing protection against the most disadvantaged and in wiping out debt. In trade promotion, Mexico has just signed a trade pact with Japan, and we haven't even signed a free trade agreement with Singapore that was all but ready months ago. Trade diplomacy diversifies risk and promotes growth. We should be the world's greatest free traders.

In diplomacy, Mexico has been a North American leader, with 63 consulate offices in the United States and with close fraternal ties to the huge Hispanic population in the south-western United States. Canada, too, has a large body of well-connected expatriates in the United States, but this resource has never been organized. We need, as well, a people-to-people diplomacy linking school boards, cities and states with their respective partners. Seasoned and numerous professionals should be the first line in our diplomatic efforts, but right behind we should mobilize the power of shared understanding.

Since the time of Immanuel Kant, we have known that the internal characteristics of a regime — authoritarian or democratic — have had an impact on peace and war. Many Canadians, in individual capacities, have contributed to democracy abroad, but until now we have had no organized Canadian structure to undertake the effort. We should create a Canadian democratic institute, which reports to Parliament, not the government, and uses the talents of parliamentarians and retired politicians from all parties to work on democratic governance abroad. And in Afghanistan, where Canada today has both our largest aid commitment and our largest troop deployment, we should join our partners by agreeing to stay in Afghanistan, in a meaningful capacity, by undertaking a commitment to staff a Provincial Reconstruction Team.

Talk is cheap. Capacity costs money. But the best way to enhance our partnership with the United States is to be seen making a real contribution to shared goals. And once this is recognized, temporary disagreements about means will be the normal disputes of diplomacy not national crises. The recent budget forecast surpluses in the order of $4-5 billion for years to come, with debt repayment being the goal. Instead, the budget should continue to be balanced, but any surplus should be invested — with a third applied to reducing the tax burden on Canadian business, a third to investing in domestic social and physical capital, and a third to reduce the capacity gap (the lion's share going to national security needs). In particular, the Canadian Armed Forces need steady annual increases, first to maintain existing capacity and then to expand it. Virtue, yes, but a new virtuous circle where capacity matches our goals and both are supported by the Canadian public. Robert Kennedy once said sarcastically, "Canada has given us all support, short of help." Our goal for the twenty-first century should be to ensure that no one can ever say that about us again.

Notes

1. This paper was originally presented as an address to the conference "North American Integration: Migration, Trade and Security," Institute for Research on Public Policy, Ottawa, April 1-2, 2004.

12

Between a Rock and a Soft Place: The Geopolitics of Canada-US Security Relations

Joel J. Sokolsky

Introduction

Home and away

Whatever its idiosyncrasies and inconsistencies, Canadian national security policy since the country's founding has been attuned to those larger geopolitical trends that have shaped its diplomatic and defence relations in North America and beyond. Especially after 1945, internationalism was a central tenet of Canadian foreign policy. Under the banner of collective security at the United Nations and, more importantly, as part of the North Atlantic Treaty Organization's collective defence arrangements, Ottawa was prepared to actively participate in global security affairs. This included dispatching forces overseas. But while the Canadian strategic perimeter seemed to rest anywhere in the world where Western interests were at risk, the very globalization of the threat also highlighted the domestic side of defence policy and bilateral security relations with the United States. Originating in the famous 1938 exchange between Franklin D. Roosevelt and William Lyon Mackenzie-King, through the establishment of the Permanent Joint Board on Defence in 1940, the Military Cooperation Committee of 1946 and the establishment of the North American Aerospace Defence Command (NORAD), geopolitics and bilateralism seemed to go together. In the parlance of the Pentagon, geopolitics for Ottawa is a matter of playing in "away games" and "home games."

So it was during wars both hot and cold, and so it is now in the global war on terrorism. Given the new American emphasis on homeland security, and Washington's concerns about Canada as a staging area for new terrorist attacks on the US, it might be thought that for Canada the home game could take precedence. In a recent speech, Northern Command/North American Aerospace Defence Command (NORTHCOM/NORAD) Commander, US Air Force General Ralph Eberhart, noted the mission of homeland security is like playing an "away game and a home game." In war, just as in sports, it's the away game that you "want to win," but it is the home game that you "must win."[1] Now as before, however, Washington will always look to Ottawa for some overseas contributions. When it comes to Canada-US security relations, and especially the war on terrorism, Ottawa cannot avoid playing in the away game, no matter how much it may increase its important contributions to the home game in North America. Thus in setting its national security policy, Canada must take into account the totality of American national security policy and its geopolitical implications, at home and away. As has been said of globalization,[2] so too might it be said of geopolitics: for Canada, at home and away, it all comes down to relations with the United States.

This paper looks at trends in US national security policy and the implications for Canada. It begins with an examination of the Bush doctrine, particularly as it relates to America's relations with its principal allies and the dilemmas this policy poses to both allies and the United States. It then turns to an examination of what American national security policy means for Canada-US security relations in the age of terrorism.

The reality of geopolitics for Canada is that bilateral security relations with the United States do not end at the North American strategic perimeter, wherever it may be drawn. Thus the paper next turns to consider what the implications of trends in American national security policy are likely to be for Canada both abroad and at home in the North American context. As has been evident since September 11, 2001, and especially since the beginning of the Iraq war, Ottawa is facing a United States determined to do what it takes to secure the American people. This is the hard geopolitical reality that will define bilateral security relations at home and away for years to come, regardless of which party holds the White House and Congress. The argument here, however, is that because Washington will be looking to allies for contributions abroad and needs Canada for the homeland security missions, the situation also provides an opportunity for Canada to make meaningful contributions to US national security home and away, consistent with Canadian national interests. Indeed, to this extent Canada, perhaps more than other allies, finds itself in the fortunate position of being caught between a rock and a soft place.

The Diplomacy of Dispensability, the Bush Doctrine, the RMA, and American Grand Strategy

When President George W. Bush declared in his recent State of the Union Address that "America will never seek a permission slip to defend the security of our country,"[3] he was only echoing a fundamental tenet in American foreign policy, one that reaches back to the founding of the republic. But the meaning of the reaffirmation of this bedrock principle in the aftermath of the Iraq war and the bitterness it occasioned between the United States and some of its traditional allies were unmistakable. Allies, no matter how old or new, are dispensable. America, which had championed and shaped a vast global network of security arrangements during the Cold War and which had worked to sustain those ties in the post-Cold War era, was prepared, if necessary, to turn its back on its own creations and go it alone if it had to in the war on terrorism. Others countries and international organizations were welcomed and encouraged to join, but more as followers whose services could be useful rather than as allies whose power and political support were indispensable to the pursuit of American national security interests. As the president observed, "There is a difference, however, between leading a coalition of many nations, and submitting to the objections of a few."[4]

The Democratic presidential nominee, Senator John F. Kerry, has taken the Bush administration to task for its policy of "unilateral pre-emption" that was not helping win the war on terrorism and has alienated allies. Yet, at the same time, Kerry feels compelled to stress to his audiences that he too would never let the United Nations or other countries prevent US military action "to protect our security, our people, and our vital interests."[5]

Indeed, Charles Kupchan argues that what has taken place is a profound shift away from the internationalist "center" that began with Franklin Roosevelt and continued into the post-Cold War era. This shift goes beyond the impact of September 11. It is the result of "secular changes in American politics, not just the idiosyncrasies of George W. Bush and his neo-conservative advisers, [which] are pushing the country away from liberal internationalism." He observes that "populist politics — and its aversion to submitting America's will to the preferences of others — [is] animating Democrats as well as Republicans."[6]

The Bush administration's 2002 national security strategy is explicit in its call for the use of force, including in a pre-emptive mode,[7] and the Iraq war seems to confirm this post-September 11 disposition. But as Seyom Brown has argued, "US government officials have been exhibiting a surprising willingness to use military force as an instrument of foreign policy" since before the fall of 2001. This willingness to employ the military instrument, he argues, is more than a response to the terrorist threat and the need to protect the homeland, and goes beyond the more traditional approach that restricted the use of force to situations where the lives of citizens or vital national interests are at risk. Rather, America now seems prepared to utilize its unmatched military capabilities "broadly as an instrument of diplomacy," in order "to counter a wide range of threats to the country's far-flung global interests." And such disposition, observes Brown, shows "every sign of persisting into the future."[8]

It is clear that in this post-September 11 world, US national security strategy is fundamentally driven by how seriously Americans, and their governmental leaders across the political spectrum, now view the dangers to their country and the importance of armed force in meeting those threats. The result of this is that when it comes to the use of force, allies, even old ones such as Canada, are increasingly regarded as dispensable. To a large extent, this is made possible by the vast discrepancy in military power between the US and its traditional partners as a result of the revolution in military affairs (RMA) that is transforming the American armed forces and reducing their need to rely on the contributions of others. But this is not simply a matter of favouring unilateralism over multilateralism. America still cultivates military ties and indeed has sought out new ones. All those willing to join "coalitions of the willing" are welcome. Some old and new allies can and do make a contribution to American operations, even if it is only in the provision of basing and overflight rights. But while the United States still prefers to act with others when it uses military force, Americans now have far less patience for the diplomacy of alliance and a far greater disposition to dispense with not only the military contributions of allies, but also their good will and counsel.

While the Bush administration has been explicit in its unilateralist approach, the Clinton administration was no less determined to maintain America's freedom of action — a freedom that with the fall of the USSR appeared unlimited. As Michael Mastanduno observed in 1997, the Clinton administration "followed a consistent strategy in pursuit of a clear objective — the preservation of the United States' pre-eminent global position."[9] There was, though, a difference in tone on Washington's part and, therefore, in the receptivity on the part of old allies and adversaries to America's unipolarism. Far from alienating other countries in the 1990s, the United States seemed able to maintain its traditional ties and forge new ones as it sought to engage itself across the globe, especially in Europe. This was unilateralism with a smile.

Even before September 11, the United States under the Bush administration seemed prepared to reject even the "pretense of concert" and adopt an approach to the world and its allies that was undermining the old sense of unity. It was unilateralism with an attitude. "The election of George W. Bush dramatically accelerated...go-it-alone tendencies. Bereft of its internationalist wing — Bush *père*, Henry Kissinger and Brent Scowcroft wield influence but no longer power — the Republican Party has veered towards a tough unilateralism shaped by neo-conservative ideology and the populism of the party base in the heartland."[10] As Ivo Daalder and James Lindsay have observed, from the beginning the Bush administration made it clear that it would judge the rest of the world "by American standards and not the other way around." The president was "polite but cursory" with other world leaders and showed a distinct "lack of concern for their interests and their advice."[11]

Not surprisingly, what looked like a widening in the value gap between Europeans and Americans was more of a "Bush gap," an unease among some European leaders and their populations not with America but with "Americanism" as it is expressed and practised by the current administration in Washington. The Clinton administration was no less anxious to perpetuate US dominance, but it somehow made others believe that "preserving the unipolar moment" was in everybody's interest. The Bush administration, although after September 11 the beneficiary of the greatest outpouring of sympathy for the United States since the Kennedy assassination, has managed to alienate some of America's oldest allies and their populations. "The September 11 terror attacks, far from pushing the administration back to the centre, stoked an angry vulnerability that put extremist voices in command. The moderates, Colin Powell most prominent among them, have been embattled ever since."[12] This was unilateralism with a vengeance.

Washington's interest in its new Eastern European and former Soviet Republic "partners" was an early indication of the diplomacy of dispensability, as the relative importance of the older allies declined and the US focused its attention on countries which not only bordered on Russia, but would also provide easier access for the projection of US military force into the Middle East and Persian Gulf. As the US drew down its forces in Western Europe, it began to look for new facilities farther east. And some of these countries, such as Hungary and Poland, anxious to secure US protection, would oblige. Indeed, after September 11 the United States was able to use former Soviet air bases in Kyrgyzstan, Tajikistan and Uzbekistan.[13]

In the new war on terrorism, the Bush administration adopted the language of a crusade of good versus evil, of "those for us" and "those against us," of those who loved freedom and democracy and those who did not. This rhetoric served to rally and unite the American people in support of the president's bold responses. "But it was alien to most foreigners...not accustomed to the blunt language and locutions of west Texas, many people outside the United States saw Bush's words as proof that their views did not matter."[14]

However, it is not entirely evident that this is just a temporary phenomenon that will disappear with a change in attitude (or regime) in Washington. It is easy to paint a caricature of a super-patriotic, gun-toting, bible-thumping president leading his country on a series of dangerous cavalry charges around the world to explain US actions. But as amusing as this may be, it obscures rather than clarifies the fundamentals of American policy. As Owen Harries has argued, many in Europe, and elsewhere, still do not understand the America they think they know so well. And one of the things they do not yet fully understand is the profound impact that September 11 has had on the United States and American national security policy. As he observed a year ago,

> after the outrage of September 11, I do not believe that the United States could have reacted in any way other than as she did. But doing so will carry a cost. The long-term significance of what happened...may be that it forced America decisively along a course of action that — by emphasizing her military dominance, by reducing her to use her vast power conspicuously, by making restraint and moderation virtually impossible, and by making unilateralism an increasing feature of American behaviour — is bound to generate widespread and increased criticism and hostility toward her.[15]

The Bush administration makes no apologies for the emergence of this gap. To be sure, its doctrine as elaborated formally in *The National Security Strategy of the United States of America* (September 2002) is replete with calls for international cooperation and repeatedly expresses the intention of he United States to work with friends, allies, and global and regional organizations in what is now the central concern of American foreign policy — the war on terrorism. However, far more than previous statements, this document is explicit in declaring that not only will America act alone if necessary to counter threats to its security, it will act pre-emptively. In dealing with terrorism, the US will "disrupt and destroy terrorist organizations" with particular emphasis on "identifying and destroying the threat before it reaches our borders."

> While the United States will constantly strive to enlist the support of the international community, we will not hesitate to act alone, if necessary, to exercise our right of self-defense by acting pre-emptively against such terrorists, to prevent them from doing harm against our people and our country.[16]

In the case of the threat from weapons of mass destruction (WMD) and terrorism, deterrence may be insufficient. Here an appeal is made to the international law which recognizes that "nations need not suffer an attack before they can take lawful action to defend themselves against forces that present an imminent danger of attack." The Bush administration argues

> The United States has long maintained the option of pre-emptive actions to counter a sufficient threat to our national security. The greater the threat, the greater is the risk of inaction — and the more compelling the case for taking anticipatory action to defend ourselves, even if uncertainty remains as to the time and place of the enemy's attack. To forestall such hostile acts by our adversaries, the United States will, if necessary, act pre-emptively.
>
> The United States will not use force in all cases to pre-empt emerging threats, nor should nations use pre-emption as a pretext for aggression. Yet in an age where the enemies of civilization openly and actively seek the world's most destructive technologies, the United States cannot remain idle while dangers gather.[17]

Commenting on the Bush doctrine, financier and ardent Bush critic George Soros recently argued that "September 11 introduced a discontinuity in American foreign policy in terms of the unabashed pursuit and use of American power. The abnormal, the radical and the extreme have been redefined as normal." In the past, the US has secured its dominant position "by not abusing its power."[18] This may be so in terms of the post-Second World War liberal internationalist tradition, but an argument can be made that, in reacting to the terrorist threat, the Bush administration has done nothing more than reach back toward the fundamentals of American national security policy. "America," Robert Kagan observes, "did not change on September 11. It only became more itself."[19]

Indeed, there is much in the response to the terrorist attacks that is reminiscent of what Walter McDougall has termed the "bedrock" of American foreign policy unilateralism abroad and the protection of liberty at home.[20] And historian John Lewis Gaddis, in an interview on his recent book, *Surprise, Security and the American Experience*, argued that the doctrine of pre-emption is well founded in American history. He refers to John Quincy Adam's policy after the British attack on Washington during the War of 1812, the various challenges from Native Americans, pirates, and the continued threat to the "weakly governed states" along the US borders from other European powers. "And so America achieved its security through territorial expansion — by filling a perceived vacuum before hostile powers could do so." For Gaddis such invasions were "pre-emptive." This policy was followed for most of the nineteenth and twentieth centuries. After the Second World War, with America an atomic superpower, it fell by the wayside as the United States exercised a liberal internationalist leadership that enhanced its own security and which, because it secured others, had the support and consent of allies. The immediate post-Cold War era produced no grand strategy, but in responding to the events of September 11 the Bush administration, according to Gaddis, has undertaken "a decisive and courageous reassessment of American grand strategy, emphasizing "democratization of the Middle East and the need to prevent terrorists and rogue states from getting nuclear weapons." While Gaddis thus supports the Bush doctrine, including the war in Iraq, he allows that some members of the administration did not "give enough weight to how frightening it can be if you have that much power and then you deploy it, and you deploy language foolishly."[21]

The rhetoric associated with the Bush doctrine has caused misgivings among allies, especially America's traditional partners. But it is evident that, consistent with the new diplomacy of dispensability, the administration is not overly concerned since, as the National Security Strategy points out, "In leading the campaign against terrorism, we are forging new, productive international relationships and redefining existing ones in ways that meet the challenges of the twenty-first century."[22] For Gaddis, this is a necessary and bold move that rejects the constraints of an outmoded international system that was really nothing more than a "a snapshot of the configuration of power that existed in 1945."[23] Here too, America has reached back to the foundations of its diplomacy. In his famous farewell address, George Washington, warning against entangling alliances, nonetheless said that the United States could "safely trust to temporary alliances for extraordinary emergencies."[24]

Jeffrey Record, in his scathing critique of the Iraq war, argues that the foolish language associated with the Bush doctrine is not just talk. Rather it represents an ill-conceived approach to the global war on terrorism.

> The administration has postulated a multiplicity of enemies, including rogue states; weapons of mass destruction...proliferators; terrorist organizations of global, regional and national scope, and terrorism itself. It also seems to have conflated them into a monolithic threat, and in so doing has subordinated strategic clarity to the moral clarity it strives for in foreign policy and may have set the United States on a course of open-ended and gratuitous conflict with state and non-state entities that pose no serious threat to the United States.
>
> Of particular concern has been the conflation of al-Qaeda and Saddam Hussein's Iraq as a single undifferentiated terrorist threat. This was a strategic error of the first order because it ignored critical differences between

the two in character, threat level and susceptibility to US deterrence and military action.[25]

This global approach led directly to the decision to launch a pre-emptive, indeed, preventive, war against Iraq, even though deterrence would be a more "credible policy for dealing with rogue states seeking to acquire WMD."[26] Record suggests that the reason Iraq was attacked despite the fact that North Korea is a more dangerous proliferator was simply because Washington wanted to employ the military instrument. As former British Foreign Minister Robin Cook, who resigned over the decision to go to war, said: "The truth is that the US chose to attack Iraq not because it posed a threat but because the US knew Iraq was weak and expected its military to collapse."[27]

American confidence in, and reliance upon, military power has grown in the post-Cold War era, as a result of the widening discrepancy in the size and technological level of the US armed forces versus those of its allies. This has allowed the United States, as Barry Posen has termed it, "command of the commons — command of the sea, space and air,"[28] thereby allowing Washington to project its power almost at will anywhere in the world.

In 2002, the United States accounted for 39.1 percent of all worldwide military expenditures. At nearly US$330 billion, this was more than the total combined military expenditures of the next 10 countries including Russia, China, France and the United Kingdom. The ability of the United States to spend this much is a result of its overall economic superiority. Even though at 3.3 percent the US ranks 48th in the world in terms of gross domestic product devoted to defence expenditure,[29] it produces 23 percent of "gross world product (GWP)." This gives it "more than twice as many resources under the control of a single political authority as either of the next two most potent economic powers — Japan with 7 percent of GWP and China with 10 percent."[30]

In every category of military power — land, sea, air — American forces are far larger than those of the next ranking military powers. For example, the United States Marine Corps "has as many personnel as the combined land and air forces of the United Kingdom, and the US Navy operates 40 special-purpose combat ships for amphibious operations, roughly the same number of major surface combatants as the entire Royal Navy."[31] In addition to quantity, the United States has an overwhelming lead in technological development. "In 2001 the US Department of Defense budgeted nearly as much money for military research and development as Germany and France did for their entire military efforts."[32] Since the end of the Cold War, American military technological development has focused on the RMA. If current US grand strategy draws upon the revolutionary traditions of the time of "the shot heard around the world," it is also based upon a technological revolution that is geared toward giving the American military the capacity to shoot anywhere in the world.

The RMA seeks to apply the most advanced technologies, particularly those in microprocessing, to warfare and, in so doing, transform the nature of war, how militaries fight wars and how they are organized. The RMA dramatically increases the speed, accuracy and lethality of weapons. Utilizing air- and space-based surveillance and reconnaissance, communication and command technologies the RMA enhances the "situational awareness" of deployed forces, further allowing them to strike with greater effect while reducing the fog of war. It has been argued that the RMA is well suited, if not specifically tailored, to a national security strategy that seeks to maintain American global hegemony without the need to permanently base large numbers of American forces overseas. It provides for dominance of the sea, air and space, and thus facilities the rapid projection and application of overwhelming and extremely accurate force against any possible adversary anywhere in the world. And by substituting technology for manpower, it seeks to reduce American casualties. It also seeks, through its accuracy, to reduce collateral damage to civilians and infrastructure. The RMA also lends itself to preemptive and preventive uses of force as a means of eliminating potential threats early on, and thus further enhances American dominance and hegemony.[33]

While the quantitative and qualitative superiority of American military power can reduce the need to rely upon allies (even other major military powers), in bringing force to bear against adversaries, it is also an important element in maintaining and expanding US global military ties. With its command of the air, sea and space, the United States is able to quickly come to the assistance of countries whose security is deemed important and where a common threat is faced. And in the absence of conflict, the US regional combatant commands, such as Central Command and Pacific Command, maintain regular military ties and conduct joint exercises with numerous countries as means of further establishing American influence.[34] In this sense, US military power permits the cultivation of new security relationships with partners who heretofore have not been close but who are now prepared to collaborate for specific purposes. For example, "In mid-2002, as the United States geared up for war against Iraq, the Indian navy escorted two supply ships through the Strait of Malacca between Malaysia and Indonesia."[35]

To the extent that the US can increase the number of participating governments in its coalitions of the willing, it enhances the legitimacy of its actions in the eyes of the international community. And, more importantly, the greater the number of foreign contributors, the easier it is to sell the operation at home where the Congress and the people are concerned with international legitimacy and burden sharing.

While it will be very difficult for partners to match the sophistication of the RMA-transformed US forces, this has not prevented the US from seeking and cultivating cooperation. Indeed, as part of the effort to extend and secure American influence, Washington makes great efforts to include contributions from other countries in its operations, regardless of their level of military sophistication and however these contributors wish to describe their contributions. As Secretary Rumsfeld noted with regard to the war in Afghanistan, "our policy of accepting help from any country, on a basis comfortable for its government, and allowing that country to characterize how it is helping (instead of our creating that characterization for it), is enabling us to maximize both other countries' cooperation and our effectiveness against the enemy."[36]

At the same time, it is clear that Washington will be the architect and the leader of any coalitions. And that above all, allied contributions will not yield any influence over the conduct of the war or its objectives. "Wars," Rumsfeld also notes, "should not be fought by committee. The mission must determine the coalition, the coalition must not determine the mission, or else the mission will be dumbed down to the lowest common denominator."[37] Thus while contributions are welcomed, the United States will not adjust the conduct and objectives of the mission to accommodate the particular wishes of contributors. It will be up to each allied government to explain its participation to its citizens. If allies press too hard for influence over the mission, even for the purposes of justifying participation to their publics, they may become dispensable.

The Dilemmas of Dispensability

Although the US may command the commons, threats will originate in what Posen calls the "contested zones," the littoral seas and especially the land and urban areas ashore. This is where the objectives of the RMA-transformed forces are likely to be found and where dominance of the seas, air and space will provide America with great advantages. But the advantages are not absolute, particularly since victory and follow-on stabilization in these areas will draw heavily upon ground forces. Here, allies can make a contribution by supplying not only infantry, a category in which the US is experiencing some scarcity relative to its global tasks, but also coastal maritime forces such as minesweepers.[38]

With regard to postconflict stabilization, the US may well have to turn to allies. "Americans," notes Record, "have historically displayed a view of war as a substitute for politics and the US military has seemed congenitally averse to performing operations other than war."[39] But Alexander F. Kagan sees the current US aversion to stabilization efforts as a function of the RMA transformation with its emphasis on "shock and awe" as a means of securing US interests. He points out that in both Afghanistan and now in Iraq "the US has been far less successful in winning the peace than it was in winning the war...Why has the United States been so successful in recent wars and encountered so much difficulty in securing its political aims after the shooting stopped?"

> The reason for this fact lies partly in the vision of war that President Bush and his administration brought into office and have implemented in the past two wars. This vision focuses on destroying the enemy's armed forces and his ability to command them and control them. It does not focus on the problem of achieving political objectives. The advocates of a "new American way of war,"...Rumsfeld and Bush chief among them, have attempted to simplify war into a targeting drill. They see the enemy as a target set and believe that when all or most of the targets have been hit, he will inevitably surrender and American goals will be achieved.
>
> War is not that simple, however. From the standpoint of establishing a good peace it matters a great deal how, exactly, one defeats the enemy and what the enemy's country looks like at the moment the bullets stop flying. The US has developed and implemented a method of warfare that can produce stunning military victories but does not necessarily accomplish the political goals for which the war was fought.
>
> To effect regime change, US forces must be positively in control of the enemy's territory and population as rapidly and continuously as possible. That control cannot be achieved by machines, still less by bombs. Only human beings interacting with other human beings can achieve it. The only hope for future success in the extension of politics that is war is to restore the human element to the transformation equation.[40]

There is, though, the view that the burden of postconflict stabilization and nation building can be shifted to allies, while the US military preserves its primary warfighting role. This has been called the "superpowers don't do windows" dimension of the US grand strategy.[41] With the US infantry already stretched, Posen suggests that after military success in parts of the contested zones, allied ground forces can be "useful in the post-conflict peace-enforcement missions necessary to secure the fruits of any battlefield victory."[42] Here, Washington would view allies as less dispensable and would be more willing to invest time and effort in allied diplomacy. This is what is being urged on the Bush administration by liberal internationalists in the US who are concerned about the rising cost of the operation in Iraq. And in the case of Haiti, Washington appears eager to enlist as many countries as possible in order to reduce the level and duration of the US deployment.

The diplomacy of dispensability is the result of trends in US national security policy that preceded the Bush administration and the war on terrorism and reach deeply into the heart of traditional American political culture. It is also a reflection of the present public mood. To be sure, as Kupchan notes, "The proclivities of the Bush White House are unique in important respects. Should the influence of neo-conservatives weaken, or the Democrats take back the White House, the current penchant for unfettered primacy would abate." But he also warns

> Nonetheless, it would be illusory to presume that a change of personnel or party will bring back America's liberal brand of internationalism. A weakening of internationalist leadership is taking place across the political spectrum. The generation of Americans that fought in the Second World War and oversaw the building of the West is retiring from politics, to be replaced by a generation that lacks the experience that formed the foundation of liberal internationalism. Moreover, the terrorist threat is here to stay and will continue to favour the extremes, not the moderate centre.[43]

At the same time, the so-called moderate centre has itself shifted because American unilateralists and liberal multilateralists share the view that to be secure the United States must remain militarily dominant and free to use that power alone and unfettered if necessary. To a certain extent, the differences are rhetorical, with the neoconservatives of the Bush administration simply placing more emphasis upon the need to be prepared to act unilaterally and dispense with allies if necessary, and its moderate and liberal critics calling for more diplomatic efforts to enlist allies under the American banner before US power is applied. Between the policy implications of David Frum and Richard Pearle's *End to Evil: How to Win the War on Terrorism*[44] and Zbigniew Brzezinski's *The Choice: Global Domination or Global Leadership*,[45] it is only a matter of tactics rather than strategy.

Brzezinski is highly critical of the Bush administration's "bullying behavior and its search for 'domination' over its allies." He wants a more cooperative approach with allies, but insists that everyone must let America lead: "The acceptance of American leadership by others is the *sine qua non* for avoiding chaos." As William Pfaff observes, "That doesn't seem very far off from the Bush administration position, just more courteous, and less shrill."[46] It is indeed only the difference between unilateralism with a smile and unilateralism with an attitude, or a vengeance.

Senator Kerry has said that he supports American leadership — not where the United States "hoards all the power," but rather where responses to crises are "shared" and "internationalized," and where unilateralism is renounced.[47] Yet, as noted above, he has also made clear that he would not give allies or international organizations a veto on US actions. Indeed, noted American historian and fellow on the Council of Foreign Relations, Walter Russell Mead observes that differences between Republican President George W. Bush and Democratic presidential nominee Senator John F. Kerry, "are moving toward a merge" in their basic foreign policy approaches. In a second term, "the Bush administration would try to get more foreign support, and a Kerry administration would sometimes have to go it alone."[48]

Both President Bush and Senator Kerry want to see an expanded UN presence, and would like other countries to contribute forces to help stabilize the situation, including a possible commitment by NATO. For both, it is a matter of fostering international legitimacy and relieving the burden on American forces. Yet for both, although the Democrats will not be as explicit, a greater UN and/or NATO role and the participation of other countries will be "multilateralism" in name only. With over one hundred thousand US troops likely to remain, with over three thousand "administrators" shifting to the new US embassy in Baghdad, with billions spent and with over seven hundred soldiers killed, the American government and public are not going to surrender control over the future of Iraq to any international organization.

In Iraq and in other locations, the US will act alone if necessary. As Ivo H. Daalder, who served on the Clinton National Security Council staff and is now with the Brookings Institution, argues: "The world is not going to be terribly different under a Bush and a Kerry presidency."

> The United States is the most powerful country in the world, and therefore the use of American power is going to be indispensable in getting anything done. Kerry will find, if he doesn't know it now, that in order to get anything done, whether it is through the UN or through NATO, that the US is going to have to lead...using power, using coercion.[49]

Thus even if the more pronounced unilateralist tendencies of the present administration are to be tempered by new leadership, and the United States seeks to cooperate more with allies, the basic objective of sustaining American global military hegemony would not change.

This is because the global character of the terrorist threat has only reinforced the attractiveness of a United States grand strategy that seeks to preserve hegemony through the ability to widely use American military power. Should there be substantial identifiable progress in the GWOT, for example the capture of Bin Laden or stabilization in Iraq and Afghanistan, the preservation of US hegemony will continue to provide the framework for American national security policy for years to come.

This strategy does make allies more dispensable than they have been in the past, in the sense that Washington no longer feels it must invest heavily in the diplomatic efforts necessary to accommodate the wishes of foreign governments, especially those of older allies on a long-term basis. Rather it will seek out new partners in ad hoc coalitions of the willing where needed, relying upon temporary inducement where necessary and seeking ways to do without support if it is not forthcoming. It will bypass and work around the coalitions of the worried and the wary. Thus, despite Turkey's refusal to allow American forces to invade Iraq from its territory, Washington was able to obtain support from its new allies among former Soviet republics.

While older allies, especially the major industrial democracies, may fret about this situation, they confront a fundamental dilemma. It is that while American military might has made them increasingly dispensable as allies, the power and the willingness of the United States to use it, unilaterally if it chooses, also remains indispensable to their own security and prosperity. In seeking to protect itself by sustaining its dominant military position in the world, the United States makes possible the stable international environment that all but eliminates any major challenge to the security of the West. And in so doing, it provides the basis for global trade and the whole panoply of commercial relations. For this reason, "the world's consequential powers have grudgingly supported US hegemony."[50]

Yet the situation may not be all that unattractive to some allies. Confronted with the enormity of the United States' military power, allies may "feel helpless and/or useless,"[51] while the American diplomacy of dispensability further relieves their political leaders of the incentive to keep pace quantitatively and qualitatively. Most will maintain armed forces, and may offer Washington small contributions for politically symbolic purposes. But they now have more reasons to devote more of their resources to their domestic needs, and other foreign policy instruments, while still enjoying the benefits of US military power.

In some sense it is the "Canadianization" of allied defence policy. As David Haglund and Stéphane Roussel have observed, this is the essence of "Canadian strategic culture," which has been to undertake less of an effort than its great-power partners want it to, but not so little as to be eliminated altogether from their strategic decision-making.[52] Thus the question in regard to military ties with the US becomes not *How much is enough?* but *How much is* just *enough?* It is not so much a matter of being free riders, but of being easy riders.

This highlights the dilemma of the diplomacy of dispensability for the United States. America prefers to act with others. It does seek out allied contributions as part of its policy of maintaining its dominant global position and for the purposes of establishing the legitimacy of its use of power abroad and securing support at home. But the greater its relative military power becomes, the less inclined it may be to engage in the often difficult and frustrating diplomacy of alliance. This will increase the incentive of allies to contribute very little. Washington may thus increasingly find itself bearing the burden of collective defence or acting alone when intervention is called for, while the very allies it shuns continue to reap the benefits of US military power.

The prospect of such a future may just be enough to convince American leaders, across the political spectrum, that while there may be no going back to the liberal internationalism of the twentieth century, neither is the diplomacy of dispensability an entirely wise approach to United States national security in the twenty-first.

But in the final analysis, there can be no long-run comfort for the allies in the dilemmas of dispensability. And thus if Washington is going to be persuaded to dispense with the diplomacy of dispensability, its allies must help out, including in Iraq.[53] A Vietnam-like US failure there will have reverberations throughout the world. It will further encourage terrorists and other enemies of the West and strike at the very heart of the global stability upon which the industrial democracies depend for their peace and prosperity. More importantly, it will likely engender a real unilateralism, one in which the United States, discouraged by war and angry with its allies for not coming to its assistance, will continue to exercise its great power alone, but will be more selective when it comes to dealing with regional crises that threaten others more immediately. All of this will make an already disorderly world order more chaotic, especially for those who lack the power or the will either on their own, or multilaterally, to cope with it. While the diplomacy of dispensability has its dilemmas and problems, the solutions do not rest with coalitions of the worried, the wary and the weak.

Dispensable but Not Invisible: Canadian-American Cooperation Overseas

Although much has changed with the end of the Cold War, the passing of the post-Cold War era and the advent of the war on terrorism, what Charles Doran observed in 1984 still remains valid: the United States has tended "to look at the Canada-US relationship through the lens of global politics, whereas Canada tends to look at global politics through the lens of its relations with the United States."[54] For Canada, this means that in the GWOT an unavoidable litmus test for support of the United States will be participation in American-led overseas coalitions in the "contested zones." Ottawa cannot avoid playing in the away game, no matter how much it may increase its important contributions to the home game in North America. Overseas, Canada may not always be conspicuous by its presence, but, as the Iraq war showed, it is conspicuous by its absence, to the detriment of its security relations with the United States.

Canada should be prepared to support and contribute to American-led multilateral operations in the war on terrorism. It should not do so on the expectation that military contributions will yield influence in Washington. Ottawa should be under no illusions about being able to significantly influence American decisions there. As with other allies, the United States will not alter its policies in order to secure Ottawa's participation. Canada should do so because American unilateralism, however it may from time to time be dressed up as multilateralism, offers the best chance for victory in the war on terror and world order. And Ottawa should do so because its own national interests, especially economic, rest on maintaining good relations with Washington. When it comes to real security and defence spending, the moment of truth is essentially a bilateral one.

This does not mean, however, that Canada requires a major augmentation and RMA-like transformation of the Canadian Forces in order to contribute to the GWOT and international stability. Those who make this argument ignore not only domestic realities regarding the likely size of defence budgets and the willingness of the Canadian public to support large sustained operations abroad, but also the nature of the US national security doctrine described above. In seeking support among allies for its overseas operations, Washington is primarily looking for a measure of international legitimacy as well as something to show Congress and the American people that allies are contributing and that the US is not bearing all the burden.

But it is more than this. Canadian participation in US-led operations overseas tells the Americans that Canada is not just concerned about its economic well-being, and thus is prepared to do whatever it takes to assure the US that it is not a strategic liability in order to convince Washington to keep the border open. Sending forces overseas tells the United States that Canada takes America's global security obligations seriously and sympathizes with the problems and dilemmas the United States faces in seeking to secure its own and, by extension, Western interests. United States Ambassador to Canada Paul Cellucci is correct in saying that for the United States, security trumps economics.

Given the highly symbolic, but by no means unimportant, character of Canadian contributions to US-led operations overseas, Canada is actually in a very advantageous position. It can, as the recently released *Securing an Open Society: Canada's National Security Policy* rightly stated, "be selective and strategic when it comes to the deployment of our armed forces," asking beforehand "which efforts would be of the greatest relevance to our national interests" and whether "we have the capacity to meaningfully contribute to a successful outcome";[55] in other words, as before, gauging *how much is* just *enough*.

Contrary to recent predictions about the current and even more impending meltdown of the CF,[56] Canada has, in fact, the capability to make useful contributions to American overseas operations. In a strategic environment dominated by the overwhelming RMA-transformed forces of the United States, the CF still maintains a credible overseas capability.

As already recognized, Canada must also seek to implement elements of the RMA if the CF is to serve the national security interest overseas, where interoperability with American forces is important, and at home, where the greater surveillance and situational awareness made possible by RMA technologies will be no less applicable. Any redistribution of resources within the defence budget toward modernization of existing equipment should seek to exploit the advantages of the RMA that can enhance the effectiveness of a relatively moderate-sized armed force such as Canada will always possess. Here, progress has been made, particularly with regard to maritime forces and some elements of the land force, such as the LAV IIIs and Coyotes, and any savings derived from reduction in some defence activities, such as from limiting the size and duration of overseas deployments, should be channelled into further exploitation of the RMA.

Given existing budgetary realities that will limit both expenditures on new technologies and the number of platforms that can be modernized, the CF will inevitably lag behind the United States in the RMA. However, as Eliot Cohen has argued, "transformation in one area of military affairs does not...mean the irrelevance of all others." Historically, new kinds of weapons have existed and been employed alongside older systems.[57] This is certainly the case today, including within the US-led coalitions and, indeed, within the US military itself. As is readily apparent in Afghanistan and in Iraq, winning in the contested areas, and long-run success in the GWOT, will very much depend upon being able to carry out postconflict stabilization and restoration activities as well as other forms of operations other than war. Military units are essential for this and can be quite effective even if they are not on the cutting edge of the RMA.

Within the CF the RMA is uneven, but it is being implemented. The navy has a small, but modern and interoperable, high seas fleet (supported by the air force's long-range maritime patrol aircraft), which will be enhanced in coming years by the addition of helicopters and new supports ships that will include the capability to transport troops and equipment. The Canadian Army, though it will remain small, should be able to deploy units in the range of five hundred to one thousand personnel. With its current Coyotes, light armoured vehicles and the proposed purchase of a mobile gun system to replace the Leopard tanks, it will have the kinds of capabilities that will be useful to coalition operations in a range of likely operations. This is especially true in the kinds of urban and small unit stability actions such as the CF are presently undertaking in Afghanistan.

Though the new support ships should increase Canada's ability to get forces overseas and sustain them, the CF, with a small air transport fleet and in the absence of a heavy strategic airlift capacity, will still lack the capability for rapid deployment. However, as is evident in Afghanistan and in other likely scenarios, the GWOT is going to be a long struggle and coalition operations will involve months if not years. The CF need not be Federal Express; it does not absolutely, positively, have to be there overnight to make a contribution. It does not have to match the American capabilities in this regard. Indeed, what Washington will likely be looking for from Canada is the ability to come in after US forces and assist with the more protracted, but no less essential, tasks of stabilization.

Even without heavy long-range strategic lift, the air force's fleet of C-130s should be able to provide support for forces overseas. It may not be possible, given budget realities, for the air force to acquire a new fighter for force projection roles, but with the American command of the air, such a capability will not undermine the CF's ability to maintain a presence overseas. Nor should one discount the other overseas capabilities that Canada can offer in terms of institution building, military training and aid to police forces in areas of former conflict, such as Canada is already doing in Iraq.

The point of this is simple but needs repeated re-emphasis, because so much criticism has been directed against CF capabilities and so many dire predictions have been made about what will happen to Canada's military capabilities over the next 10 years. Given the legacy capabilities that will be around for the next decade and longer, and in light of planned and feasible modernizations, the CF will be able to contribute effectively to international security operations without having to implement a radical, and essentially unattainable, wholesale American-style RMA transformation of its capabilities, doctrine or organization.

In sum, the United States will continue to welcome whatever contributions Canada can make to the overseas dimensions of the GWOT. Just because the United States will still bear the overwhelming bulk of the burden and will not surrender ultimate strategic and political control does not mean that such assistance will be invisible in Washington, as Prime Minister Martin discovered in his recent conversations with congressional leaders.[58]

North American Homeland Security and Defence: Capabilities Will Matter More Than Organizations

In the global war on terrorism and in any foreseeable future security environment, Washington will *expect* Canada to contribute to overseas operations, but it will *demand* that Canada attend to Canadian homeland security and defence. The difference for Canada is as follows: in overseas operations Ottawa will have greater flexibility in terms of missions and the level and duration of commitments, while at home there will be no choice. There is no question of invisibility in Washington, as the Americans have already taken, and continue to take, ample note of Canadian deficiencies in terms of its domestic and counterterrorism efforts.[59]

It is not anti-Americanism but realism to point out that, as in the Cold War and so too in the GWOT, the United States does not help Canada defend itself against threats to Canada; it helps Canada defend itself against threats to the United States. Canada, while sharing concern about the threat, is also wary about having to defend itself against American help. In the post-September 11 international security environment, the nature of this reality is confronting Canada's leaders as never before. Yet as Canada seeks to respond to American demands about improving its own domestic security efforts, it is very important to keep in mind that American homeland security and defence efforts are still in a state of flux, uncertainty and conceptual confusion, even as new vast bureaucracies and military commands are put into place.

In July 2002, President Bush stated that "there is an overriding and urgent mission here in America today, and that's to protect our homeland. We have been called into action, and we've got to act." The 2002 *Homeland Security Act*, which created the Department of Homeland Security (DHS), prohibits the DHS from engaging in the "military defense of the United States," and stresses that "warfighting" activities are strictly the responsibility of the Department of Defense (DOD). But as a recent study by the Lexington Institute on the whole of the National Guard pointed out, neither the *Homeland Security Act* nor the 2002 *National Strategy for Homeland Security* details the Pentagon's role in homeland security. Currently, US military doctrine attempts to make distinctions between homeland defence and homeland security and to limit its role in the latter. Homeland defence is "protecting US borders and waters against military threats," whereas homeland security is described as "detecting, preparing for, preventing, protecting against, responding to, and recovering from terrorist threats or attacks within the United States."[60]

United States Northern Command (NORTHCOM) states explicitly that "job one" is defending the homeland. But its mission statement goes further to say that its "mission is homeland defense and civil support, specifically: conduct operations to deter, prevent, and defeat threats and aggression aimed at the United States, its territories, and interests within the assigned area of responsibility; and, as directed by the President or Secretary of Defense, provide military assistance to civil authorities including consequence management operations."[61]

In drawing a distinction between defence and security, DoD officials often point to the *Posse Comitatus Act*, a 1879 statute put in place in response to abuse by Union forces in the reconstruction-era South, that generally prohibits the armed services from undertaking law enforcement tasks in the US. More generally, it is argued that a distinction can be made between law enforcement and war. Homeland security, being closer to law enforcement, must be carried out under the rule of law, with all its restrictions, while homeland defence is close to warfighting and is thus exercised under the law of war, where any measure, including pre-emption, may be taken to defend the United States. Thus, for example, if a ship is suspected of carrying terrorists intent on sneaking into the US to perform acts of terrorism, the Coast Guard and other law enforcement agencies would be used. But if a commercial vessel is believed to be carrying a cruise missile with a biological weapon, then the US Navy could be directed to simply destroy it.

But as the Lexington study observes,

> The distinctions between "defense" and "security" seem to have little practical or legal utility. Also, it is not clear that America's enemies will clearly recognize these boundaries and limit their methods and operations so that they can be countered either by the military or other federal agencies as appropriate. In fact, they may seek to use these artificial distinctions to find gaps and seams in US security that can be easily exploited...In practice, many federal agencies will play a role in countering terrorist threats. Attempting to create artificial distinctions between these missions contributes little. In short, the Defense Department's effort to draw a bright red line between it and the Department of Homeland Security is the least useful of its current initiatives.[62]

One might ask why — if the security against terrorist threats to the American homeland is so important to have resulted in such a massive reorganization of the US government, including the establishment for the first time in its history of a military command specifically for the defence of the United States and its people at home — the DOD seems intent on drawing distinctions between security and defence. Given the sweeping nature of the *USA Patriot Act*, it would seem that the law could be made to serve the interests of domestic security however the Congress deems is required.

It would appear that the deeper reason for the DoD trying to make the distinction is the fact that the US military has not fully bought into the homeland security *or* the homeland defence mission. The individual services are still overwhelmingly committed to the force projection warfighting role. This is a disposition reinforced by the fact that two overseas wars are now going on. The argument for this is that the United States, traditionally, has always sought to meet and defeat its enemies as far away from American soil as possible, a historical orientation very much reinforced by the Bush doctrine.

Although the war on terrorism began in the American homeland and led to the largest reorganization of the American government in US history to protect the republic at home, homeland security, and in particular homeland defence, is not the focus either in policy or in the allocation of human and fiscal resources to the GWOT. Indeed, there is growing concern in the US that homeland security is not being adequately funded or organized. As Jeffrey Record points out, while "Operation Iraqi Freedom and its continuing aftermath will not affect funding of the relatively cheap counterterrorist campaign against al-Qaeda...those costs most assuredly will impede funding of the woefully underfunded homeland security requirements."[63] A recent study by the RAND Corporation noted serious delays in the modernization of the United States Coast Guard necessary to meet the new demands of homeland security,[64] while another study has noted deficiencies in the training and organization of National Guard units for homeland roles.[65]

The problem here is the nature of the threat. To be sure, it makes absolute strategic sense to go after terrorist bases abroad, as was done in Afghanistan, using the global force projection might of the American military as only it can do. And the American military needs to use its command of the commons to intercept and destroy threats destined for US shores or airspace, anywhere in the world.

But even if the US is successful in capturing and killing terrorists in their overseas bases, or destroying aircraft and ships in distant locations, the American homeland would still be vulnerable to terrorist attack, either by groups and individuals who will seek to enter the United States, or, as in the case of the September 11 attacks, by the "enemy within," by those already in the US.[66] Moreover, as the Hart-Rudman Commission on National Security in the 21st Century pointed out in January 2001, an attack on the American homeland is "most likely to occur when the United States is involved in a conflict overseas."[67] Thus in the GWOT, the advantages of meeting America's enemies abroad rather than before they immediately threaten the homeland are limited. Success overseas does not remove the homeland threat, and it may actually be exacerbated as groups sympathetic to terrorists eliminated abroad seek to take the war to American soil. In addition, to the extent that overseas operations draw upon military units, such as the National Guard and Coast Guard, which would be required to come to the aid of civil authorities in a major terrorist attack, such operations may actually reduce the DoD's ability to meet its homeland roles.

In a certain sense, the global war on terrorism is closer to international policing than international war, although warfighting capabilities are needed. As Peter Andreas has pointed out, the emphasis is now not upon the traditional defence of borders against armed attack by organized military forces, but upon the policing of frontiers against "clandestine transnational actors," those nonstate actors who "operate across national borders in violation of state law to erode law enforcement efforts." In the past, these individuals have been associated with organized crime, drug trafficking and illegal immigration. The focus now is on transnational terrorists. In response, borders have not been redrawn but "recrafted," in the sense that the border control problem has involved a thickening of policing efforts that extend beyond the formal national boundary, or point of entry, to take more distant approaches. This has involved the application of capabilities and procedures, such as satellite surveillance and intelligence that had formerly been associated with the military defence of borders. The objective is to prevent the entry of terrorists, while securing the borders for the continued free movement of commerce and legitimate travelers. For America and thus for its allies this border policing has become the new "high politics" of international relations.[68]

Given its wealth and power, the United States may well still be able to afford two militaries, one for overseas operations and one for homeland security and defence. But Canada cannot. (Although, as noted, given the strain resulting from the Iraq war on National Guard units, so necessary for homeland defence and security roles, even America may be not able to afford such a division.) It is an approach that must be avoided in the setting of Canadian defence policy, especially the determination of the proper roles, missions and equipment for the CF.

Canada need not follow the US in this doctrinal confusion. Indeed, *Canada's National Security Policy* adopts a more comprehensive perspective, reaffirming the overall priority of "protecting Canada and Canadians at home" and integrating the military's role into this larger goal. It avoids the kinds of artificial distinctions that are present in US policy and thus provides a good basis for pursuing a broad-based security collaboration with the United States. It draws, moreover, on the traditional "aid to the civil power" role of the CF.[69]

Ottawa should, therefore, continue to vigorously pursue all aspects of continental security, taking the initiative where possible in order to try and shape US homeland and continental security policy in directions favourable to Canada. Work is already being done by the Binational Planning Group to explore new dimensions of bilateral cooperation. New institutional arrangements, such as an expansion of NORAD to include maritime and land surveillance, may be useful.

But capabilities will matter more than organizations. To plan for an expansion of NORAD's mandate to include maritime collaboration and cooperation in the utilization of land forces in response to emergencies that may spill over the border is meaningless, unless the capabilities exist and a measure of interoperability is achieved. If Canada is to play a greater role in continental security, and indeed if new arrangements under NORAD are to be more than subsidiaries to NORTHCOM completely dominated by the US, then Ottawa must sustain and improve the capacity of the CF to contribute to homeland security and defence.

Here too, the task should not be beyond the capacities of the Canadian Forces. For in the new era of air defence that focuses on threats from within North American airspace, the existing Canadian contribution to NORAD, including the CF-18 fleet, should suffice.

The most significant thing Canada can do to assure the US that it shares America's concerns is to agree to the integration of ballistic missile defence into NORAD. The United States has indicated it will proceed with the deployment of an initial, limited BMD system, whatever Canada decides. A Canadian decision not to give NORAD a BMD role would likely compel the United States to re-evaluate the future of the command. It might be possible to give the air-defence role to NORTHCOM and establish new cooperative measures with Canada while moving the BMD role to another US command. But the broader implications of a Canadian decision not to agree to a missile defence role for NORAD will be profound. It will signal that Canada is not prepared to compromise over an issue that is important to the United States. For the sake of its own and more immediate national interests in fostering bilateral security co-operation, Canada must simply put aside its arms control difficulties regarding the possible weaponization of space and agree to this role for NORAD.

This appears to be what is happening. Consistent with expectations, the new Martin government has moved on the issue, with the minister of national defence and US Secretary of Defense Rumsfeld exchanging letters on 15 January 2004 "stating the interest of both nations in negotiating an agreement on the ballistic missile defence of North America." After the negotiations, Ottawa will assess its position and make a decision "on whether Canadians' interests have been met and will take a decision on Canadian participation in BMD." In the meantime, an interim amendment to the NORAD agreement will give Canadian personnel access to US BMD planning information.[70] The minister of defence has also suggested that if an agreement were reached Canada might make "some of its geography in the north available in lieu of a major cash contribution," allowing the US to station "rocket launchers and radar stations as part of a continental defence program."[71]

Plans to make greater use of the Canadian Army reserve units in Canada to respond to possible terrorist attacks will further enhance domestic security and potential expansion of bilateral cooperation on the ground in the event of an emergency. At sea, the Canadian navy will remain primarily focused on overseas operations, but it too has a long tradition of providing for defence of the ocean approaches and aiding the civil powers in coastal defence, and it is moving to bring its surveillance and other capabilities to the assistance of the homeland maritime defence role. The four submarines can also provide extended coastal surveillance.

But more is required if the navy is to fully contribute to North American maritime defence and work closely with the US Coast Guard. The navy should forego replacement of the four Tribal Class destroyers in favour of building a modern coastal defence fleet with short- and longer-range vessels similar to those being built under the USCG's Deepwater project. If, as was announced on May 7, 2004, "The assertion of Canadian sovereignty in our waters always has been and will continue to be the navy's top priority,"[72] then an emphasis on coastal defence and security should be the navy's first priority in future acquisitions as well.

The capacity of the CF to contribute to the protection of the Canadian homeland, and in so doing provide the capabilities that will make an expansion of bilateral security cooperation possible, is bolstered by the myriad other measures the Canadian government has taken and will take as part of the national security policy. Indeed, in contrast to the United States, the overwhelming bulk of Canadian spending on security is going to homeland protection and law enforcement tasks. This includes everything from "smart borders" to immigration reform to enhancement of domestic intelligence capabilities to greater port security to the establishment of the new Department of Public Safety and Emergency Preparedness.

In addition, one of the factors that has made bilateral security cooperation work in the war on terrorism, and work well, is that there already existed an extensive, multifaceted network of collaboration at the bureaucratic levels between the law enforcement, intelligence, health and public safety agencies of Canada and the United States.[73] Indeed, to a certain extent, the present efforts of NORTHCOM and the Binational Planning Group to investigate the state of bilateral cooperation for consequence management purposes is an exercise in the re-discovery of what many bureaucrats and agencies on both sides of the border already know and have been practising for decades. These long-standing functional ties, which have been so important to bilateral relations, provide Canada with further advantages in making significant contributions to American homeland security.

It is not clear that the current US policy toward the GWOT accords winning the home game equal priority with winning the away game. But given geopolitical and thus bilateral realities, an approach that does emphasize the importance of the home game is one that the government in Ottawa would be well advised to continue to follow.

Conclusion

Confronting realism with realism and finding a soft place

Canadians cannot avoid the harsh geopolitical realities of how American global military power is exercised and the impact of this on bilateral security relations. This was so in the past and is even more so in the age of terrorism. But the best response to realism is realism, which, although not openly acknowledged, has always been the Canadian response. We should therefore continue to shun the "unrealism" of those who say that our independence and sovereignty, and world peace, will be at risk if Canada collaborates at home and away with the United States. Equally, however, we need to continue to avoid the unrealistic, the romantic realism, of those who predict a total meltdown in bilateral relations unless Ottawa dramatically increases military spending, fully embraces the RMA and mounts major overseas operations. Both approaches offer radical solutions that are impossible from an international relations standpoint and unsustainable domestically, because both assume that Canada is now between a rock and hard place. It is not. As such, neither approach would serve the national interest.

To be sure, we have seen the "rock" — unbridled American unilateralism, the ignoring of Ottawa's good will and counsel, and allegations that Canada is not doing enough to fight terrorism, all of which have resulted in strains in bilateral security relations. But in a world in which US hegemony and the GWOT will be the reality for years to come, we have also seen the "soft place." This is the opportunity Canada now has to continue to improve bilateral security relations, by sustaining its contributions overseas and contributions to American homeland security as no other ally can. Why is this a soft place? Because regardless of reservations Ottawa might have about how the Bush administration has pursued US national security, American leadership still provides the kind of global stability that Canada needs for its own security and prosperity.

It is also a soft place for Canada because the kind of efforts needed for national security at home will make a direct contribution to the security of the American people and their economic well-being. A Canadian approach to national security and defence that places a high priority in policy and posture on the defence of the Canadian homeland — and by extension of North America — will best serve not only Canada's vital interests but also those of the United States.

And finally it is a soft place because the military, and especially nonmilitary, security spending needed to makes contributions at home and away is well within the fiscal capacity of the government to accommodate. This is provided, of course, that the government means it when it says it will be "selective" in deploying forces overseas. There will be demands that it not be so from groups who want to impress the broader international community by taking a greater role in multinational peacekeeping and country-building efforts and from those who want to impress the United States by sending more combat forces to US-led coalitions. Both will demand larger overseas deployments and generate new demands for costly equipment. Given limited defence budgets, this will only draw resources away from more immediate and important domestic and continental defence roles.

Taken together, Canada is not confronted with insurmountable problems here. It has the capacity to contribute to the global war on terrorism both at home and away, and thus support international stability in the context of US leadership in a manner consistent with Canada's interests and values. Moreover, if Canada takes advantage of this opportunity, then bilateral security relations will be not only a soft place for Ottawa but also a comfort zone for Washington, providing the latter with an assurance of the kind of allied support Americans do not always recognize and appreciate, but which they do need.

Notes

Parts of this paper are drawn from Sokolsky (2004a). Other parts are drawn from Sokolsky (2003a) and Sokolsky (2003b). I would like to thank Ms. Sara Moller for her research assistance with these works. Parts also appeared in Sokolsky (2004b). Earlier versions were presented as papers at the Atlantic Council of Canada Spring Conference, Toronto, 21 May 2003; the conference on "Canada and the Globalizing World," Seventh Biennial Conference, Russian Association of Canadian Studies, Institute of the USA and Canada, Moscow, 25-27 June 2003; and as an MIT Security Studies Program Seminar (Sokolsky 2004c).

1. Eberhart (2004).

2. Higginbotham (2004).

3. Bush (2004).

4. Bush (2004).

5.Healy (2004) A12.

6. Kupchan (2004).

7. United States (2002).

8. Brown (2003) 1.

9. Mastanduno (1997) 51.

10. Kupchan (2004).

11. Daalder and Lindsay (2003) 189.

12. Kupchan (2004).

13. Posen (2003) 17.

14. Daalder and Lindsay (2003) 189.

15. Harries (2002).

16. United States (2002) 6.

17. United States (2002) 15.

18. Soros (2003) 64-65.

19. Kagan, Robert (2003) 85.

20. McDougall (1997).

21. Secor (2004).

22. United States (2002) 7.

23. Secor (2004).

24. Washington (1796).

25. Record (2003) v.

26. Record (2003) 41.

27. Record (2003) 31.

28. Posen (2003) 7.

29. See: International Institute of Strategic Studies (2003).

30. Posen (2003) 10.

31. Posen (2003) 18.

32. Posen (2003) 10.

33. On the RMA see: Sloan (2002), Gongora (2000), O'Hanlon (2000), Hundley (1999).

34. See: Priest (2002).

35. Hamilton (2003).

36. Rumsfeld (2002) 31.

37. Rumsfeld (2002).

38. Posen (2003) 44.

39. Record (2003) 45-46.

40. Kagan, Frederick (2003).

41. Hillen (1997).

42. Posen (2003) 44.

43. Kupchan (2004).

44. Frum and Perle (2003).

45. Brzezinski (2004).

46. Pfaff (2004) A15.

47. Pfaff (2004) A15.

48. Stockman (2004) A1, A8.

49. Stockman (2004) A8.

50. Posen (2003) 46.

51. See Moller and Shepherd (2004) 10.

52. Haglund and Roussel (2002) 14.

53. See: Friedman (2004).

54. Doran (1984) 139.

55. Canada (2004a) 50.

56. See, for example: Bland (2004).

57. Cohen (1996) 51.

58. Fagan (2004).

59. Delacourt and Harper (2004).

60. Delacourt and Harper (2004).

61. U.S. Northern Command.

62. Carafono (2004) 9.

63. Record (2003) 43.

64. Rust (2004).

65. Carafono (2004) 15.

66. Chalk and Rosenau (2004).

67. As quoted in Tulak et al. (2003) 133.

68. Peter Andreas (2003) 78-80.

69. Canada (2004a).

70. Canada (2004b).

71. Salot (2004).

72. Canada (2004c).

73. Higginbotham (2004).

References

Andreas, Peter. 2003. "Redrawing the Line: Borders and Security in the Twenty-First Century." *International Security* 28 (fall): 2.

Bland, Douglas (ed.). 2004. *Canada Without Armed Forces?* Montreal, Kingston: McGill-Queen's University Press.

Brown, Seyom. 2003. *The Illusion of Control: Force and Foreign Policy in the 21st Century.* Washington, DC: Brookings Institution Press.

Brzezinski, Zbigniew. 2004. *The Choice: Global Domination or Global Leadership.* New York: Basic Books.

Bush, George W. 2004. *State of the Union Address*, January 20. http://www.whitehouse.gov/news/releases/2004/01/20040120-7.html

Canada. 2004a. *Securing an Open Society: Canada's National Security Policy.* Ottawa: Privy Council Office. www.pco-bcp.gc.ca

———. 2004b. "Canada and Ballistic Missile Defence: Discussions with the United States on Possible Canadian Participation in the Ballistic Missile Defence of North America." Department of Foreign Affairs and International Trade. www.dfait-maeci.gc.ca/department/focus/bmd-en.asp

———. 2004c. "Marine Security Contribution Program Announced," news release no. GC 005/04, May 7. Transport Canada. http://www.tc.gc.ca/mediaroom/releases/nat/2004/04-gc005ae.htm

Carafono, Dr. James Jay. 2004. *Citizen-Soldiers and Homeland Security: A Strategic Assessment.* Arlington, Virginia: Lexington Institute. http://www.lexingtoninstitute.org/defense/CitizenSoldiers.pdf

Chalk, Peter, and William Rosenau. 2004. *Confronting the "Enemy Within": Security Intelligence, the Police, and Counterterrorism in Four Democracies.* Santa Monica, CA: The RAND Corporation. http://www.rand.org/publications/MG/MG100/MG100.pdf

Cohen, Eliot. 1996. "A Revolution in Warfare." *Foreign Affairs* 75 (March/April): 2.

Daalder, Ivo H., and James M. Lindsay. 2003. *America Unbound: The Bush Revolution in Foreign Policy.* Washington, DC: Brookings Institution Press.

Delacourt, Susan, and Tim Harper. 2004. "PM's Message to Bush: 'We'll Defend Ourselves.'" *The Toronto Star*, April 30. http://www.thestar.com

Doran, Charles. 1984. *Forgotten Partnership: U.S.-Canada Relations Today.* Baltimore, Maryland: The Johns Hopkins University Press.

Eberhart, General Ralph. 2004. Quoted in Sgt. 1st Class Doug Sample, "Defending the Homeland Is a 'Must Win' Game," American Forces Information Service news article, 26 Feb. http://www.defenselink.mil/news/Feb2004/n02262004_200402263.html

Fagan, Drew. 2004. "PM Plans to Push UN to Sidelines." *The Globe and Mail*, April 30. www.globeandmail.com

Friedman, Thomas. 2004. "Axis of Appeasement." *The New York Times*, March 18, A33.

Frum, David, and Richard Perle. 2003. *An End to Evil: How to Win the War on Terror.* New York: Random House.

Gongora, Thierry, and Harald von Riekhoff (eds.). 2000. *Toward a Revolution in Military Affairs? Defense and Security at the Dawn of the Twenty-First Century.* Westport, Connecticut: Greenwood Press.

Haglund, David, and Stéphane Roussel. 2002. "The Contradictions of Canadian Strategic Culture: 'Imperial' Commitments Within a 'Democratic Alliance'." Paper presented to the Biennial Meeting of the American Association of Canadian Studies, Portland Oregon, November.

Hamilton, Robert A. 2003. "Navy's Top Officer Calls for a Global Naval Force: Clark Outlines Strategy for Combating Terrorism." *The New London Day*, October 28. http://www.theday.com

Harries, Owen. 2002. "Understanding America." Lecture given at the Centre for Independent Studies, Australia, April 3. www.cis.org.au/events/CISlectures/2002/Harries030402.htm

Healy, Patrick. 2004. "Kerry Speech Blasts Bush's 'Unilateral Pre-emption.'" *The Boston Globe*, February 28.

Higginbotham, John. 2004. "21st Century Diplomacy: Understanding Trans-Governmental Networks in Canada-U.S. Relations." Presentation at Harvard University, 21 April.

Hillen, John. 1997. "Superpowers Don't Do Windows." *Orbis* 41 (spring).

Hundley, Richard O. 1999. *Past Revolutions, Past Transformations: What Can the History of Revolutions in Military Affairs Tell Us About Transforming the U.S. Military.* Santa Monica, California: RAND National Defense Research Institute.

International Institute of Strategic Studies (IISS). 2003. *Military Balance 2003-4* 103:1.

Kagan, Frederick W. 2003. "War and Aftermath." *Policy Review* 120 (August and September). http://www.policyreview.org/aug03/kagan.html

Kagan, Robert. 2003. *Of Paradise and Power: America and Europe in the New World Order.* New York: Alfred A. Knopf.

Kupchan, Charles A. 2004. "America Searches for Its Centre." Originally published in the *Financial Times*, February 10. Available at http://www.cfr.org/pub6758/charles_a_kupchan/america_searches_for_its_centre.php

Mastanduno, Michael. 1997. "Preserving the Unipolar Moment: Realist Theories and U.S. Grand Strategy after the Cold War." *International Security* 21 (spring): 4.

McDougall, Walter A. 1997. "Back to Bedrock: The Eight Traditions of American Statecraft." *Foreign Affairs* 76 (March/April): 2.

Moller, Richard, and Mark W.S. Shepherd. 2004. "Sovereignty Suicide: Canada's March into the American Empire." A paper presented at the Sixth Annual Society for Military and Strategic Studies Conference, University of Calgary, February 27-28. Conference proceedings available at the Centre for Military and Strategic Studies, University of Calagary. http://www.stratnet.ucalgary.ca/news_views/papers/smss-proceedings-2004.pdf

O'Hanlon, Michael. 2000. *Technological Change and the Future of Warfare.* Washington, DC: Brookings Institution Press.

Pfaff, William. 2004. "How Kerry Differs on Foreign Policy." *The Boston Globe*, March 8 .

Posen, Barry R. 2003. "Command of the Commons: The Military Foundation of U.S. Hegemony." *International Security* 28 (summer): 1.

Priest, Dana. 2002. "A Four-Star Foreign Policy?" In *American Foreign Policy,* edited by Glenn P. Hastedt. 8th ed. Guilford, Connecticut: McGraw-Hill/Dushkin.

Record, Jeffrey. 2003. *Bounding the Global War on Terrorism.* Carlisle, Pennsylvania: Strategic Studies Institute, U.S. Army War College.

Rumsfeld, Donald H. 2002. "Transforming the Military." *Foreign Affairs* 81 (May/June): 3.

Rust, Emily. 2004. "Study: Coast Guard Plan Inadequate: New Demands Since 9/11 Demand More Equipment, RAND Says." CNN.com International Edition, April 26 . http://www.cnn.com/2004/US/04/26/coast.guard.study/index.html

Salot, Jeff. 2004. "Canada May Host U.S. Missiles." *The Globe and Mail,* February 23.

Secor, Laura. 2004. "Grand Old Policy." Interview with John Lewis Gaddis, *The Boston Globe,* February 8.

Sloan, Elinor. 2002. *The Revolution in Military Affairs.* Kingston: McGill-Queen's University Press.

Sokolsky, Joel. 2004a. "The Diplomacy of Dispensability: The Bush Doctrine, Military Intervention and the Allies," paper presented to the conference "Military Intervention and Peacekeeping in the Twenty-first Century: New Challenges and Their Societal Consequences," at the Munk Centre for International Relations, University of Toronto, March 11-13.

———. 2004b. "A Hard Bilateral 'Moment of Truth': Canada and the United States in the Age of Terror." In *Independence in an Age of Empire: Assessing Unilateralism and Multilateralism,* edited by Graham F. Walker. Halifax: Centre for Foreign Policy Studies.

———. 2004c. "Realism Canadian Style: The Chrétien Legacy in National Security Policy and the Lessons for Canada-U.S. Relations." MIT Security Studies Program Wednesday Seminar Series, February 25, 2004. http://web.mit.edu/ssp/spring04/sokolsky.htm

———. 2003a. "The Power of Values or the Value of Power? America and Europe in a Post-9/11 World." *Columbia International Affairs Online,* case study (August). www.ciaonet.org

———. 2003b. "Between Venus and Mars: Canada and the Transatlantic Gap." *Connections Quarterly Journal* 2 (September): 3. www.ciaonet.org

Soros, George. 2003. "The Bubble of American Supremacy." *The Atlantic Monthly* 292 (December): 5.

Stockman, Farah. 2004. "Kerry Faces PR Fight Over Foreign Policy," *The Boston Globe,* April 26.

Tulak, Arthur N., Robert W. Kraft, and Don Silbaugh. 2003. "State Defense Forces and Homeland Security." *Parameters* 33 (winter 2003-04):4. http://www.carlisle.army.mil/usawc/parameters/03winter/tulak.pdf

US Northern Command. "Defending the Homeland is Job #1." www.northcom.mil

United States. 2002. *The National Security Strategy of the United States.* Washington, DC: The White House. http://www.whitehouse.gov/nsc/nss.html

Washington, George. 1796. *Washington's Farewell Address.* The Avalon Project, Yale Law School. http://www.yale.edu/lawweb/avalon/avalon.htm

Notes on Contributors

Thomas S. Axworthy is an adjunct lecturer in policy studies and is chairman of the Centre for the Study of Democracy at Queen's University. He is also the executive director of the Toronto-based Historica Foundation of Canada. From 1989 to 2000 he was executive director of the Charles R. Bronfman Foundation in Montreal. He taught at Harvard University from 1984 to 2003, most recently as adjunct lecturer in public policy at the Kennedy School. From 1974 to 1984 he served in several capacities in the Canadian government, principally as Prime Minister Pierre Trudeau's senior policy advisor and principal secretary. He co-authored, with Trudeau, *Towards a Just Society: The Trudeau Years* (1990). Dr. Axworthy served in the offices of the minister of national revenue, the minister of state for urban affairs, and the leader of the Opposition (as senior policy advisor). His publications include four books and numerous articles, including "Limping into the Millennium: American Liberalism in the Year 2000" (1995), and a draft proposal for the United Nations entitled "A Universal Declaration of Human Responsibilities" (1997).

Douglas Bland is a professor and holds the Chair in Defence Management Studies at Queen's University's School of Policy Studies. His research is concentrated in the fields of defence policy-making and management at the national and international levels, the organization and functioning of defence ministries, and civil-military relations. He has published books, articles, and reports and lectured in these fields in Canada, the United States, Europe and

South Africa. Dr. Bland is a graduate of the Canadian Army Staff College, the NATO Defence College in Rome, and holds a doctorate from Queen's University. He was a NATO Fellow in 1992-93. His published works on defence management include *The Administration of Defence Policy in Canada 1947-84* (1987); and the first two volumes of an annotated collection of government statements on defence policy, *Chiefs of Defence: Government And The Unified Command of The Canadian Armed Forces. Canada's National Defence, Volume 1 — Defence Policy* (1995), and *Volume 2 — Defence Organization* (1998). In 1999, he initiated the defence management monograph series, The Claxton Papers; the most recent volume is entitled "Canada without Armed Forces?"

Ann M. Fitz-Gerald is an associate professor at Cranfield University's Department of Defence Management and Security Analysis at the United Kingdom Defence Academy. She is also the director of Cranfield's Centre for Managing Security in Transitional Societies. She holds degrees in commerce, war studies, political science and defence management. Her Ph.D. thesis examined the disparities within multinational military forces and the impact these disparities have on sustainable peace and development strategies. Dr. Fitz-Gerald has carried out extensive fieldwork in conflict resolution, the security-development nexus and security sector reform in Haiti, the Balkans, Northern Ireland, West Africa, Afghanistan and Iraq in both research and practitioner capacities. She is widely published and is co-editor of *Providing Security for the People: Security Sector Reform in Africa*, and author of the recently published book *Security Sector Reform in Sierra Leone*. She is currently the director of the Global Facilitation Network for Security Sector Reform, a contract funded by the UK government, which involves an international team of experts to facilitate all policy development and capacity-building in security-related areas.

W. Don Macnamara, a retired air force brigadier general, has taught international business and strategy in undergraduate and MBA programs at the Queen's University School of Business since his retirement in 1988, and is now the associate director of the Queen's Public Executive Program . He is also a senior fellow in the Queen's Centre for International Relations and current president of the Conference of Defence Associations Institute.

Sean M. Maloney is the strategic studies advisor to the Canadian Defence Academy and teaches in the Department of War Studies at the Royal Military College. He served in Germany as the historian for the Canadian Army's NATO forces and is the author of several books, including *Canada and UN Peacekeeping:*

Cold War by Other Means 1945-1970 (2002) and *Operation KINETIC: The Canadians in Kosovo 1999-2000* (forthcoming). He has conducted extensive field research on Canadian and coalition military operations throughout the Balkans, the Middle East and South West Asia.

Danford W. Middlemiss is a professor of political science at Dalhousie University and a former director of that university's Centre for Foreign Policy Studies. Specializing in Canadian defence policy, the politics and processes of defence procurement, civil-military relations and similar subjects, Dr. Middlemiss has lectured at the National Defence College, the Canadian Forces Maritime Warfare Centre, and at the Canadian Command and Staff College. Among his recent publications are *Maritime Interests, Conflict and the Law of the Sea: A Conference Report* (co-editor with Fred W. Crickard and Susan J. Rolston), and *Canadian Defence: Decisions and Determinants* (co-author with J. J. Sokolsky).

Hugh Segal is president of the Institute for Research on Public Policy, a position he has held since 1999. He teaches at the Queen's University's School of Policy Studies and School of Business, where he is, respectively, the Ivey Foundation Fellow and a professor of public policy. He is the chair of the Walter and Duncan Gordon Foundation, a governor of the Kingston General Hospital, and sits on the board of the Atlantic Council of Canada. He is a past governor of the University of Ottawa, from where he graduated in 1972, and former vice-chair of the Institute of Canadian Advertising and the Advisory Committee on Confederation for the Province of Ontario. During fifteen years in the public sector he was associate secretary of Cabinet for federal-povincial affairs in Ontario and chief of staff to the prime minister in Ottawa. He was awarded the Order of Canada in 2003, and holds a doctorate in laws (honoris causa) from the Royal Military College of Canada.

Joel J. Sokolsky is dean of arts and a professor of political science at the Royal Military College of Canada (RMC). He is also a research fellow with the Institute for Research on Public Policy. He received his Ph.D. in government from Harvard University. Dr. Sokolsky has taught at Johns Hopkins University, Dalhousie University, Duke University and Bridgewater State College. His most recent works include "Sailing in Concert: The Strategy and Politics of Canada's. Naval Interoperability," *The Soldier and The State in the Post-Cold War* (co-editor), and "Realism Canadian Style: National Security and the Chrétien Legacy." He has been the recipient of several scholarships and awards, including two NATO Fellowships and two Fulbright Scholarships.

Denis Stairs is currently McCulloch Professor in Political Science at Dalhousie University in Halifax. Between 1971 and 1975 he was director of Dalhousie's Centre for Foreign Policy Studies. He is also a former president of the Canadian Political Science Association. He served for six years on the Social Sciences and Humanities Research Council of Canada and was Dalhousie's vice-president (academic and research) for 1988 to 1993. A fellow of the Royal Society of Canada, Stairs writes primarily on Canadian foreign policy, Canada-US relations, the policy-making process as it applies to the conduct of Canada's foreign affairs, and the impact of domestic constituencies on foreign-policy decisions.